# Cyber-Physical Systems

# and

# Metrology 4.0

S. Yatsyshyn
B. Stadnyk
*Editors*

# Cyber-Physical Systems and Metrology 4.0

International Frequency Sensor Association Publishing

S. Yatsyshyn and B. Stadnyk, *Editors*
Cyber-Physical Systems and Metrology 4.0

Published by IFSA Publishing, S. L., 2021
E-mail (for print book orders and customer service enquires):
ifsa.books@sensorsportal.com

Visit our Home Page on http://www.sensorsportal.com



ISBN: 978-84-09-26899-3
e-ISBN: 978-84-09-26898-6
BN-20210129-XX
BIC: TBM
BISAC: TEC022000

The Monograph is printed upon the decision of the Scientific Council of Institute of Computer Technologies, Automation and Metrology of Lviv Polytechnic National University (Protocol No.1, 15.09.2020).

# Contents

# Foreword

It has been 5 years since the monograph '*Cyber-Physical Systems. Metrological Issues*', 2016 was published. A lot has happened during this time. The world has accelerated even more. The editors of this book have involved 30 authors whose scientific achievements for the last 5 years cover a significant IT segment.

The basics of development of CPSs, built on the implementation of smart sensors and actuators, can be divided into 2 areas: a) Creation and development of information-measuring subsystems, as a single-out structural component; b) Studies of the functioning of these subsystems, as MIs that are responsible for CPS specifics. Ensuring the metrological unity of measurements with compliance to the dimensionality and reproducibility of units requires the introduction of 'intrinsic' standards of physical quantities, the work of which is based on the fundamental constants of the matter, f. e. the standards of electrical resistance based on the Quantum Hall effect and the computer-based equipment providing the frequency stability up to $10^{-16}$ in CPS's operation. Modern information-measuring technologies are implemented as real hardware and software measurement tools and as virtual tools attention to which is paid below. The CPS potential can be strengthened due to the emergence of new branches, which include biometrics, biosynthesis within which the mRNA-vaccine can be considered as the necessary component of the health-care CPSs, etc. The authors foresee the CPS development in the directions of the sensors' and actuators' data fusion, their accumulation and processing, especially in large arrays, the noise analysis behavior prediction, and also embedment of means of self-checking, self-adjustment, and middleware self-implementation.

We express deep gratitude to the leading experts in the field - Dr. Sc. V. Babak, Ukraine, and Dr. Sc. T. Fröhlich, Germany, for comments and recommendations while writing the manuscript, and especially to Dr. S. Y. Yurish, Director of IFSA Publishing, Barcelona, who took on the main burden of promptly bringing the manuscript to the scientific society.

Lviv, Ukraine, January 2021 *Editors*

# Authors

1. Yuriy Bobalo, Dr. Sc., Prof.
2. Mykola Mykyichuk, Dr. Sc., Prof.
3. Bohdan Stadnyk, Dr. Sc., Prof.
4. Svyatoslav Yatsyshyn, Dr. Sc., Prof.
5. Yaroslav Lutsyk, Dr. Sc., Prof.
6. Tetiana Bubela, Dr. Sc., Prof.
7. Ihor Mykytyn, Dr. Sc., Prof.
8. Zenoviy Kolodiy, Dr. Sc., Prof.
9. Pylyp Skoropad, Dr. Sc., Prof.
10. Roman Baitsar, Dr. Sc., Prof.
11. Yevhen Pokhodylo, Dr. Sc., Prof.
12. Vasyl Yatsuk, Dr. Sc., Prof.
13. Serhiy Prokhorenko, Dr. Sc., Prof.
14. Myroslava Prokhorenko, Ph.D., As.-Prof.
15. Orest Kochan, Dr. Sc., Prof.
16. Yuriy Khoma, Ph.D., As.-Prof.
17. Ihor Likhnovsky, Ph.D., As.-Prof.
18. Ivan Pytel, Ph.D., As.-Prof.
19. Roman Ivakh, Ph.D., As.-Prof.
20. Iryna Petrovska, Ph.D., As.-Prof.
21. Vasyl Parakuda, Ph.D., As.-Prof.
22. Viktor Semerak, Ph.D., As.-Prof.
23. Olha Lysa, Ph.D., As.-Prof.
24. Olha Oleskiv, Ph.D.
25. Andriy Midyk, Ph.D. Student
26. Tetiana Fedyshyn, Ph.D. Student
27. Anastasiia Riznyk, Ph.D. Student
28. Serhiy Lazarenko, Ph.D. Student
29. Oleksandr Shpak, Ph.D. Student
30. Oleksandr Kosterov, Ph.D. Student

# Acknowledgment

The scientific results, presented in Chapter 5 of this book, were obtained in the frame of Joint Ukraine – Germany R&D project number 0119U103134, 27.09.2019 - 31.12.2020, financially supported in Ukraine by the Ministry of Education and Science.

# List of Abbreviations

AC – Active Current
ADC – Analog to Digital Converter
AE – Acoustic Emission
AMC – Active Measurement Converter
ANN – Artificial Neural Network
BIPM –British Institute of Physical Measures
CCAUV – Consultative Committee on Acoustics, Ultrasound, and Vibration
CMF – Coriolis Mass Flowmeter
CNT – Carbon Nano Tube
CPS – Cyber–Physical System
CRD – Coordinate–Rotary Device
DAC – Digital to Analog Converter
DC – Direct Current
DNN – Deep Neural Network
DoE – Degree of Equivalence
DSC – Differential Scaling Calorimetry
DSP – Digital Signal Processing
ECG – ElectroCardioGram
EEG – ElectroEncephaloGram
EFS – Energy Fluctuation Spectrum
EIS – Electrical Impedance Spectroscopy
EMS – ElectroMyoStimulation
FET – Field–Effect Transistor
FPGA –Field–Programmable Gate Array
FPT – Fixed–Point Temperature
FSC – Filamentous Single–Crystal
HLA – High Level Architecture
IEC – International Electrotechnical Commission
IRS – Ionizing Radiation Source
ISO – International Organization for Standardization
KCRV – Key Comparison Reference Value
KNN – K Nearest Neighbor
LabVIEW – Laboratory Virtual Instrumentation Engineering Workbench
LAN – Local Area Network
LBDS – Lviv Biometric DataSet

LDA – Latent Dirichlet Allocation
LDA – Linear Discriminant Analysis
LR – Logistic Regression
MCU/PC – Microcontroller Unit/Personal Computer
MEMS – MicroElectroMechanical System
MG – Metallic Glass
MI – Metrological Instrument
MNP – Magnetic Nano Particle
MLP – Multilayer Perceptron
MOSFET – Metal–Oxide–Semiconductor Field–Effect Transistor
MSE – Mean Square Error
NPL – National Physical Laboratory
OA – Operational Amplifier
PCA – Principal Component Analysis
PhT – Phase Temperature
PID – Proportional Integral Derivative
PPG – PhotoPlethysmoGram
QHE – Quantum Hall Effect
R&D – Research and Development
RF – Radio–Frequency
RMS – Root Mean Square
RPS – Reference Power Supply
RQUIT – Reduced Quantum Unit of Temperature
SS – Signal Source
STS – Source of Test Signal
SVM – Support Vector Machine
TTPW – Thermodynamic Triple Point of Water
UC – Ultrasonic Converter
UME – Ultrasound Medical Equipment
VC – Vector Converter
VR – Voltage Repeater
VST – Vector–Scalar Transducer
VT – Vibration Treatment
WS – Working Substance
WSN – Wireless Sensor Network

# Introduction

The purpose of this book is presentation and consideration of main trends in the field of metrology of Cyber-Physical Systems which are becoming a key element of everyday life. In the first place, it destined for engineers, lecturers, students, persons who are not acquainted enough with the specificity of Cyber-Physical Systems and their Metrology but are interested in it. The authors tried to highlight the emergence and development of these systems, combined with the study of their metrology maintenance. Authors also presented their achievements respectively in the set of 15 publications in the *Sensors & Transducers* journal, under the common notion "Metrological Array of Cyber-Physical Systems": starting from Vol.186, Is.3, 2015, pp.1-11, Part 1, Challenge of Modernity, and pp.12-17, Part 2, Checked Instrument Based on Quantum Resistance Standard; finishing with Part 15. Approach to the Creation of Temperature Standard on Basis of Fundamental Physical Constants in Vol.199, Is.4, 2016, pp.62-66.

Day after day the globalization of industry, agriculture, transport, healthcare, and so on becomes more total. Certainly, it contributes to the continual development of Internet technologies, one manifestation of which is the occurrence of Cyber-Physical Systems. Objective quantitative information on the progress of technological processes is obtained by measuring their parameters. Measurement should be considered as a holistic process that starts from the perception and transformation of object measurement data to its processing, storage, transmission, and application for developing retroactivity in controlled technological objects. Therefore, in current conditions, one of the most important CPSs' parameters is their general and metrological reliability. The latter defines the validity and significance of the obtained information that becomes especially important for CPSs operation.

Indeed, the manufacturing CPSs should not cause environmental damage, greater from the acceptable standards. The problem of preventing environmental and technogenic accidents and disasters should be noted also. Exploration of this book would give academic researchers and practitioners a novel insight into the complex problem of CPSs conjugated with unceasing development of metrology science, and powerful tools to analyze the arising daunts.

The presented monograph includes several arrays of metrological studies based on studying:

- Verifying and validating the metrological tasks for parameters determining the controlled equipment, process, materials through the development, implementation, and realization of specific metrology and standardization methods, instruments, and equally the same as the facilities, etc. that is successfully described by the terms "metrological hardware, software, and middleware". Then, the hardware is necessary to provide at the design stage and to install at manufacturing this component of CPS or at its operation. Software is usually installed while training the hardware. Middleware has to be transferred via Internet base stations, and should be installed automatically and provided the set of automatic calibration actions, measurement, and characterization of objects in-situ. Calibration could be performed remotely under the condition of code access to CPS LAN with implemented appropriate software and middleware.

- Aspects of metrological reliability, including its prediction, particularly of CPS integrated metrological subsystems remote nodes. This includes not only microwave research essential for the operation at optical wavelengths but low and ultralow frequency methods in which it can be detected successfully the hidden and latent defects of complicated systems.

- The necessary metrological reliability of information and measurement system that, as a separate subsystem is the part of CPSs, in practice, constantly supervise the measurements. Reliable measurement information of required accuracy or trueness can be obtained only through technically informed choice of measuring instruments and includes the following data: availability of measured or monitored parameters of the object; tolerance for deviations of these parameters and allowable measurement uncertainties; the allowable probability of false and unidentified rejections for each of monitored and controlled parameters and the values of confidence for them; distribution laws of measuring (controlled) parameters and their measurement errors that can arise while using the measuring instruments; measuring conditions: mechanical loads (vibration, shock, acceleration, etc.), climatic impacts (temperature, humidity, pressure, etc.), and so on.

We are very grateful to our reviewers for well-wishing criticism.

## Chapter 1

# Measuring Instruments for Cyber-Physical Systems

**Mykola Mykyichuk, Yaroslav Lutsyk, Ihor Mykytyn, Olha Oleskiv and Iryna Petrovska**

### Definition

- **Cyber-Physical Systems** are the systems of collaborating computational entities which are in intensive connection with the surrounding physical world and its on-going processes, providing and using, at the same time, data-accessing and data-processing services available on the Internet. This definition is similar to the definition of the work [1].

### Introduction

Nowadays, the CPS area is expanding from academic research and laboratory development to real projects. The rapid perception and advancement of the CPS in developed countries where they have received state support are quite simple; CPSs are critical to ensuring national security and are an essential part of the modern technological revolution. The development of the information industry did not reduce production but only moved it to third countries. As a result, the US and EU population became dependent on these countries. To regain production independence, especially in the postCOVID'19 period, a process was reversed to outsourcing is arising, and the CPSs can solve. Although, of course, the scope of CPSs is much broader, since these systems help in creating qualitatively new healthcare, transport, energetics, and more. Several technical prerequisites led to the emergence and rapid development of CPSs [2].

The first one is the increase in the number of devices with built-in microprocessors and storage media: sensor networks operating in technical infrastructures; medical equipment; smart homes and more. The second is an integration that maximizes impact by combining individual components into larger systems: Internet of Things, World Wide Sensor Net, smart living environments, and the defense systems of the future. The third is to limit the cognitive abilities of humans that evolve more slowly than machines. It emerges a moment when people become no longer be able to cope with the amount of information needed to make decisions; so, some parts of the action must be transferred to the CPS by removing the human from the control loop (humans out of the loop). Here, we can consider the competitions of special chess programs and artificial intelligence systems in which the latter win. At the same time, in some cases, CPSs can enhance human analytical capabilities, so there is a need to create new interactive, high-level systems that leave people in the control loop (human in the loop). A key factor in the work of the CPS is the model used in the management system: how much it coincides with reality depends on efficiency. Considering in this context classical embedded systems, it should be admitted that they are products of XX century engineering based on a simplified conception and, respectively, on simplified models of the properties of nature and the environment. Today, simplified models lead to man-made disasters when conditions not provided by the model are created. One of the main reasons for this state is the insufficient equipping of the production chains with the sensors caused by the deficient computing power of the control subsystems. Then the smart sensors and smart actuators as mandatory components of CPSs with their computing powers are essential. The afore-mentioned approach is a counterpart of an Industry 4.0 – platform. The latter pays considerable attention enhancing quality and raising throw-out by accelerating manufacturing processes and making them more accurate. In such a way, multisensors' technology becomes insufficient for powering the production output; it needs to introduce new technologies based on the contemporary scientific practices. One bright example seems to be the closed-loop technology [3] which is a process within the operation cycle that integrates both manufacturing and measurement operations, following inspection; it raises the level of machining accuracy to an entirely new level, ensuring the highest possible standard for the particular production line. Another example of the continuous development in this branch is the implementation of new technics for calibration, using the Internet or similar network to provide an efficient service and permanent communication with the client. Here, the calibration results are sent directly to the client data base, applying Metrology 4.0 [4].

The CPSs are currently divided in next-generation vehicle control, distributed computation and smart production systems, smart grids, biometrics, health-care, social and environmental systems, their verification and validation, etc. Some of them are considered below in each Chapter of Monography.

- Smart manufacturing includes various production systems, which, however, are based on the use of smart sensors and smart actuators that perform certain technical activities within a given cycle. Their metrological support may involve several measuring and control means that have to be reprogrammed, with the remote access and installable Middleware. Equally important is ensuring the continuity of production, without interruptions to metrological checks, etc., which are scheduled and further burden manufacturers with the need to partially dismantle production facilities. Therefore, particularly responsible areas of smart manufacturing should include code-operated measures [2] and the built-in standards of physical quantities, f. i. of electric resistance [5]. The latter are discussed below.

- Smart health-care systems have certain requirements for the development of medical implants. The classic metrology challenge is the study of a person-adapted, smart cancer-treatment system. Human cells are selectively irradiated with microwaves so that, due to the action of the energy field, diseased cells are destroyed at a certain temperature and healthy cells remain intact. Selective absorption and irradiation power are realized through feedback, for example, on the base of temperature nanosensors [6].

CPSs are continuously complicated comparing to traditional objects, with the changeable over time their structure and composition, and the components and connections between them are not 100% reliable. There are some problems in the creation of the CPS: for example, concepts such as reliability, including metrological reliability, determination, failure period, etc. It is becoming increasingly important to access the metrological risks of their components avoiding cycle interruptions or shutdowns. The time factor, absent in traditional programming, but certainly inherent in the CPS's operation, is gaining weight here.

Typically, CPS involves data tracking, computing, and activating. They combine traditional information technology from sensor data passing to data processing in computation, as well as operational technologies for control and management.

A substantial feature of CPS is the combination of information and operational technologies along with time constraints. To ensure the implementation of smart programs, you must create an application model of the system, have the appropriate metrology and software, for instance, based on single-board microcontrollers and microcontroller kits. Besides, compatibility between heterogeneous components and systems is required, including, first of all, the harmonization of their timings and their certification algorithms. In this regard, developments in the field of metrology (calibration, complex product quality assessment, model-based diagnostics) and in the field of software and middleware as well as in the area of their metrological verification for MIs (further – MIs) are required (Fig. 1.1).

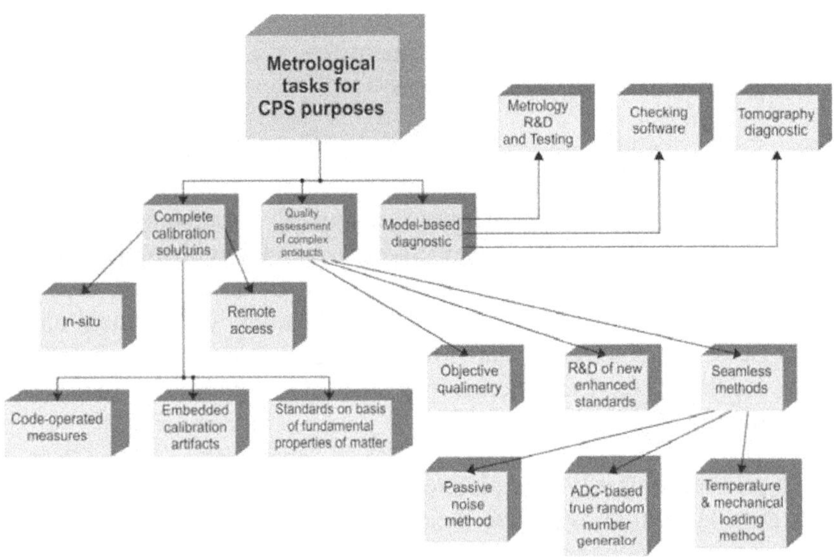


**Figure 1.1.** Metrological challenges for cyber-physical systems.

Already today we can state the existence of many purely metrological problems of the CPSs. These include the following:

1. The existing set of standards in different countries may be considered insufficient to successfully describe the means and tools of modern scientific technology [7].
2. Leading R&D centers using advanced production equipment cannot be tested in certified laboratories due to their complexity and delivery problems.
3. Unique technology often requires self-controlled and advanced specific metrological equipment for quality assurance.
4. Existing standards lost by several orders of their value characteristics exactly when transferring them to the end-user; it is considered the normal metrological practice that cannot be considered sufficient for CPS.

Their solution can be achieved by understanding the main trends in the development of metrology at the present stage. These include:

- The transition to smart sensors, devices, and systems;
- Replacement of traditional measures with measures based on fundamental physical constants (intrinsic measures);
- Development of systems: self-checking, self-verification, self-calibration, self-adjustment, etc.;
- Changing the MIs to similar ones that are based on the frequency-dependent properties.

Smart measuring instruments are the prerequisite for CPS design as they constitute the essential units of information and measurement subsystems. There is a set of smart MIs which is divided into the following subsets: smart sensors, smart transducers, their grids, etc. that can be joined together in modern WSNs. The emerging field of cheap and easily deployed sensors offers an unprecedented opportunity for a wide spectrum of various applications. When combined, they offer numerous advantages over traditional networks. These include a large-scale flexible architecture, high-resolution data, and application-adaptive mechanisms as well as a row of metrologically specific features and performance (self-check, self-validation, self-verification, self-calibration, …, self-adjustment).

For instance, special attention according to the opinion of Carnegie Mellon University scientists can be given to verification the WSNs in full compliance with well-known methodology. Timing verification has to guarantee the execution of assigned tasks in real-time WSNs within the planned duration. In this area, our efforts are directed in developing the schedule-ability techniques for multi-core platforms, where new challenges of shared resources, such as memory, different and changeable algorithms of signal processing and storing, transmission and aggregation have to be considered. Often it would be needed for decision-making in real-time systems, especially in health care and life safety.

## 1.1. Smart Sensors and Their Grids

Main milestones in everyday work aiming at the insurance of reliable WSN reliable operation lie in the direction of functional and probabilistic verifications. We provide the software and middleware development aiming to reach predetermined behavior. The easiest way to achieve this may be demonstrated on the example of widespread wireless fire detectors networks. They are characterized by some special algorithms directed on as fast as possible and accurate triggering and actuating the automation of a higher level. So, it becomes the necessity of researching and implementing the original operation algorithms for fire sensors and also the checking algorithms for periodic real-time software examination. Considering their structural complexity (presence of smoke and heat-sensitive elements, various principles of elaboration of the received signals, their drift of characteristics, and pollution of translucent elements, etc.) development of such algorithms is a daunting task. Herein, human life may be the price of a bug. Equally important seems to be probabilistic verification that is assigned to boost the probability of reaching WSN declared goals (estimation of their chances being achieved). This can be fulfilled in two known ways: a) probabilistic model based on a Markov approach, b) the similar one based on Monte Carlo simulation techniques.

Significant attention should be given to the collaborative autonomy of WSNs designed from the autonomous changeable groups of dissimilar sensors. We try to optimize performance and scalability of them by designing the decentralized operating environment. It may be open- and closed- sourced; the latter is proposed for WSNs that perform mission-oriented tasks. The new challenge of CPSs operation is their ability of rapid adaptation to a continuous environment alters even unto malfunctions. "Self-adaptive system is the system, capable of changing its behavior and structure, intending to adapt without human intervention to changes in itself and its operating environment" [8] while considering the WSN's approaches. We obligate to develop not only overall adaptation strategies including predictions of environment evolution but also the specific metrological proactive adaptive mechanisms caused by mentioned changes. If it is the MIs, the arrangements, in particular, are bound with the change in metrological characteristics and as the consequence of CPS dysfunction. Additionally, due to modeling, the adaptations can be performed proactively outpacing the happening drifts and other kinds of performance changes. As a result, it spares the time necessary for an adaptation strategy that has to be fulfilled.

Smart sensor, according to generally accepted industry definitions, combines a sensing element, analog interface circuit, ADC, and bus interface; all in one housing [9]. Except for it, a smart sensor includes the microprocessor that conditions the signals before transmission to the control network. Also, it filters out undesired noise and compensates for errors before sending the data. Making the grade against the newest generation of smart sensors, however, means that additional functionality must be included, such as self-testing, self-identification, self-validation, or self-adaptation. Of particular interest and importance to designers are such smart sensor capabilities as self-calibration and self-diagnosis, the ability to use signal processing, and multi-sensing capabilities.

Smart Temperature Sensor is the analog or digital primary thermosensitive transducer combined with a processing unit and a communicating interface [10] and able to perform a row of smart metrological functions due to installed metrological software. This is an intelligent temperature sensor with many specialized algorithms provided in the design or installation stage, i.e. a sensor with such embedded algorithms that are necessary to provide an implementation of following specialized metrology functions. Namely, such functions include, f. i. the ability to realize automatic switching of sub-range of measurement, depending on input signal value; automatic self-validation, self-check, self-diagnostics, etc.; the introduction of adjustments when the action of impact factor takes place; linearization of metrological characteristics; compensation of cold-junction temperature for thermocouples and so on.

The high-mentioned autonomous smart sensors being spatially distributed are frequently combined in-network assigned to monitor preset or environmental conditions (T, p, etc.) and to cooperatively send received and processed data by the network to the end-user. The set of such sensors is named WSN if specified sensors are interconnected and distributed in space. Furthermore, the most modern WSNs can operate bi-directionally which means that the control of smart sensors becomes enable as well as their software and middleware updating.

### 1.1.1. Sensors Network and Middleware

Each WSN consists structurally from a large number (up to $10^3$) of nodes which are the individual sensors that can radio communication with one or several neighboring units. The most common WSN is the fire alarm sensors network each branch of which has up to 26 sensors that were caused by limiting the length of microcontroller register [11]. The topology of every WSN may differ: star, a cluster tree, mesh, up to advanced multi-hop mesh WSN. The propagation technique between the hops of the network can be routing or flooding. Nowadays, the problem arises to adapt the traditional network topologies to contemporary communicating conditions. Its significant segment is described in [12] and considered below.

Early sensor networks used simple twisted shielded–pair implementations for each sensor. Later, the industry adopted multidrop buses (e.g., Ethernet). Now we can apply web-based networks (e.g., the World Wide Web) implemented on the factory floor. Three classic network topologies (point-to-point, multidrop, and web) are well-known. In the more reliable first one, each sensor node requires a separate twisted shielded–pair wire connection. In a multidrop network, each sensor node puts its information onto a common medium; this requires careful attention to protocols in hardware and software. Afterward, few networks have provided frequency-modulated signals on wires to carry multiple sensor readouts on the appropriate channels. Furthermore, early WSNs were simple radio-frequency implementations of this topology. They applied RF modems to convert the RS-232 signal to a radio signal and back again. Due to simple FM coding and interference impact, the reliability of such networking does not satisfy principally. Only when it was selected the particular frequency ranges of network operation the complete wireless local area networks have emerged. Remote data acquisition systems of similar topology are frequently implemented with in-field data concentrators and radio transmitter aiming the hosts, where the signals are demultiplexed into original sensor signals.

Once the industry began the migration to multidrop buses, problems associated with digitization began to emerge [13]. Especially, it occurs with point-to-point systems, where a single clock of the host could be used to timestamp when the analog signals from multiple sensors were acquired. With distributed smart sensors required to implement a multidrop network, synchronization of their clocks becomes the critical issue and important design parameter. Somewhat different problems are inherent in a web topology in which all nodes are potentially connected to all other nodes. Some of them can be eliminated by applying the repeaters and routers to make virtual connections.

Significant interest has emerged to ad hoc networks. They are multi-hop networks consisting of wireless autonomous hosts, where each host may serve as a router to assists traffic from other nodes [14]. Wireless ad hoc networks cover a wide range of network scenarios, including sensor, mobile ad hoc, personal area, and other networks. Research activities in the specific field of WSNs include sensor training, security through smart node cooperation, sensor area coverage with random and deterministic placement, object location, sensor's position determination, energy-efficient broadcasting, and activity scheduling, routing, connectivity, data dissemination and gathering, path exposure, tree reconfiguration, and topology construction. Network topology involves not only the sensor nodes but also the base stations and cross-layers for normal operation. Cross-layering becomes quite important for WSNs. So cross-layer can be used to make the optimal modulation to improve the transmission performance, such as data rate, energy efficiency, etc.

Following [14] the ideal wireless sensor is networked and scalable, smart and programmable, capable of fast data acquisition, reliable and accurate over the long term, and requires minimal maintenance. Selecting the optimum sensors and wireless communications link requires knowledge of the application and problem definition. Battery life, sensors update rates, and sizes are all major design considerations. Examples of low data rate sensors include temperature, humidity kinds. Examples of high data rate sensors are the sensors of strain, acceleration, and vibration. Contemporary examples of extra high data rate sensors are thermal and optical imagers, remote control sensors, and others.

Advance in microelectronics has resulted in the ability of sensors, based on digital electronics, and radio communications to develop the integrated units. Such sensors acquire the ability to communicate with each other with the help of wireless data routing protocols. The WSN preferably has to consist of a base station ("gateway") that can communicate with a few wireless sensors via radio-waves. Data is accumulated at the sensors' node, compressed, and transmitted to the main or the intermediate base-station.

We are not considered the architecture of WSNs below as it is traditional (star, mesh, hybrid star - mesh). Similarly, it is not the special purpose to study ordinary network standards (IEEE802.11x that is used for LANs at high bandwidth data transfer; IEEE802.15.1 and 2 that are known as Bluetooth that is of lower power and applies for personal area network; IEEE802.15.4 that is specifically designed for the requirements of WSNs, here the worldwide license-free band – 2… 4 GHz – is applied; ZigBee which alliance encompasses the IEEE802.15.4 specification and expands on the network specification and the application interface; and one of the most modern standards is IEEE1451.5 that aims the operating of smart sensor working groups in their wireless option).

If a centralized architecture is used in a sensor network and the central node fails, then the entire network will collapse, however, the reliability of the sensor network can be increased by using distributed control architecture. Distributed control is used in WSNs for the following reasons: sensor nodes are prone to failure; for better collection of data; to provide nodes with backup in case of failure of the central node; resources have to be self-organized.

A smart grid sensor is a small, lightweight node that serves as a detection station in a sensor network. Smart grid sensors enable the remote monitoring of equipment such as transformers and power lines and the demand-side management of resources on the energy smart grid [15]. Smart grid sensors can be used to monitor weather conditions and power line temperature, which can then be used to calculate the line's carrying capacity. This process is called dynamic line rating and it enables power companies to increase the power flow of existing transmission lines. Smart grid sensors can also be used within homes and businesses to increase energy efficiency [16]. According to NanoMarkets, companies like GE, LG, and Whirlpool have already announced their commitment to building home appliances that are smart-enabled. Smart grid sensors will link these appliances with smart meters, providing visibility into real-time power consumption. Power companies can use this information to develop real-time pricing and consumers can use the information to lower their power consumption at peak times [17].

A middleware layer is a novel approach to fully meeting the design and implementation challenges of WSN technology. We are considering the WSN middleware as a software infrastructure that glues together the network hardware, operating systems, network stacks, and applications. A complete middleware solution should contain a runtime environment that supports and coordinates multiple applications, and standardized system services such as data aggregation, control, and management policies adapting to target applications, and mechanisms to achieve adaptive and efficient system resources use to prolong the sensor network's life. Middleware should provide low-level programming models to meet the major challenge of bridging the gap between hardware technology's raw potential and the necessary broad activities such as reconfiguration, execution, and communication.

The relevant middleware projects for WSNs were studied in [18]. In cases where physical contact for replacement or maintenance is impossible, wireless media is the only way for remote accessibility. Hence, middleware should provide mechanisms for the efficient processor and memory use while enabling lower-power communication. Major WSN properties are studied below. Scalability is the first one and is defined as follows: if an application grows, the WSN has to be flexible enough to permit this growth anywhere and anytime without affecting network performance. Efficient middleware services must be capable of maintaining acceptable performance levels as the network change its dimensions and topology. The latter is subject to frequent changes owing to factors such as malfunctioning, device failure, mobility, and so on.

Most WSNs' applications are real-time phenomena; so, middleware should provide real-time services to adapt to the changes and provide consistent data. Application knowledge's design principles dictate another important and unique property of WSN middleware. Middleware must include mechanisms for injecting application knowledge of WSN's infrastructure. This lets developers map application communication requirements to network parameters, which enable them to fine-tune network monitoring. WSNs are being widely deployed in domains that involve sensors information for example, in healthcare and rescue areas. The untethered and large deployment of WSNs in harsh environments increases their exposure to malicious intrusions and attacks such as a denial of service. Also, the wireless medium facilitates eavesdropping and adversarial packet injection to compromise the network's functioning. All of these factors make security extremely important. Furthermore, sensor nodes have limited power and processing resources, so standard security mechanisms, which are heavy and resource consumption, are unsuitable. These challenges increase the need to develop comprehensive and secure solutions that achieve wider protection while maintaining desirable network performance. Middleware efforts should concentrate on developing and integrating security in the initial phases of software design, hence achieving different security requirements such as confidentiality, authentication, integrity, freshness, and availability [19-20].

### 1.1.2. Specifics of Smart Measuring Instruments

Analog interface is a set of MIs that are an integral part of measuring channel between the primary measuring transducer and ADC. It performs the following functions: scale transformation of measurement signals, their filtering, temperature compensation of thermocouples cold junctions, galvanic separation of transferring and receiving parts, linearization of characteristics and initialization of passive MTs, measuring signals multiplexing, measurement and service information transmission.

Smart sensors are supplied with digital information transmissive means by equipping them with built-in digital controllers to match the universal network interface or by combining the technology of analog and digital transmission in a single measuring channel. By the structure, all smart sensors are divided into 4 groups: sensors of centralized and decentralized types, as well as sensors with digital and analog buses. By correction methods, the analog interfaces with smart sensors are divided into the groups: with manual error correction, with auto-correction of errors in the analog-digital form, and with digital correction of errors.

The further development of R & D must work regarding the assessment of accuracy in the field of metrology in particular while single measuring the multivariable values by using multiparameter smart MIs, for instance, CMF [21]. They can be considered flexible tools and equipment to be able to quickly re-engineer existing procedures. Resultantly, by choosing the right measurement technology solution we can efficiently supply both development and production with relevant data.

CMF transformation function is the dependence of the liquid mass or volume on its flowing velocity through the specified cut. It is determined by the comparison of the time characteristics of two identical sensors in the CMF input and output. More phase difference of mentioned characteristics corresponds to a faster-controlled environment flowing. So, there is a dependence on the hydrodynamic regime of the current environment flowing through the CMF, its viscosity, the temperature, etc. Within the current metrological conception, the CMF consists of transducers with appropriate sensors and peripheral devices and the microprocessor unit of received signal processing. The CMF sensors determine the flow velocity, temperature and provide information in form of output signals to the microprocessor that carries out the function of the brain of the measurement device and system in total providing access to the display, main menu and output device of processed information for the interaction with other systems, for instance, the filling system. Peripheral devices provide monitoring, warning signalization, and other functions, for instance, periodic processes management and the function of liquid density more accurate determination, etc. The CMF transfer function error temperature component is the error specified by the temperature regime of liquid/gas flow. It depends on the temperature of the control environment (on the temperature dependence of the liquid flowing regime through the CMF); on the CMF body outlet temperature that is provided with the help of temperature detecting means for detecting a temperature of the inner tube, and temperature means for detecting a temperature of the outer tube; temperature correcting means for compensating an instrumental error according to a change of temperature of the inner tube; temperature difference correcting means for compensating an instrumental error according to a difference between a temperature of the inner tube and a temperature of the outer tube.

The next example of the necessity to develop smart MIs could be industrial tomography systems. Their development consists of the development and implementation of methods and algorithms to process the results of measurements which would enable a limited number of gauges exactly and quickly obtain high-quality images of the distribution of the studied variables. Mainly it would be the spatial distribution of the electrical conductivity of the medium. Improvement of the exactness and rate of tomography measurement has been reached [22] by reducing additive gauge errors through the use of the different measurement method; by elaborating the particular methodology for calculating the sensitivity matrix without of methodical errors inherent in finite-difference methods and based on only a unitary solution of the direct task; ensure the stability and convergence of the iterative procedure of the spatial distribution determination.

## 1.2. Verification of Measuring Instruments and Their Software

There exists a particular metrological aspect concerning software usage. The problem consists not only of the bugs or errors of the software. They are detected at the stage of software development by the developer or even the user. It seems to be more complicated due to arising inaccuracies even incorrectly executed software for certain MI. The latter is predominantly caused by software-hardware platform mismatch while they're developing. Their detection is a rather complex issue, the failure of which can lead to rather large errors in certain ranges of the MI. The solution consists in implementing the metrological kind of software verification.

Presented here approach substantially differs from generally accepted one. The latter is described as following. Commonly as the software metric is considered a standard of measure of a degree to which a software system or process possesses some features. Since quantitative measurements are essential in all areas, the goal is obtaining objective, reproducible and quantifiable measurements, which may have numerous applications in schedule and budget planning, cost estimation, quality assurance testing, software debugging, software performance optimization, and so on.

Our specific measurement consists in the evaluation of MI's performance reliability, trueness, and other metrological properties, due to the quality of the certain kind of metrological software, or the software linked with metrological features of MI.

The design and development of MIs involve the reduction of analog units and the expansion of digital ones due to the advances of last. These include the digital microcontrollers, programmable logic arrays, and more. The trend of modernity consists in growing the "weight" of software in measurement means. Inconsistency of software to the measuring tasks of certain MI, accidental, or intentional alteration of its functions can lead to incorrect measurement results. It is therefore advisable to conduct verification of software to derive its impact on the metrological characteristics of MI and the possibility of further running the studied software as a consisting part of the device, tools, or metrological mean [23].

Under MI's software, we mean a set of programs and procedures designed to register, aggregate, process, display and save, and post the measurement results. Such software as a functional part of MI is delivered jointly with hardware [24-25]. According to [25] software, that may affect metrological characteristics of MIs, concerns: a) programs and program modules that participate in the processing of measurement results; b) involved in the calculations the software parameters that affect the results; c) programs and program modules that carry out the presentation of measurement data, its storage, and transmission, software update, and identification, secure software development [26] and data protection; d) components of protected interface for data exchange between software modules of CPS units.

The last 2 items are appropriately attributed to overall not metrological verification of software since no measuring or computing actions that affect the measurement results have been fulfilled. These points characterize the correctness of the software functioning in general. So, we have the expertise to incorporate security practices – authentication, authorization, and auditing – into each phase of the verification, from software design and implementation to testing and deployment [27]. Secure software engineering has become an increasingly important part of software quality, particularly due to the development of the Internet. While IT security measures can offer basic protection for the main areas of our IT systems, secure software is also critical for establishing a completely secure CPS's environment. Every single software MI developer must care about security because users need to be able to trust the proposed software including the MI software and middleware. The entire cover cycle of the aforementioned wares development includes requirements engineering, trust & threat modeling, secure coding, security testing, and security response to code protection [28].

MI software metrological verification raises the problem of choosing the appropriate methods of software and middleware assessing, testing, and certifying. The result of the metrological validation must be the confirmation or negation of studied ware to the requirements noted in normative documents. Procedures and methods of checking software, and determining its disadvantages are considered below. Software study includes first of all the fulfillment of the procedures of unambiguity ensuring the operating functions for generated data. The selection of the procedures is defined by regulation requirements, as well as by software developer or the user desires to confirm its compliance with target specification.

To validate the type of MI, test procedure should envisage the identification and evaluation of software impact on metrological characteristics as well as prevent unauthorized software reconfiguration and interference that can degrade the trueness of received measurement results. Therefore, the developed under tests, project description of MI type should contain additionally to metrological characteristics the description of the software, identity, impact assessment, and the level of protection against unintentional and intentional changes. Certification of MI software and middleware is the research that aims at the determination of characteristics, features, identification data, and confirmation of requirements compliance. Under the methodology of software certification to determine one or more characteristics (analysis of documentation and source code, functional inspection under controlled conditions, etc.) the tests are carried out [29].

Distinguish two kinds of certification: general and metrological. The object of general appraisal is the complete software, the study of which is conducted to justify the application of algorithm (program) within specific tasks. To assess the impact of software on the data inexactness, the metrological certification examines merely that software which is an integral part of complete software of the particular MIs. Software certification is mostly voluntary except the software that performs especially responsible functions, where the lack of quality, mistakes, or failures can severely disrupt or are dangerous for life and health (aviation, nuclear power, management of authorities and banking systems, etc.) [30].

It is important to choose the correct verification method for MI software and middleware. There are considered the possibility and cost of its implementation, the checking quality, and more.

Method of comparative testing with applying the reference software is used for certain software which helps to identify its features being checked. As the reference software can be studied: attested and/or certified software, the functionality of which is similar to tested software features; specially developed software with functions that are identical to functions of software that has to be checked; software for assignments of computing tasks (e.g. spreadsheets, software for mathematical and statistical calculations, etc.).

To develop reference software, somebody may resort in cases when the software under test is not too complicated and its implementation algorithm is rather simple. Reference software should not reproduce all the functionality of the software that is tested and could only contain functions and parameters that affect MI metrological characteristics. In certain cases, the peculiarities of graphic user interface and functions that do not participate in the handling of measurements (e.g., display functions, data storage, etc.) are not taken into account. The given method enables us to consider the metrological characteristics and maximum peculiarities of software that is tested. The drawback of this method is that often the complexity of software implementation makes it inexpedient due to the high costs of developing the reference software.

In the absence of reference software, the priority is given to the method of comparative trials by using output data models and comparative tests with generating the "standard" data. The latter is recommended for certification of data processing algorithms of measurement results. The method enables us to evaluate the algorithm possibilities by comparing the processed results of the output data model obtained with the help of the mentioned algorithm, towards the specified parameters of this model. The method of output data models is a kind of generation method of "standard" data, not just data generated by specially developed programs; only data are not generated by the specially developed program, and programmatically they are set at the input of software which is tested. Output data models are selected in a way that they fully regard the measuring tasks covering the greatest range of possible values. These may include data that completely cover the range of possible values; data close to the largest and smallest values as well as several intermediate values; specific values of input variables – points of the sharp rise or rupture of derivatives, zero, single, and extremely small numerical values of variables, etc.

If values of a variable depend on values of another variable, the test is carried out for certain combinations of these variables, such as the equality of two variables, or large and small their difference, or zero values of variables. The test method of output data models is easier to implement than the classic method of generating "standard" data. However, the development of this method requires a priori information about software algorithms and their program realization, which is not always known.

Method of generating "standard" data, as the method of output data models, is applied alternatively to the reference software method, or in the case of inability to check the particular functions of implemented software. One of the prerequisites of the method of generating "standard" data is the availability of a priori information about the corresponding measuring task. "Reference" data is produced by a specially developed generating program, i.e. the generator of "standard" data based on specified output data. The last realize in one of the programming languages or by using the standard mathematical (statistical) software package. Initial data for testing, including for the generation of "reference" data, are formed considering the properties of software algorithms that are checked. Method of generating "standard" data is an alternative one to the method of using the reference software. On the other hand, the development generator of "standard" data is advisable as it is cheaper than other methods of implementation [31].

If there are several program realizations of the same measurement algorithm at the absence of the reference software, it is expedient to carry out checks by comparing method implementation. According to it, the same set of "standard" data is submitted to inputs software products, and a comparison of relevant test results is performed. The comparison method is simple and requires no extra programs. However, programs with the same functions are quite rare.

When testing on an analysis of the source code of the software, the following items are checked: conformity of algorithms structure to the rendered documentation; correctness of recorded algorithms to the chosen programming language; matching the selected algorithms up measuring tasks (detection of unstable algorithms).

When checking the compliance of algorithms structure to the rendered documentation, the algorithms block diagram can be composed and compared with the algorithms described in the documentation. In the event of differences between the structures of algorithms, the additional analysis of block design elements is conducted. When checking the correctness of algorithms to the chosen programming language, the compliance of code to programming regulations, the presence of uncertain variables and operators, the correct organization of cycles, and so on are established. The conformity of selected algorithms with measuring tasks can be estimated by mathematical analysis of implemented software algorithms. It can be explored algorithms of realized different characteristics, in particular, may be executed an optimal analysis of numerical methods for solving measurement tasks. This method provides an opportunity for a detailed assessment of metrological software. However, to implement a test method on analysis of source code, the experts in the software industry and metrology must be engaged simultaneously. So, costs for implementation of it are much higher compared to other methods.

Middleware, being downloaded from the Internet and installed in MIs, is the software that provides services to software applications beyond those available from the operating system of the programmable block of MI. It is often can be presented as "software glue". Middleware makes it easier for software developers to fulfill communication and input/output, so they can concentrate the attention on their problems. Middleware is the software layer that lies between the operating system and such an application as WSNs. As a result, it supports complex, distributed business software applications.

Middleware includes Web servers, application servers, content management systems, and similar tools that support application development and delivery. It is especially integral to information technology based on Extensible Mark-up Language, Simple Object Access Protocol, Web services, Web 2.0 infrastructure, and Lightweight Directory Access Protocol, which is commonly used for communicating and managing in distributed applications.

In simulation technology, middleware is generally applied in the context of high-level architecture to some distributed simulations. This is a layer of software that lies between the application code and the run-time infrastructure. Middleware mainly consists of a library of functions, and enables a row of applications — simulations or federates— to page these functions from the common library rather than re-create them for each application. Wireless networking developers can use middleware to meet the challenges associated with WSNs. Implementing a middleware application allows developers to integrate operating systems and hardware with a wide variety of various applications that are currently available. For instance, radio-frequency software toolkits provide middleware to filter noisy and redundant raw data.

### 1.2.1. Metrological Verification of Software

This point has always evoked the lively discussion and often misunderstanding and rejection of the proposed concepts of metrological verification of MI software. Why is there such a situation? And why, in our opinion, should be given essential attention to the abovementioned problem.

First, this issue is on the border between metrology and digital (microcontroller and microprocessor) technology, which in recent decades has rapidly burst into the measuring equipment [32-35]. Almost every modern MI is not projected without a microcontroller. In many cases, a personal computer becomes a part of MI (LabVIEW, etc.). In this case, the main processing of the measurement results is fulfilled on a PC. Second, no one ever metrological principles have been applied to digital equipment and software since the basic necessity of it is absent. Nowadays, taking into account the wide integration of microprocessors with software to MI, which is characterized by the metrological characteristics, the logical question arises: how to take into account software inexactness and how it should pass metrological verification of measuring methods to obtain valid results of verification? Third, software vendors conduct their testing and just do not understand the goal of software metrological testing or more correctly verification. Meanwhile, there exists a particular difference between software testing and metrological verification, noticed by the authors.

Software testing is the process of technical study designed to identify information about software quality relative to the context in which it is used. The testing technique includes both a process for finding errors or other bugs and examinations of software components for a specific assessment, such as:

- Compliance with the requirements of designers and developers;
- Correct answer for all possible inputs;
- Performance of functions at an acceptable time;
- Practicality;
- Compatibility with current software and operating systems;
- Compliance with customer's tasks, etc.

Software testing provides objective, independent information about software quality, risks, failures, refuses, etc. [36]. As can be seen from the above list of tasks performed through testing, none of them provides a quantitative assessment of the results of calculations conducted using the software, but only provides a qualitative assessment.

The metrological verification of software is significantly different from testing. And the main difference lies in the fact, that is, first of all, gives a quantitative estimate, namely, allows determining the error of the software. Software, conditionally, is the separate block of MI and participates in the calculation of the measurement result. This approach allows us to apply metrological principles concerning software, substantially improve, simplify and accelerate the process of metrological verification of modern MI, and also to unify the metrological verification of MI software.

Fourth, the CPSs are emerging on the market for production and services [37-39], the main elements of which are practically ready for use. And if not today, shortly such systems will become the main production units of the world economy, which, depending on the task, will form a robotic complex with the corresponding software. And here are two questions. What software should use CPS so that they can perform different tasks? And by what criterion CPS should choose the software, including, for the involved MI? On the above questions, we will try to answer in the following.

### 1.2.2. Key Moments in the Development of Measurement Technology

First, let's make a brief excursion in the near past of measuring technology. Note the key occasions that will help to understand the issues of the metrological verification of MI software, that for today hung up in the indefinite state and one way or another must be resolved.

In the 19$^{th}$ century, as a result of researches in the electric phenomena, the first galvanometers appeared devices for measuring direct and variable current. This time physicists from around the world were beginning to develop new methods of electrical measurements: Lenz has suggested the ballistic method; Christie suggested the bridge method and Poggendorff suggested the compensation method. At the end of the 19$^{th}$ century, two scientists D'Arsonval and Deprez have created the first high-sensitivity galvanometer. In a few years, a physicist Dolivo-Dobrovolsky developed devices that later became the basis for modern voltmeters, ammeters, and wattmeters. It should be noted that in the first steps of the development of electrical MI, no one paid special attention to the need for metrological verification of developed MI, in the context as we understand it for today. However, in parallel, there were developed standard instruments and exemplary measures that provided reproduction, storage, and transmission of the unit of physical size, and later were used as means of metrological verification.

At that time MI was mainly used in scientific researches. However, the rapid development of the creation and production of new materials and complex products that needed a certain technological process led to the fact that MIs began to be widely used in production. The quality of the manufactured products depended on the accuracy of maintaining certain technological parameters such as temperature, pressure, etc., which were measured by MI. Therefore, the accuracy of the measurement of physical quantities influenced the quality of production. Therefore, the metrological verification of MI was necessary, which allowed ensuring the proper condition of the measuring technique. Verification of MIs is the set of operations performed in order to confirm the compliance of MIs with metrological requirements.

Let us consider the most common analog MI. A block diagram of metrological verification of such MI is presented in Fig. 1.2.

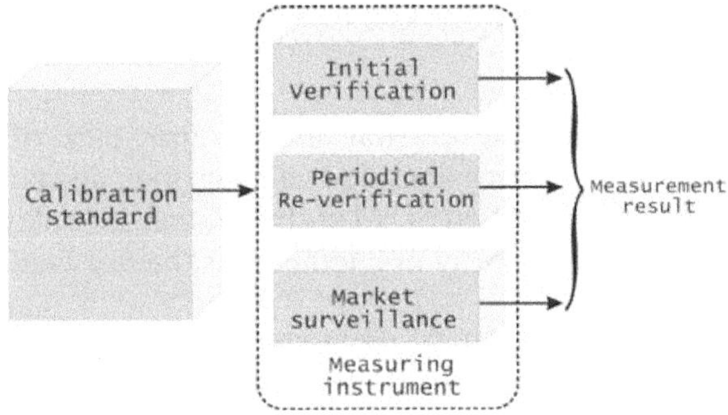


**Figure 1.2.** Block diagram of metrological verification of MI.

The known physical quantity is given to the MI input. A measurement result is obtained at the output. When the measurement error is derived, it becomes possible to compare its value with the value of the MI class. Then we receive the possibility to conclude about the suitability or unsuitability of the MI for a certain application.

Development of electrical MI element base including the transistors, amplifiers, specialized microcircuits, and more, and the simplification of electrical MIs, lead to the emergence of a wide range of non-electrical MIs. Their distinguished features of the design are the next:

- Primary transducer, which converts the measured physical quantity into an electrical one, mainly in current and voltage;
- MI of aforesaid electric value.

Here is arising up an interesting circumstance, which in the future would allow understanding the necessity of MI software metrological verification. What does it mean? Nothing hinders the metrological verification of such MI according to the structural scheme presented in Fig. 1.2: for instance, primary transducers are checked in such a way.

Nowadays, the metrology is continuously altering due to the digital transformation of contemporary society (Fig. 1.3) which consists in the instant conformating the different bodies (manufacturers, customers and users, laboratories and market surveillances, etc.) involved in the particular MI quality assessment.

Why the functional blocks of MIs have to be checked separately?

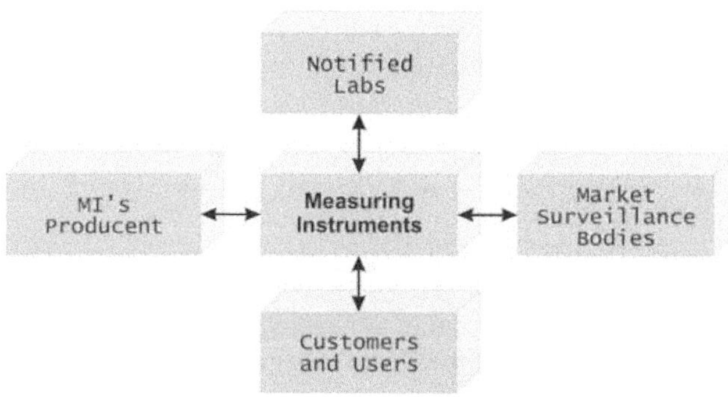


**Figure 1.3.** Digital transformation of Metrology for contemporary society.

First, because the standards of electric quantities, mainly, are inherent in lower uncertainties compare to the standards of non-electric quantities.

Second, the duration to verify non-electrical MI is longer than of electrical MI. For example, during metrological verification, the transition from one voltage to another one takes the seconds, and the similar transition in temperature would take from ten minutes to an hour.

Third, the separation allows unify the process of metrological verification of electrical MI: mainly the output signal of the primary transducers of non-electrical quantities is unified, f.i., 0-10 V, 0-20 mA. So, for convenience and speedy metrological verification, functional blocks of such MI are checked separately.

The last step for today in the development of measuring technique has been fulfilled with the dissemination of microcontrollers. What has essentially changed with the emergence of these two elements of digital technology? This allowed moving practically all mathematical processing of the measurement results from the analog part to the digital and significantly simplifies the analog part and, as a consequence, reduce the errors of MI. This can be demonstrated on the example of measuring the RMS voltage, which is determined by the formula:

$$U_{skz} = \sqrt{\frac{1}{T}\int_0^T u(t)^2\,dt} \qquad (1.1)$$

A block diagram of analog MI of the RMS voltage is presented in Fig. 1.4.

**Figure 1.4.** Block diagram of the analog measuring instrument of the RMS voltage.

To find the RMS voltage $U_{skz}$ is necessary to bring the input signal u(t) to the square using an analog multiplier, integrate and take the square root. The error of such transformation is at best 0.5 – 1 %.

The use of microcontrollers allowed calculating the RMS voltage in digital form by the formula:

$$U_{skz} = \sqrt{\frac{1}{n}\sum_{i=1}^{n} u_i^2}\,. \qquad (1.2)$$

The block diagram of the digital MI of the RMS voltage is presented in Fig. 1.5.

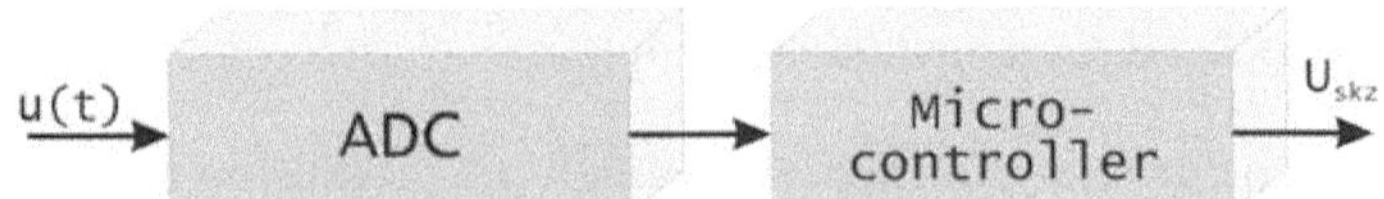


**Figure 1.5.** Block diagram of a digital measuring instrument of the RMS voltage.

Due to such changes in the structure of the MI succeeded in significantly reducing the measurement error. Comparing Fig. 1.3 and 1.4, it is possible to conclude that the main data processing and obtaining the result of measurement occurs in the microcontroller software. And here is another metrological curiosity. In our case, almost 100% of mathematical processing digital data obtained with the use of ADC, and the calculation of the measurement result is carried out using the software. Despite this, the metrological characteristics of the ADC are normalized, and the software remains in the metrological sense incognito and conceals your metrological characteristics.

### 1.2.3. In-Situ Verification of Measuring Instruments and Their Software

More than once during the discussions presented in this chapter sounded assertion: the device is checked in general, and there is not a necessity to check the software separately. But is it so? Do we properly check modern MI? To answer this question, we give a small example carried out by the authors and provide some explanations. But first let's clarify that these issues are exacerbated as the further development of CPSs and the complexity of MIs, in connection with the acquisition of intelligence traits [40]. We understand them as the ability of the CPSs to self-develop, including through implemented software and middleware

Verification of classic MI (and, unfortunately, modern also) mainly implemented in a few points from the range of MI. Such a small number of verification points are mostly sufficient, since the dependence of the error on the measured value has a monotone character, for example, is monotonically increasing or monotonically decreasing. And this, in our opinion, was correct only until the main mathematical processing and calculation of the measurement result began to be conducted using the software. What does it mean?

Fig. 1.6 shows the dependence of relative error $\delta_{fs}$ of amplitude calculation of spectral harmonics by the function fft () from signal frequency for data of type double (floating point numbers) and data of type int (integer numbers, which complies with 16-bit ADC performance).

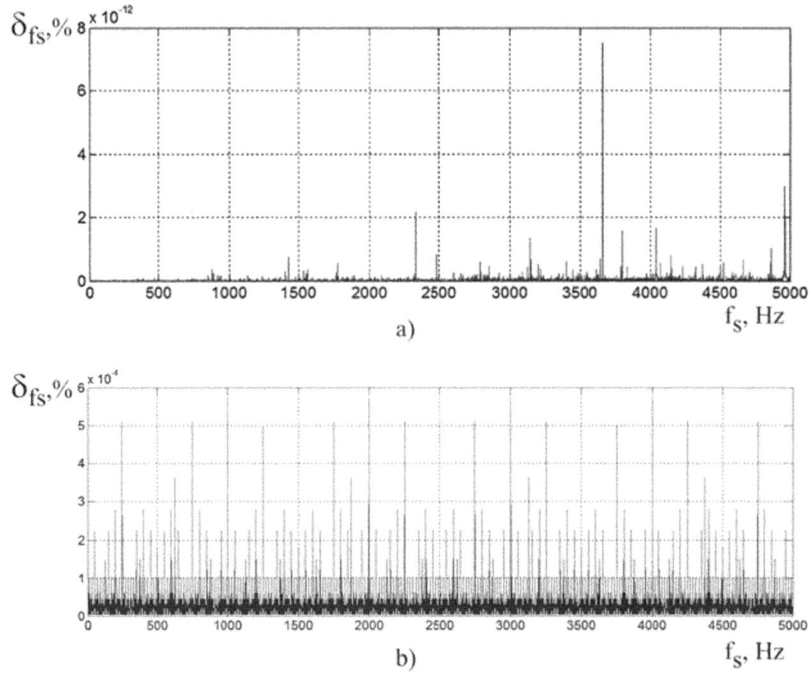


**Figure 1.6.** Dependence of relative error $\delta_{fs}$ of amplitude calculation of spectral harmonics by the function fft () from signal frequency for a) Data of type double; b) Data of type int, which complies with 16-bit ADC performance.

As can be seen, the relative error of the amplitude calculation of spectral harmonics has a pseudo-random character. There is no clearly expressed monotonically increasing or monotonically decreasing dependence. In such a case, taking only a few points for verification of MI, whose software makes the function fft (), it is possible to get the improper results of metrological verification. The probability that we get the maximum error of the function fft () is sufficiently small. Therefore, the verification of such MI must be done at all points of the range of the measured value, in this case by frequency. Certainly, such verification on the structural scheme of metrological verification (Fig. 1.2) will be sufficiently long. And there is no necessity to check the analog part at every point in the range of the measured value since the above errors relate only to the software that is used to calculate the values of spectral harmonics. So, this suggests that it is necessary to conduct the metrological verification of software separately. This will allow us to make a quick verification in autonomy mode for each measured value and define the maximal error of measuring result, which is determined using the software.

Another argument against the metrological verification, which is put forward by specialists of digital technology is that taking into account modern possibilities, they can provide practically any computational error (within reasonable limits). This can be achieved by increasing the data bus of microcontrollers and microprocessors. And this is certainly true. However, the specialists of measurements are always limited by the ADC bit depth. And regardless of size, the data bus of the microcontroller or microprocessor, the determining influence on the calculation error of measurement result with the software will have the ADC bit depth. And the influence of bit on the calculation error of measuring result is quite large.

According to the results of the research (Fig. 1.6), the transition from the data type double (floating-point numbers) to the data type int (integers) taking into account the size of the ADC (in our case, 16 bit) leads to a significant increase in the error of amplitude calculation of spectral harmonics by the function fft (). For data type double, the maximum relative error of the function fft () is $7.5 \cdot 10^{-12}$ % (Fig. 1.6, a), and for data type int it is $5.7 \cdot 10^{-4}$ % (Fig. 1.6, b). Thus, the error increases by almost 8 orders of magnitude.

Therefore, for the quantitative estimation of error of MI software, that predefined by the limited nature of ADCs bit, it is expedient to conduct metrological verification of MI software using digital data standards. And every software would have to get some metrological characteristics, which in some way connected with the metrological characteristics of the ADC.

The next debatable question is the necessity of metrological verification of MI software on the stage of exploitation. Since the software for MI does not change during exploitation, then his verification should be done only at the design and development stage.

However, today the theoretical and practical foundations of CPS are being created (except for emergencies). Depending on the put task CPS will form an original robotic computer system. For this purpose, components of the CPS and related software of general-purpose and software that will be used for MI of robotic equipment will be involved from the existing park. For CPS to perform varied tasks, they must use flexible software, which will mainly be obtained through cloud-based technologies. That is, with the statement of the new task will be used another MI software. And it suggests an idea that in such case software, which will calculate the result of the measurement using a certain algorithm, must undergo metrological verification. This will guarantee the correct work of the robots involved in the CPS, and thus will ensure the safety of human life.

Besides, there is another question: after what criterions the CPS choose software, and also for the involved MI. Which software parameter should characterize it so that the CPS correctly chooses the software that is optimal for the task? And there is a simple answer to this question: for MI software, which is used in the CPS, such characteristic is the error of software computing the measurement result. That is, the software will be characterized by not qualitative, but a quantitative characteristic, which enables quickly to compare variants of the same type software and choose the optimal, in obedience to the put task of measurement. Certainly, the producers of MI software must conduct metrological verification and ascribe declared metrological characteristics of software. On the other hand, the CPS must conduct a metrological verification of software and check whether the received metrological characteristics are under the declared. If the result is positive, then the software can be used. Otherwise, another software is proposed and studied. During the operation should be considered even parameters such as fluctuations in the electrical resistance of the circuit or grounding [41], which can stochastically effect through the chip technique to the CPS's operation.

The aforesaid concerns the in-situ verification. But to conduct remote metrological control at the site of CPS's exploitation we propose to use differential measurement method and code-controlled impedance measures [42] considered below in Chapter 4.

## 1.3. Conclusions

1. Implementation of Cyber-Physical Systems is impossible without easily deployed Smart Measuring Instruments that suppose an opportunity for a wide spectrum of various applications. Numerous advantages include a large-scale flexible architecture, high-resolution data, and application-adaptive mechanisms as well as a row of

metrologically specific features (self-check, self-validation, self-adjustment, etc.) which unprecedentedly improve their performance.

2. Predictable ways of Smart Measuring Instruments development consist in further studying the Smart Sensors Grids with improved parameters that are reached not only by software & middleware enhancing and metric introduction, but also by ensuring their scalability, flexibility and several other characteristics as well as by providing the specific metrological checks, for instance, of software, or by model-based diagnostics of Smart Instruments reliability.

3. Recently considering science, the metrological verification of MI software is used extremely rarely. In the nearest future, due to the implementation of CPSs in production and services, the issue of metrological verification of MI software will need to be resolved and regulated by law.

## References

[1]. Rajkumar et al., CIRP Encyclopedia of Production Engineering, *Springer*, 2010.

[2]. Cyber-Physical Systems. Metrological Issues, Editors S. Yatsyshyn, B. Stadnyk, *IFSA Publishing*, Barcelona, 2016.

[3]. Industry 4.0: The Future of Metrology (http://www.ogpuk.com/news/industry-4-0-future-metrology/)

[4]. R. Benitez, C. Ramirez, J. A. Vazquez, Sensors calibration for Metrology 4.0, in *Proceedings of the II Workshop on Metrology for Industry 4.0 and IoT (MetroInd4.0&IoT)*, Naples, Italy, 4-6 June 2019, pp. 296-299.

[5]. J. Weis, K. von Klitzing, Metrology and microscopic picture of the integer quantum Hall effect, *Royal Soc. Publ.*, October 2011.

[6]. D. Lockwood, Nanosensor reveals temperature variation in the muscles of live creatures, 2016 (https://cen.acs.org/articles/94/web/2016/11/Nanosensor-reveals-temperature-variation-muscles.html).

[7]. The little big book of metrology, *NPL*, 2011.

[8]. A. Platzer, Logical Foundations of Cyber-Physical Systems, *Carnegie Mellon University,* (http://symbolaris.com/logic/lfcps.html).

[9]. C. Mathas, Smart Sensors – Not Only Intelligent, but Adaptable, *Electronic Products*, 09-29-2011, https://www.digikey.com/en/articles/smart-sensors---not-only-intelligent-but-adaptable

[10]. S. Yurish, Sensors: Smart vs. Intelligent, *Sensors & Transducers*, Vol. 114, Issue 3, March 2010, pp. 1-6.

[11]. B. Deb, S. Bhatnagar, B. Nath, A topology discovery algorithm for sensor networks with applications to network management, *CiteSeerX*, 2002, (http://citeseerx.ist.psu.edu /viewdoc.1886).

[12]. W. Manges, *Oak Ridge National Laboratory* (http://archives. sensorsmag.Com/articles /0500/72/WirelessSensorNetworkTopologies).
[13]. Ad-hoc Sensor Networks, *Brunel University London* (https://www.brunel.ac.uk/electronic-and-electrical-engineering/research-and-phd-programmes/Research-areas/Ad-hoc-Sensor-Networks).
[14]. Sensor technology handbook, (Ed. by Jon S. Wilson), *Elsevier*, 2005.
[15]. Smart grid sensors, *IoT Agenda* (http://searchnetworking.techtarget.com/ definition/smart-grid-sensor)
[16]. A. Singh, Smart grid sensor, *Int. Journ. Comp. Eng.,* Vol.2, Issue 7, 2012, pp. 930-963.
[17]. G. Bitelli, P. Conte, T. Csoknya, E. Mandnnici, Urban energetics applications and Geomatic technologies in a Smart City perspective. *Int. Rev. of Appl. Sc. and Eng.,* Vol. 6, Issue 1, pp. 19-29 (https://akjournals.com/view/journals/1848/6/1/article-p19.xml).
[18]. H. Salem, M. Nader, Middleware: middleware challenges and approaches for wireless sensor networks, *IEEE Distributed Systems Online 1541-4922*, Vol. 7, No. 3, March 2006.
[19]. What is Middleware?. *Middleware.org. defining technology*. 2008. Retrieved 2013-08-11.
[20]. S. Hadim, N. Mohamed, Middleware challenges and approaches for wireless sensor networks, *IEEE Distributed Systems Online*, Vol 7, Issue 3, 2006, Retrieved March 4, 2009.
[21]. M. Kazahaya, A mathematical model and error analysis of Coriolis mass flowmeters, *IEEE Transactions on Instrum. and Measurement,* Vol. 60, Issue 4, 2011, pp. 1163 – 1174.
[22]. G. Bolton, K. Primrose, An overview of electrical tomography measurements in pharmaceutical and related application areas, *AAPS PharmaSciTech*, June 2006.
[23]. Procedure for certification of software of measuring instruments (http://www.uazakon.com/).
[24] WELMEC 7.1, Edition 2 Information document. Development of software requirements. Vienna, 2005, 48 p.
[25]. O. Velychko, V. Gaman, O. Hrabovskii, T. Gordiyenko, Features of testing of the built-in software of measuring instruments, *Collection of Scientific Works ODATRYA*, № 1 (16) 2020, pp. 36-41.
[26] T. Zimmergren, Embrace a Secure Software Development Lifecycle (SDLC) for Azure, Jan. 2020, https://zimmergren.net/author/zimmergren/
Microsoft secure software development, The 3rd Conf. on minimization of software vulnerabilities in its development.
[27]. What is CSSLP (certified secure software lifecycle) (www.searchsecurity.techtarget.com /.../CSSLP-certifie...).
[28]. ISSECO, the International Secure Software Engineering Council (www.isseco.org/).
[29]. G. Zlygosteva, S. Muravyov, A generalized model of the test procedure for measuring software, *Bulletin of the Tomsk Polytechnic un.*, Vol. 318, Issue 4, 2011, pp. 62-67 (In Russian).

[30]. E. Kovalevskaya, Metrology, quality and certification of software, *Moscow International Institute of Econometrics, Computer Science, Finance and Law*, Moscow, 2002 (In Russian).
[31]. T. Fröhlich, O. Oleskiv, I. Mykytyn, Problems of Metrological Verification of Software for Modern Measuring Instruments, in *Proceedings of the 59th Ilmenau Sc. Col.,* Technische Universität Ilmenau, 11 – 15 September, 2017 (https://core.ac.uk/download/pdf/224746145.pdf).
[32]. Guidance for the Management of Computers and Software in Laboratories with Reference to ISO/IEC 17025/2005, *Eurolab. Techn. Rep.*, No 2, 2006 (http://www.eurolab.org/documents/2-2006.pdf).
[33]. Z. Kolodiy, B. Stadnyk, S. Yatsyshyn, A. Kolodiy, Integral power of flicker-noise in real systems, in *Proceedings of the Int. Conf. on Noise and Fluct. (ICNF),* June 02-06 2015, Xian, China, 2015, pp. 1-3.
[34]. Development of Software Requirements: Informative Document, WELMEC 7.1, Issue 2, 2005 (http://www.vniims.ru/009lab/docs/welmec_7_1.pdf).
[35]. Software Guide (Measuring Instruments Directive 2004/22/EC). 2011: WELMEC 7.2, Issue 5, 2012.
[36]. C. Kaner, Ja. Falk, H. Q. Nguyen, Testing Computer Software, New York, *John Wiley and Sons, Inc*., 1999.
[37]. R. Rajkumar, I. Lee, L. Sha, J. Stankovic, Cyber-physical systems: the next computing revolution, in *Proceeding of the Design Automation Conference (ACM),* 2010, pp. 731-736. (https://www.cs.virginia.edu/~stankovic/psfiles/Rajkumar-DAC2010-Final.pdf).
[38]. Technologie-Konzept (http://www.its-owl.de/technologie-netzwerk/strategie/technologie-konzept/).
[39]. E. A. Lee, Cyber-Physical Systems: Design Challenges (http://www.cs.virginia.edu/~son/cs851/papers/CPS.challenges_ISORC08.pdf).
[40]. O. Ivakhiv, T. Dominyuk, B. Stadnyk, Intelligent measurement multiplex system, in *Proceedings of the 2nd IEEE Intern. Workshop on Intel. Data Acq. and Adv. Comp. Syst.: Techn. and Appl. (IDAACS' 2003)*, September 8–10, 2003, Lviv, Ukraine, 2003, pp.13–15.
[41]. Z. Kolodiy, B. Stadnyk, A. Kolodiy, Energy Spectrum of Stochastic Signals Caused by Variations of Electrical Resistance, *Autom. Control and Comp. Sc.,* Vol. 52, No. 4, 2018, pp. 311–316.
[42]. V. Yatsuk, T. Bubela, Y. Pokhodylo, Y. Yatsuk, R. Kochan, Improvement of data acquisition systems for the measurement of physical-chemical environmental properties, in *Proceedings of the 9th IEEE Int. Conf. on Int. Data Acq. and Adv. Comp. Systems: Technology and Applications*, 21-23 September 2017, Bucharest, Romania, 2017, 6 p.

# Chapter 2

# Standards of Physical Quantities and Other Measuring Instruments

**Yuriy Bobalo, Svyatoslav Yatsyshyn, Bohdan Stadnyk, Anastasia Riznyk, Vasyl Parakuda, Oleksandr Kosterov, Oleksandr Shpak, Myroslava Prokhorenko, and Serhiy Prokhorenko**

## Definitions

- **Quantum thermometry** means thermometry based on the application of temperature quanta aiming the provision of the measurements through implementing a quantum value determined solely by the fixed quantum constants (defined previously based on the fundamental constants of matter, that is, the Planck's constant and the Boltzmann's constant).
- **Nanothermometry** is the branch of thermometry that is serviced and used in its everyday practice by the nanoscale thermometric measuring instruments, for example, by the liquid-in-nanotube thermometers.

## 2.1. Standards of Physical Values Based on Fundamental Constants of Matter

At the end of the $20^{th}$ century, due to intensive research in Nanotechnology and Metrology 4.0, most (6 of the $7^{th}$, except for temperature in Kelvin degrees) basic units of the SI system, were successfully identified through fundamental physical constants of the matter [1]. Such units include m, A, kg, sec, mol, cd. Besides, a similar result was obtained for some additional values of the system SI. Since these standards are received in such a way and are based on the received particular units of minimal size that cannot be changed, they pertain to exclusively metrologically reliable and stable.

### 2.1.1. Quantum Standards of Electric Resistance and Electric Voltage/Current

With the discovery of the Josephson and the quantum Hall effects, two electrical quantum standards became available. These quantum effects play a major role in the modernization of the SI. By fixing conventional values for the von Klitzing constant $R_{KI}$ and the Josephson constant $K_J$, the worldwide consistency of the electrical measurements has improved considerably. The QHE together with the Josephson effect have played an important role in the replacement of the last before end material artifact in the SI, the kilogram. The research and development of a primary ac resistance standard based on the QHE are of great importance in metrology [2]. It simplifies and, as a consequence, improves the measurement capabilities in the field of impedance measurements which are considered an integral part of Metrology 4.0 and are discussed below. Despite the successful application of the QHE in metrology, our understanding of the results attraction is still incomplete. In particular, the effect allows us to obtain only one fixed value of electrical resistance. Traditionally, you need at least two values to build a scale.

Therefore, we study below the possibilities of using the achievements of metrology to extend the scale of measurable resistances, taking the value corresponding to the constant Klitzing as a nominal value. Namely, it is about increasing and decreasing the dimensions of the measured values of resistances with the use as a reference value while maintaining the accuracy characteristics.

Further work should also be carried out to gain a better understanding of the processes which are responsible for the obtaining the value of the quantum of temperature needed the last in the replacement really of the last material artifact in the SI that is $1^0$C as one-hundredth of the difference between boiling and freezing points of water.

- **Standard of Electrical Resistance Based on von Klitzing Constant**

Not only mechanical properties and melting points of substances, and also their electrical characteristics change in nanoscale. In 1998, it turned out that the resistance of nanoresistors did not depend on their size and substance, but was determined by only two fundamental physical constants:

$$R_0 = \frac{h}{2e^2}, \quad (2.1)$$

where h is the Planck's constant, e is the charge of an electron and is equal to 12.9 kΩ. The value of $R_0$ was called the quantum of electrical resistance, implying that in the nanoscale the resistance of all resistors is the same. An example of a nanoresistor characterized by ballistic conductivity is the carbon nanotube. This is long cylindrical frame structures with a diameter of one to several tens of nanometers and a length up to a centimeter, consisting only of carbon atoms. The current is transmitted in the wire by electrons. Such nanotubes pass electrons as light waves pass through light lines. Thus, in nanoscale, electricity is converted to optics, and Joule heat is dissipated only at the nanotube boundaries, where it connects to a bulk conductor. Let's try to derive a formula that relates the quantum of resistance to the fundamental physical constants. Since the nanotube has ballistic conductivity and Joule heat is not released in it, it can be assumed that its length is less than the free path of the electron conduction. Suppose that a voltage U is applied between the sections A and B of the nanotube (Fig. 2.1), and its current is I.

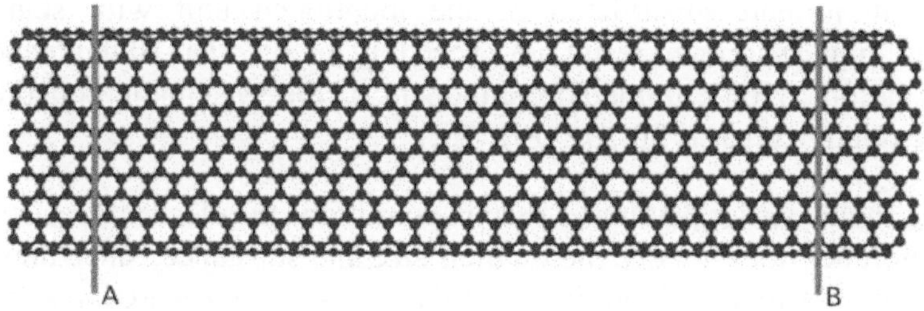


**Figure 2.1.** Schematic representation of a carbon nanotube.

Since energy is not dissipated, the change in electron energy between sections A and B is E = eU. This energy change occurred over a time interval t equal to the time of flight between the mentioned sections. The Heisenberg uncertainty ratio imposes some restrictions on the changes E and t: whence it follows that $U \approx \frac{h}{e\Delta t}$.

Let us now estimate the current in the nanotube. The latter is a one-dimensional quantum structure. In it, as in a helium atom, may be taken only two electrons with different spins. This means that the current I between sections A and B is equal to $I = \frac{2e}{\Delta t}$ . From the last relations it is easy to obtain the formula for the required resistance - see equation (2.1). Since the nanotubes do not heat, they are capable of transmitting currents of enormous density - more than $10^7$ A/cm$^2$. If carbon nanotubes had normal (non-ballistic) conductivity, then at such currents their temperature would increase to 20 000 K, which is much higher than their combustion temperature.

So, we accept that for the metrological purposes (creation of the quantum standard of electric resistivity), the specimen based on QHE is considered. A row of issues [3-4] is devoted to the aforesaid problem. Below we discuss the problem of implementation of the high-precision reference standards just as it was regarded by participants of the Royal Society Discussion Meeting, for instance by [5]. We consider the Ukrainian state standard of electrical resistance. This standard is proposed to develop by applying the latest nanotechnology research including the electrical conductivity of carbon nanotube.

According to the state primary standard unit of electrical resistance, the physical values of an electrical resistivity that is realized by the standard are equal to 1 Ω; 100 Ω. State primary standard provides storage, verification, and supervision of the mentioned unit with standard deviation measurement result not exceeding $3 \cdot 10^{-8}$ Ω at 10 independent observations. Non-eliminated component of the systematic constituent of the relative error is $30 \cdot 10^{-8}$ (Fig. 2.2).

It was found earlier that under certain conditions the electrical resistance of nanotubes does not depend on their size and substance; since they are inherent in superconductivity. Simultaneously a measured resistance determined by the resistance transient contacts constitutes von Klitzing constant (for electron couple) $R_0 = \frac{h}{2e^2} = 12906.4037$, where h is the Planck's constant [$6.62606957(29) \cdot 10^{-34}$ J/s], e is the charge of an electron ($1.602176565 \cdot 10^{-19}$ C). So, the standard unit of electrical resistance realized by CNT is determined by only two mentioned fundamental physical constants. The latest is defined with high precision utilizing a set of different physical methods. To study Planck's constant the research methods are applied: Watt balance, installations of studies of X-rays crystal density, Magnetic resonance, Faraday constant,

Josephson constant. 2018 CODATA recommended value of weighted mean Planck's constant relative uncertainty is $u_h = 4.4\times10^{-8}$ [6]. The same concerns the charge of the electron, which uses the results of different research and its evaluation methods: Millikan's oil-drop experiments, E. Rutherford, and other scientists' investigations: relative standard uncertainty of electron charge determination is $u_e = 2.2\times10^{-8}$ [6].

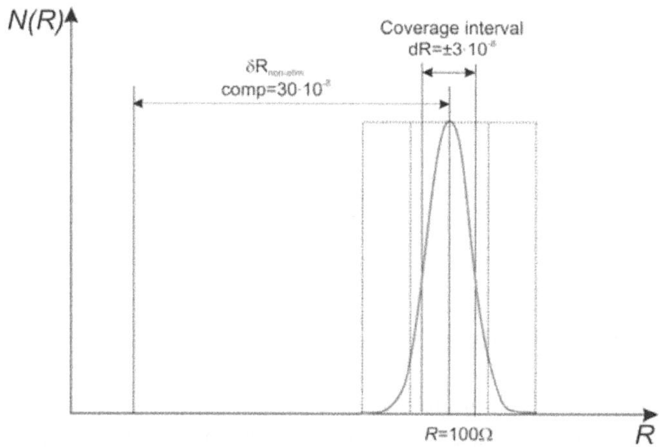


**Figure 2.2.** An instrumental error of the Ukrainian standard of electrical resistance.

It was estimated the total uncertainty resistance of CNT resistors: $u_R = \pm(u_h + 2u_e) = \pm\ 8.8\cdot10^{-8}$. Transition to the mean square error $15{,}242\cdot10^{-8}$ is made by the equation: $\sigma = \frac{u}{\sqrt{3}}$. The absolute value of the proposed standard tolerance limit is defined as $\pm\ 0.0019672\ \Omega$ (Fig. 2.3).

Thereby, the evident result of nanotechnology achievements direct implementation seems to be creating the State standard of electric resistance based on von Klitzing constant calculated value of which is determined as $(12906.4037 \pm 0.0020)\ \Omega$. Such a standard can be successfully realized on any superconducting nanotubes, for example, CNTs. Due to an array of world-wide previous researches we argue that the non-eliminated component of the systematic constituent of relative error of the proposed standard is 2-fold reduced in comparison with the existing standard.

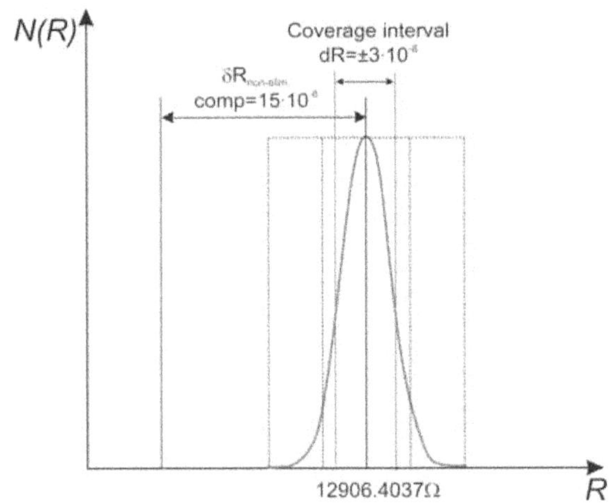


**Figure 2.3.** The instrumental error of State standard of electrical resistance, based on von Klitzing constant.

- **Josephson Voltage Standard**

A Josephson voltage standard is a complex system that applies a superconductive integrated circuit chip operating at a temperature of 4.2 K to generate stable voltages that depend only on an applied frequency and fundamental constants. It is an intrinsic standard in the sense that it does not depend on any physical artifact. It is the most accurate method to generate or measure voltage and, by international agreement in 1990, is the basis for voltage standards around the world. Standard is based on the Josephson effect, 1962, within which the electrons can tunnel between two superconductors separated by a thin insulating barrier. This is a Josephson junction. An applied dc voltage V across the barrier would generate an AC at the frequency f = 2eV/h. Conversely, an applied AC of frequency f would generate a dc voltage Vn at the quantized values Vn = nhf/2e, where n is an integer and the value of 2e/h is approximately 483.6 MHz/V [7].

Naturally, the Josephson voltage standard was initially used to improve the measurement of the constant 2e/h based on voltage values derived from the SI volt realization as maintained by Weston cells. The stability of the Josephson volt depends only on the stability of frequency (which currently is higher than $10^{13}$, owing to small differences in national standards, different values of KJ were adopted by various countries [8]. This inconsistency was corrected in 1990 when, by international agreement, the constant KJ-90 was assigned the value 483597.9 GHz/V and adopted by all standards laboratories. The uncertainty in KJ-90 is 0.4 ppm.

Standards such as the Josephson volt that depend on fundamental constants rather than physical artifacts are known as the intrinsic standards. Although the Josephson voltage standard does not realize the SI definition of the volt, it provides a very stable reference voltage that can be reproduced anywhere without the need to transfer artifacts such as Weston cells. It seems to us, the similar is happening with the quantum temperature standard considered below.

### 2.1.2. Quantum Standard of Temperature

At the end of the 20th century all major SI units, except the temperature, were expressed through fundamental physical constants. A need for a Quantum standard of temperature has been demonstrated by the issues of the 13th TEMPMEKO International Symposium [9]. Contemporary technologies persistently require the most exact and precise standards as we invariably lose accuracy characteristics by several orders while uploading them to the end-user.

At this moment "Temperature" remains the last value among 7 main units of SI that is not regulated with the help of the mentioned constants. It is caused partly by the existing preconception that temperature is the only value which cannot be quantized; partly by the absence of proper metrological instruments and mainly by the lack of development the theory and practice of appropriate "intrinsic" temperature standards based on physical laws rather than on stability of material artifacts: for instance, CNTs as such ones in Raman method of thermometry.

- **Temperature and Its Definition**

In USA, UK, Germany and other countries through several years the endeavors of elaborating the unit of temperature scale in the form of a quantum energetic unit (minimal by size a discrete value of the energy or heat energy that can be defined, established and fixed by the experimenter) have been carried out at the high methodological level. The recommended by [10] new format of a unit with a new definition is the next. The kelvin, symbol K, is the SI unit of thermodynamic temperature. It is defined by taking the fixed numerical value of the Boltzmann constant k to be $1.380\,649 \times 10^{-23}$ when expressed in the unit

J $K^{-1}$, which is equal to $kgm^2s^{-2}K^{-1}$, where the kilogram, meter and second are defined in terms of h, c, and $\Delta\nu_{Cs}$. This definition implies the exact relation k = 1.380 649·$10^{-23}$ $kgm^2s^{-2}K^{-1}$. Inverting this relation gives an exact expression for the kelvin in terms of the defining constants k, h and $\Delta\nu_{Cs}$:

$$1\,\mathrm{K} = \left(\frac{1.380\,649}{k}\right) \times 10^{-23}\ \mathrm{kg\ m^2\ s^{-2}} \tag{2.2}$$
,

which is equal to:

$$1\,\mathrm{K} = \frac{1.380\,649 \times 10^{-23}}{\left(6.626\,070\,15 \times 10^{-34}\right)\left(9\,192\,631\,770\right)} \frac{\Delta\nu_{Cs} h}{k} \approx 2.266\,6653 \frac{\Delta\nu_{Cs} h}{k} \tag{2.3}$$

The effect of the proposed definition is that the kelvin is equal to the change of thermodynamic temperature that results in a change of thermal energy $k_BT$ by 1.380 65 …·$10^{-23}$ J/K [11]. Then using $k_B$ rather than TTPW to define kelvin better reflects modern practice in determining thermodynamic temperature directly by primary methods, particularly at very high and low temperatures.

The unit of thermodynamic temperature, the kelvin, has been redefined by fixing the value of the Boltzmann's constant. The present CODATA recommended value of $k_B$ is determined predominantly by acoustic gas-thermometry results. To provide a value of this constant based on different physical principles, purely electronic measurements were performed by using a Johnson noise thermometer to compare the thermal noise power of a 200 Ω sensing resistor immersed in a triple-point-of-water cell to the noise power of a quantum-accurate pseudo-random noise waveform of nominally equal noise power. Measurements integrated over a bandwidth of 550 kHz and the total integration time of 33 days gave a measured value of $k_B$=1.3806514(48)·$10^{-23}$ J/K, for which the relative standard uncertainty is 3.5·$10^{-6}$ and the relative offset from the CODATA 2010 value is +1.9·$10^{-6}$ [3]. Efforts have led to some success. To provide a value of $k_B$ based on different physical principles, purely electronic measurements have been performed by using a Johnson noise thermometer, by Quantum Voltage Noise Source method [3, Table 1 (Summary uncertainty budget for a determination of Boltzmann constant)].

TEMPMEKO-2019 technical summary concludes that its important result is following. "… scientists can now develop new technologies and measurement methods that aren't constrained with the triple-point of water definition (previously kelvin was defined as a change in temperature equal to 1/273.16, the assigned temperature for the triple-point of water). Any fundamental equation that has both temperature and the Boltzmann constant can be the basis of a measurement technique to measure temperature directly and on the true thermodynamic temperature scale. Eventually, scientists would like to see temperature measured on the thermodynamic temperature scale, instead of the current practical ITS-90 …" [10].

The possible and evident way of defining temperature by the experimental values of energy through micro, nanocalorimeter readouts is characterized by fundamental flaws. Severe difficulties raise in the area of ultralow energies gauging [5-6]. Applying the energetic method, as the basic one, for measurement of temperature, could bring in an additional error/uncertainty, as the direct temperature measurement becomes indirect. Such kind of determination is always less accurate since a single error $\delta T$ of direct measurement is replaced by the sum of two errors: $\delta E + \delta k_B$ in the indirect method.

Therefore, we agree with and support three summarizing statements, especially the second one, provided by Graham Machine of NPL UK, in his plenary talk on the matter [10] that:
- The redefinition of the kelvin will have no immediate effect on temperature measurement and for most users, it will pass unnoticed;
- The redefinition lays the foundation for future improvements free of material and technological constraints and will enable the development of new techniques for measuring temperature, especially at extreme temperatures.

The Mise en Pratique of the definition of the kelvin will guide the world-wide dissemination of the kelvin by describing methods for measurement of thermodynamic temperature.

Nevertheless, the high-mentioned definition does not hold the base in the form of a quantum of the physical quantity to which it is assigned, that is, a quantum of temperature. It should be noted that in actuality for electrical resistance and its determination using the QHE (here one quantum of resistance is equal to $R_K$), the quantization is rather mandatory to perform the contemporary standard.

- **Investigation in Creating the Quantum Unit of Temperature**

Temperature is the statistically defined quantity, determined by the inner energy of a body of sufficient size to apply the thermodynamic consideration to this body. It seems to be one of the fittest terms among the considerable number of temperature definitions that try to identify temperature in nanothermometry. A thermodynamic notion of temperature is related to heat exchange between two systems. The necessity to characterize a state of thermodynamic systems by some specific quantity becomes obvious. So, a notion "Thermodynamic temperature" has been introduced for this purpose. The objective measurement of temperature is possible due to the transitivity of thermodynamic equilibrium. Therefore, there is a possibility to compare the object temperatures among themselves per se contact. Temperature as a physical quantity that characterizes the internal energy of bodies is not being measured directly nowadays. All usable measuring instruments transform temperature in some other physical value that could be measured.

Nowadays try to link the term "Temperature" to basic constants of microphysics, on the one hand, and to threshold sizes of nanoparticles where this notion is still applicable, on the other hand. The special significance is bestowed to the definition of the minimum particle size to which the notion "local temperature" could be adopted, i. e. the temperature at which a part of the thermodynamic system remains in a canonical state, and the energetic distribution of electrons corresponds to the exponentially falling one-parametric function [12]. We also have moved down a similar path [13] and successfully studied the dependence of temperature measurement trueness on the ratio of linear sizes of sensors and the measured objects while they are reducing to the nano-dimensions. Anything better was not achieved yet.

- **Macro Properties and Nanoproperties, Expressed by Fundamental Physical Constants**

A Boltzmann constant consideration related only to the energy of electrons scattering in the process of collision with atoms may be incomplete and therefore not quite correct. Ignoring the process of acquiring energy by electrons to which may be involved in any fundamental physical constant except Planck's constant, the obtained model would be not quite perfect. Two conjugated sides of process combine a balanced approach to the problem of temperature arising due to the heat manifestation (in case of transmission of electric current through the substance) of the conduction electrons interacting with atoms. Therefore, the availability of the Planck's constant in the relation of the Quantum Temperature Unit becomes relevant.

At the same time, there is another, equally effective way to study the macro properties of materials through their micro- and nano properties. It is indicated in the example of QHE research [14]. Here, a passing result has turned out in establishing the link between the macro property, expressed in the value of 25812.807 557 ± 0.0040 Ω with the nanoscaled characteristics of the substance, which as have been proved are the charge of the electron and the Planck's constant. Similarly, the studies [4] have envisaged the relation of one of the major electrical unit (voltage) with the same fundamental physical constants.

While studying their phenomenology we prove the possibility of the existence of Quantum of Temperature as the manifestation of substance's nano properties due to the electron-phonon interaction. Moreover, considering the phenomenology of the mentioned effects in conjunction with the similar phenomenon of thermoelectricity, which nano properties in the form of a complex of thermoelectric effects are manifested as macro quantity - integral thermopower, we can achieve a logical conclusion regarding the metrology. On the one hand, we get minimal, barely noticeable changes (the temperature jumps ~ $10^{-11}$ K) due to electron-phonon dissipation; on the other hand, while passing, for example, $10^{8}$ electrons per 1 s, we obtain caused by an integral effect (thermo-EMF) the value, sufficient for measurement (~$10^{-3}$ K).

Let's consider below the possibility of researching contemporary measure of temperature based on fundamental constants of matter with involvement of the standard of electrical resistance based on the inverse of conductance quantum [4] as well as the standard of voltage based on the Josephson junctions [15] that can produce voltage pulses with time-integrated areas perfectly quantized in integer values of h/2e. The synthesized voltage is intrinsically accurate because it is exactly determined from the known sequence of pulses, the clock frequency, and fundamental physical constants.

Thus, we consider the investigation of the electrical resistance value that is based on Klitzing constant, and of the voltage standard on the Josephson Effect for exact frequency-to-voltage conversion, combined with the clock. As the mentioned resistance we propose to study one of widespread FET constructions, namely the CNTFET with built-in CNT [16] which has to be superconductive. Source and drain have to be manufactured from dissimilar metals that form the thermoelectric pair through the CNT. The latter, being in the superconductive state, is characterized by the resistance 25812.807 557 ± 0.0040 Ω, due to transient resistance of contacts. While studying the dissipation of electric power ($I^2R = U^2/R$) on such an electric resistance in temperature measurement area:

$$E = U^2\Delta t/R_{Kl} = I^2 R_{Kl}\Delta t = N\frac{3}{2}k_B T \qquad (2.4)$$

it was shown the opportunity to estimate the change of thermodynamic temperature T. Substituting equation by $I = \frac{\Delta Q}{\Delta t} = \frac{Ne}{\Delta t}$ ($\Delta t$ is the period), we simplified the equation to:

$$\frac{(Ne)^2 h}{(\Delta t)^2 e^2}\Delta t = N\frac{3}{2}k_B T, \qquad (2.5)$$

when the electrical current is formed per unit time by N conduction electrons that transfer the energy $\frac{3}{2}k_B T$ to the atoms. From here the temperature jump $\Delta T$ when passing current I through superconductive CNT (cooling is considered to be negligible), is defined:

$$\Delta T = \frac{2hI}{3k_B e} = \frac{2hN}{3k_B \Delta t}, K \qquad (2.6)$$

Otherwise, temperature increment caused by single-electron relaxation on phonon of superconductive CNT's junction with source/drain and due to unit time application is reduced and defined only by fundamental constants of matter (h and $k_B$); it is equal to $2h\cdot 1s./3k_B = 3.2\cdot 10^{-11}$ K. Providing power supply from Johnston junctions array, it appears a possibility to pass a particular number of electrons through the standard of electrical resistance (or rather through the carbon nanotube of FET). Normalizing the resulting characteristic to unit time, we have derived the formula for the Quantum of Thermodynamic Temperature (macro property), expressed, as should be expected, in fundamental physical constants conjugated with nano properties of the substance. Reduced to single electron-phonon dissipation per unit time, value is identified as RQUT:

$$\Delta T\Big|_{\substack{\Delta t \to 1s. \\ N \to 1}} = \frac{2h}{3k_B}\left[\frac{K}{s.}\right] \cdot 1[s.], \qquad (2.7)$$

Hence, the figures [17] regarding interrelation and inter-definition of basic SI units and the principles of the study of the mentioned units through the fundamental constants of matter are modified by the obtained results (see Fig. 2.4 and Fig. 2.5).

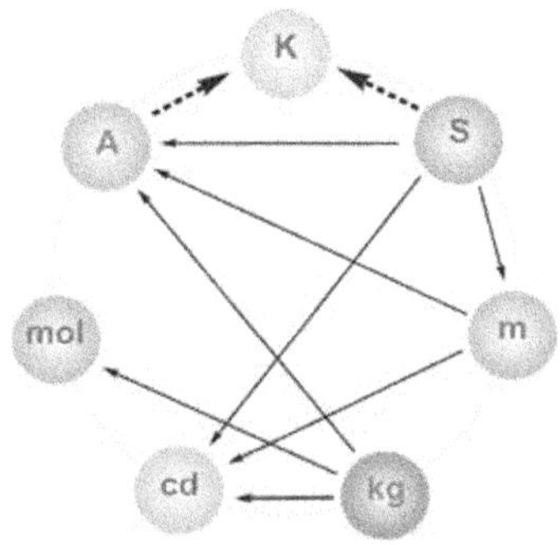


**Figure 2.4.** The interrelation and inter-definition of basic SI units: blue arrows show the revealed relationship of the studied unit *T* with the unit *I, A* (by unit *V* and unit *R*) and with unit *t, s.*

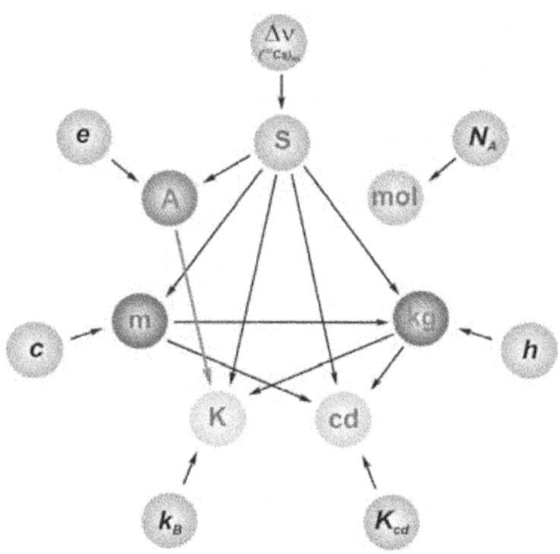


**Figure 2.5.** Principles of the SI units' study through the fundamental constants of the matter: elimination of interrelation between unit *m* and unit *T* as well as the emergence (blue arrow) of interrelation between unit *I, A,* and unit *T, K.*

In such a way the RQUIT independent of kind of matter is recommended for the creation of Temperature standard. It would be the standard based on the couple of quantum effects (von Klitzing Effect and Josephson Effect) and, having been measured against the SI system of units, has a certain value with uncertainty determined by the sum of two uncertainties [18]: of Planck's constant and of Boltzmann's constant [6] which together make its total relative uncertainty value $59.2 \cdot 10^{-8}$. The last value also includes the relative standard uncertainty of the atomic unit of time that is 5 orders of magnitude smaller ($5.9 \cdot 10^{-12}$ [6]).

- **Metrological Conception of Quantum Unit of Temperature and Possibility of Its Implementation**

The derived RQUT is equal to $3.199\ 493\ 42 \cdot 10^{-11}$ K with relative standard uncertainty $59.2 \cdot 10^{-8}$ at single electron-phonon dissipation per unit time. Note that the components of this uncertainty are determined as the combined values of the set of following appropriate methods.

For instance, to study the Planck's constant is applied the method of Watt balance, installations of studies: of X-rays crystal density, Magnetic resonance, Faraday constant, Josephson constant. CODATA 2010 recommended the value of mean Planck's constant relative uncertainty is only $u_h = 4.4 \times 10^{-8}$. Methods of the Boltzmann constant are the next. Constant $k_B$ has been determined from a measurement of the sound speed in helium gas in a quasi-spherical resonator (volume 0.5 l) maintained at a temperature close to the triple point of water (273.16 K). The acoustic velocity c is deduced from measured acoustic resonance frequencies and the dimensions of the quasi-sphere; the latter being obtained via simultaneous microwave resonance [19]. An optical (laser) method for the measurement of the Boltzmann constant which reaches an uncertainty of $2 \cdot 10^{-4}$ after a cumulative time of 61 hours is very promising [20], and other methods are considered by NIST in $k_B$ determination. So, the obtained value of the Boltzmann constant is determined as the average of several methods. Each of them is inherent in its particular systematic constituent of error/uncertainty. Therefore, obtaining the mean value is not a simple issue. Note that application the single, although the best method can contribute a component of systematic error (or unsatisfactory trueness in the uncertainty approach).

Definition of 1 A as $6.2415093 \cdot 10^{18}$e passing through a conductor's section per 1 s reveals that if an electron pump would count $10^8$ e/s or the precise ammeter would measure the electric current $6.24 \cdot 10^{-10}$ A, we receive the temperature jump: $3.2 \cdot 10^{-11}$K $\cdot$ $10^8 = 3.2 \cdot 10^{-3}$ K. At sensitivity ~43 μV/K of T-type thermocouple the measured value is ~0.14 μV. The predefined relative uncertainty $59.2 \cdot 10^{-8}$ enables to assert that this value is determined with uncertainty ~$1.9 \cdot 10^{-9}$ K.

It evidences the major advantage of the advanced temperature standard based on fundamental physical constants. From the prior known precise value of temperature jump or temperature quantum, we can come to the extremely helpful Temperature Standard that relates to the primary thermometric means.

The realization of temperature standard based on fundamental physical constants takes place on contacts of superconductive CNT, graphene, or another substance with QHE and may occur at the room temperatures [21]. For such purpose are involved the standard of electrical resistance based on the inverse of conductance quantum as well as the standard of voltage based on the Josephson junctions' array that can produce voltage pulses with time-integrated areas perfectly quantized in integer values of h/2e (Fig. 2.6). Synthesized voltage is intrinsically accurate, because it is exactly determined from the known sequence of pulses, the clock frequency f, and fundamental physical constants h and e.

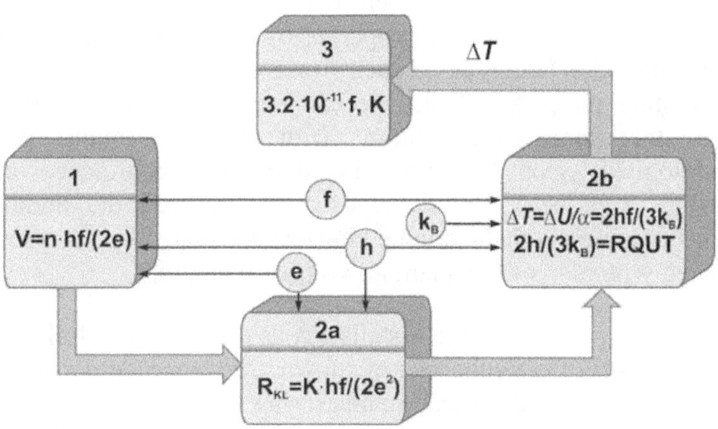


**Figure 2.6.** Scheme of Advanced Temperature Standard on Basis of RQUT and transfer of Unit Temperature to the Working Standard: 1 is the Josephson Voltage Standard; 2a is the block of quantizing the temperature based on CNT FET and simultaneously the thermocouple hot junction of which is the CNT; 2b is the block of adjustment; 3 is the Working Standard.

The mentioned, very small quantity as ΔT is quite difficult to be measured. As the standard of electrical resistance can be applied the CNTFET which source and drain have to be manufactured from two dissimilar metals (for example from Ni and Cu), that constitute the built-in thermocouple via superconductive CNT quasi-junction of ~ 0.1 μm length (Fig. 2.7).

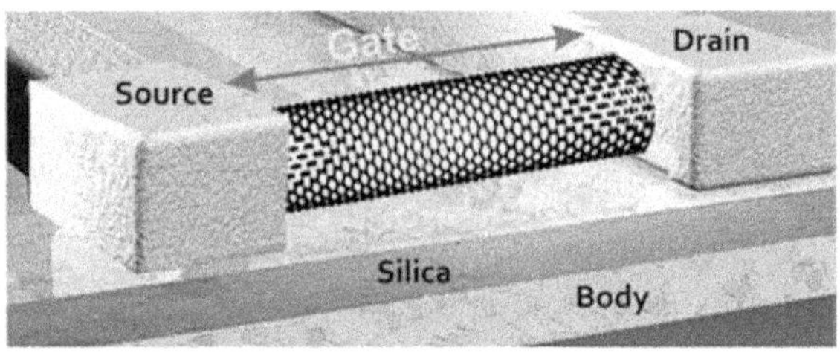


**Figure 2.7.** CNTFET with a channel width of 18 nm, with a nanotube diameter of ~ 1 nm (at a supply voltage of 0.4 V, current of 15 μA).

The source and the drain together form here a built-in thermocouple with a quasi-junction (base) from superconducting CNT of ~ 0.1 μm in length. When a nanotube is selected with superconducting properties, such a transistor is characterized by the electrical resistance of 25812.807 557 ± 0.0040 Ω due to the resistive properties of the contacts exclusively. At current I through a nanotube on such a resistance it is possible to obtain a temperature jump ΔT. Thus, we can measure by a thermoelectric method the temperature jump with the minimum possible methodological error (or with the maximum confidence in the uncertainty approach). Simultaneously is measured the number of electrons passing via the nanotube contacts controlled by the frequency measuring with help of the quantum voltage standard.

The mode of operation of the installation as a whole is as follows. The same device serves as a generator of known in advance temperature jump at the 1$^{st}$ stage and as Temperature measuring instrument at the 2$^{nd}$ stage. Firstly, the studied appliance is proposed to supply by short (~10$^{-2}$ s.) pulse voltage consequences, the effect of which is measured at the 2$^{nd}$ stage (power absence).

Further transfer of the particular temperature jump may be realized with help ordinary mean - with 2$^{nd}$ thermocouple applying with its junction located nearby the mentioned quasi-junction. At deviation of the received signal from the signal of reference thermocouple that may be caused by heat removal, the considered signal has to be powered to the required value.

Thus, in studies carried out on electrical resistance, the value of which is based on the von Klitzing constant, supplying power from an electrical standard of voltage, whose work is based on the Josephson effect, when involving a time standard, we can realize a Quantum standard of temperature.

The main application of such a standard is to improve the transfer of temperature unit size. Due to the known values of the temperature jump (when passing nano current through the CNT), which is determined with an extremely small uncertainty, unlike the classical standards, there is no need to make repeated measurements, as well as to constantly confirm the declared accuracy of studies.

Additional areas of the quantum temperature standard usage are the verification and clarification of the laws and regulations of thermoelectricity, as well as the limits of its application in the micro and nanoscale; study of the nature of thermoelectricity, including the effect of phonon drag with its influence on the Seebeck coefficient and the applied aspects of the further development of the model [22] of eddy currents in thermoelectricity.

- **I-T Converting Element and Elastic Strain Engineering**

I-T converting element is built on the FET, which gate is made from superconducting. The technology of the I-T converting element is complicated and provides coppering (of another similar) process for nanotubes free ends. Impacts of the I-T element's manufacturing defects on the quality of subsequent operations, which causes the emergence phenomenon, can be significant and essentially different. For their analysis, the classical theory of Markov processes turns out to be not quite suitable.

The technology of the I-T converting element is complicated and provides coppering (of another similar) process for nanotubes free ends. Impacts of the I-T element's manufacturing defects on the quality of subsequent operations, which causes the emergence phenomenon, can be significant and essentially different. For their analysis, the classical theory of Markov processes turns out to be not quite suitable.

Methods of design, manufacture, and operation simulation stages of similar radio-electronic elements enable to conduct formalized issue for solving the problem of optimizing parameters behind two original stochastic optimization models. Application of one of them as a target function, and the second - as constraint, opens up the possibility to solve optimization tasks, particularly in terms of providing the highest possible quality at acceptable costs, depending on the abovementioned purpose of I-T converting element and other conditions.

So, operating conditions of this element are complicated due to abruptly rising elastic strains caused the current passing via it and jumping the temperature in sections A and B (Fig. 2.1) that is, in places of CNT attachment.

- **Elastic Mechanical Strains Engineering and Nanoscale**

The elastic strains engineering can significantly improve the defining characteristics of the created standard considered in this chapter. It has revealed its effectivity while studying rather close technical mean yielding in the main unit of the mentioned standard, that is, FET [23]. Its achievements are the synthesis of nanostructures with embedded elastic effects; application of force and measurement of consequences of its action; study of mechanisms of energy dissipation; predicting the effects of these effects on specific physical properties (in this case, thermo-EMF). Although mechanical failure is a consequence of deformation to be avoided, elastic deformation can have a positive effect on substance properties. The effect of elastic deformation becomes more apparent at small sizes since micro / nanoscale materials and structures can withstand exceptionally high elastic deformations before failure. Studies of elastically-loaded nanotubes and nanosheets have shown the following: 1) Nanotubes can support significant elastic deformations, and their bending modulus increases exponentially as the nanotube diameter decreased; 2) The strains modify the electronic structure of semiconductor nano/microprobes, causing the transition of the metal-insulator at room temperature and effectively converting mechanical energy into electrical energy.

The plastic deformation is absent in the nanoscale due to the proximity of the interface surfaces to the location of possible packaging defects (dislocations, nanopores, and nanowires). As a consequence, there is only elastic deformation. The latter has 3 linear deformation components and 3 torsional deformation components. Linear deformation is with +

and -. Torsional deformations are left and right. In other words, 12 types of deformations can be calculated and reproduced on the one hand, and used for dosed and reproducible effects on thermo-EMF while measuring the temperature in the vicinity of CNT attachments. This is the content of the engineering of elastic strains for nanoscale objects of the proposed I-T converting element, which can be represented as in particular as thermocouple with thermoelectrodes.

We have studied previously the influence of elastic deformation (within the limits of up to 1%) on thermo-EMF of different bulk metals and revealed that impact is reversible (Fig. 2.8):

$$\Delta U_{\sigma} = \int_{T} \alpha_{\sigma} dT = -\frac{\varepsilon^2}{2em} \int_{T} \nabla_T E_U (T) dT \qquad (2.8)$$

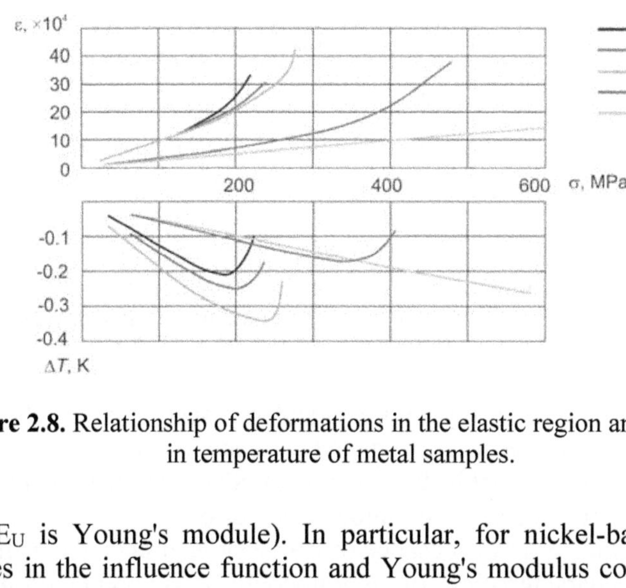


**Figure 2.8.** Relationship of deformations in the elastic region and changes in temperature of metal samples.

(here $E_U$ is Young's module). In particular, for nickel-based alloys, changes in the influence function and Young's modulus correlate with the coefficient K = 0.598. At the same time, the temperature changes of the samples were studied and shown in the lower box of Fig. 2.8.

The strains were accessed taking in account that the needed current 1 nA passing through superconducting CNT and dissipating in places of CNT attachment, provides the temperature jump ~ 1.0 K. When the temperature coefficients of linear expansion of the electrode materials (Ni; Cu) are respectively $13 \cdot 10^{-6}$ $K^{-1}$ and $16.5 \cdot 10^{-6}$ $K^{-1}$, then, at a temperature jump, there arise the thermo structural mechanical stresses.

They are caused by the differences of their expansions to the silicon substrate ($5 \cdot 10^{-6}$ $K^{-1}$) on which they are fixed (estimation for nanomaterials have been made in [24]). This can lead to additional uncertainty, which should be taken into account especially for the exact temperature standard.

Additional areas of the quantum temperature standard usage are the verification and clarification of the laws and regulations of thermoelectricity, as well as the limits of its application in the micro and nanoscale; study of the nature of thermoelectricity, including the effect of phonon drag with its influence on the Seebeck coefficient, etc. According to the 2nd law of thermodynamics (mid. of 19th cent.), Peltier heat is converted into electrical energy with the maximum possible, within the thermodynamic theory of Thomson, the efficiency of the Carnot cycle. Hence, we obtain the 1st Thomson thermoelectric ratio ($\pi = \alpha T$), which relates the coefficients of 2 thermoelectric phenomena: Peltier and Seebeck. The 2nd Thomson thermoelectric ratio ($\tau_A - \tau_B = T\, d\alpha/dT$), based on the consideration of the thermal energy balance in the thermoelectric circle, substantiated the appearance of Thomson coefficients of each of the 2 thermoelectrodes ($\tau_A$; $\tau_B$) of the considered circle. It has manifested 1st theoretically predicted thermoelectric effect, in contrast to the experimentally verified Seebeck and Peltier effects. In 1853, Thomson tested his prediction concerning his thermodynamic theory of thermoelectricity; he measured Thomson's heat with thermometers. In 1867, F. Le Roix repeated the above experiment, replacing thermometers with thermocouples, and confirmed Thomson's results [25].

With the Quantum Temperature Standard, such studies should be repeated once more, especially since they relate to nanoscale effects and are carried out on nanoobjects (nanotubes). Since Thomson deliberately realized his discovery of thermoelectric effect within the classical thermodynamics or in the absence of irreversible effects, the drift of the calibration characteristics of thermocouples is predetermined by irreversible factors of influence in the approximation of the thermodynamics of irreversible processes. So, we have an interest in the Quantum Temperature Standard appliance. While studying the Peltier and Thomson coefficients at the nanoscale with help of Quantum Voltage Standard, the I-T converting element on CNTFET structure with the source and the drain are made of dissimilar materials A and B, is recommended. Then for the previously calibrated thermocouple A / B, when carrying out the experiments, rely on the value of its thermo-EMF.

By slightly changing the CNTFET structure and making the source and drain from the same material, it becomes possible to study the low-temperature effect of phonon drag on thermopower. The latter occurs according to [26] at very small diameters (1.0... 10 nm, which corresponds to the diameter of the CNT) of the conductive material in the area of its contact with other, more massive material.

Of particular interest is the study of deviations in the temperature rises from the calculated values measured thermoelectrically within the proposed method. Otherwise, their dependence on some impact factors contributes to the identification of peculiarities of forming not only thermo-EMF at the micro- and nanoscale, as also the refinement of ambiguous statistical-thermodynamic interpretation of temperature in nanotechnologies.

## 2.2. Traditional Standards of Physical Quantities

Almost all sectors of the economy, science, and technology widely use machines, vehicles, and equipment operation of which is based on ultrasound. Ultrasonic devices can detect objects and measure distances in water and air. Ultrasound imaging or sonography is often used in medicine. Non-destructive testing of products and structures is quite an important area [27]. Industrially, ultrasound is used for cleaning, mixing, and accelerating chemical processes. The role of ultrasound for medicine and its technologies is noted. Medical ultrasound technologies have led to the creation of sophisticated ultrasound medical equipment, in particular diagnostic, therapeutic, and surgical.

However, for any type of ultrasound application, it is important to determine and normalize its intensity level. Note that no precise metrology equipment for ultrasound devices has been created so far. The guaranteed uncertainty of ultrasound measurement, established by several leading metrological centers in the world, exceeds 5-10%. Let's compare with the uncertainty of modern quantum and not only quantum standards, where it is $10^{-4}$ - $10^{-14}$ (the latter value is inherent in the frequency measurements).

### 2.2.1. Standard of Unit of Ultrasonic Pressure in Aqueous Medium

It should be noted that filling in this point of the section applies precisely to ensure the metrologically reliable operation of different types of ultrasonic equipment through the development and implementation of standards for determining sound parameters and in different environments - in water and the air. This is due to its scope (for medicine, mostly working bodies are bodies that are close to the liquid in the structure; for technology - both liquids (echo sounders) and solids).

UME should be safe for the patients and the care staff and simultaneously be effective in diagnosis and treatment. When using ultrasound in medicine, it is necessary to take into account the physical effects that occur in the tissues of the human body and affect safety. These are effects such as the heating and destruction of human tissues that occur at high levels of ultrasonic pressure therapy. The parameters of the acoustic output of the UME should be monitored both at the stage of its implementation, while assessing compliance with the requirements of the technical regulation for medical devices, at while its operation. In such a manner, the efficiency and safety of UME are determined. The list of parameters of the acoustic output of UME is standardized by the International Electrotechnical Commission. In particular, the parameters of ultrasonic diagnostic equipment are normalized per IEC 61157 [28] and the parameters of therapeutic equipment – per IEC 61689 [29].

The basis of the measurement of the parameters of the UME is the measurement of ultrasonic pressure in the patch of the emitter/transducer in the aqueous medium [28-29]. By measuring ultrasonic pressure, it can be determined other parameters of ultrasonic irradiation such as its intensity and power. Needle and membrane hydrophones are used to measure ultrasonic pressure.

An indispensable condition for the effective use of ultrasound is a sufficient level of accuracy of measurement of its parameters and traceability of the measurement results concerning SI unit standards. Until 2017, there were neither primary nor secondary standards of the ultrasonic pressure unit in Ukraine that have formed the drawbacks of the state system of technical supervision. It is fair to admit that acoustic standards are only available in a few countries (USA, Germany, Denmark, China, Japan, Poland, RF).

- **Method of Reproducing a Unit of Ultrasonic Pressure in an Aqueous Medium**

The reproduction of ultrasonic pressure in the aqueous medium is carried out by determining the sensitivity of the hydrophone (coefficient of conversion of ultrasonic pressure into voltage). Known absolute methods of calibration of hydrophones were considered and analyzed. In particular, such methods as the classical reciprocity method and its modifications, the method of optical interference, the method using the phenomenon of distortion of the sound wave during its propagation in water, the method of planar scanning of the ultrasonic beam cross-section with known acoustic power [30, 31]. According to the results of the analysis, one of the methods was recommended for the calibration of hydrophones in the ultrasonic frequency range per the IEC 62127-2 [32] was chosen for further research. That is the method of reciprocity with two converters.

The method of reciprocity with two transducers is that the hydrophone is calibrated in a known acoustic field created by an auxiliary mutual ultrasonic transducer. The latter is fulfilled in the form of a flat circular active surface able to generate an acoustic field that is quite similar to the field previously computed for a flat piston source. Also, the auxiliary converter must be reciprocal, i.e. linear, passive, and inverted [31]. The inverted converter can operate as a hydrophone and as an emitter of the ultrasonic acoustic signal. The main characteristic of the hydrophone is its sensitivity in the free field that is, the ratio of the idling voltage of the hydrophone to the sound pressure in the free field of a plane wave [32]. The main characteristic of an emitter is its "electrical current" transfer characteristic describing the ratio of sound pressure in the sound wave to the current at the inlet of the emitter. For any system in which a reciprocal converter operates as a hydrophone and in the next step as an emitter, the ratio of its free-field sensitivity to its current-transfer characteristic is equal to the reciprocity factor. For flat sound waves, this factor is a function of the effective surface area of the auxiliary transducer and the acoustic resistance of the medium (water) [33].

The determination of the hydrophone sensitivity by the reciprocity method with two transducers occurs in two steps. In the first one (see Fig. 2.9), the transfer characteristic of the auxiliary UC is determined by the self-reciprocity method.

The acoustic signal emitted by the auxiliary transducer is reflected from the unit R, located perpendicular to the radiation axis; then it enters the UC, which is now in acoustic reception mode. At the output of the auxiliary transducer is measured the voltage U1 generated by the reflected ultrasonic signal and the current I1 that passes through the transducer during irradiation. In the second measurement step, the hydrophone is placed in the acoustic field of the auxiliary UC. The excitation current of the UC at the first and the second steps of the measurement remains unchanged and is equal to I1. Then is measured the voltage U2 arising at the output of the hydrophone under the action of the auxiliary transducer of ultrasonic vibrations. During measurements, the hydrophone must be in a free field, i.e. in a sound field where the impact of the reflected signals is negligible. The hydrophone shall be positioned at such a measuring distance from the surface of the auxiliary transducer, which is three times bigger the length of the near-field of the auxiliary converter.

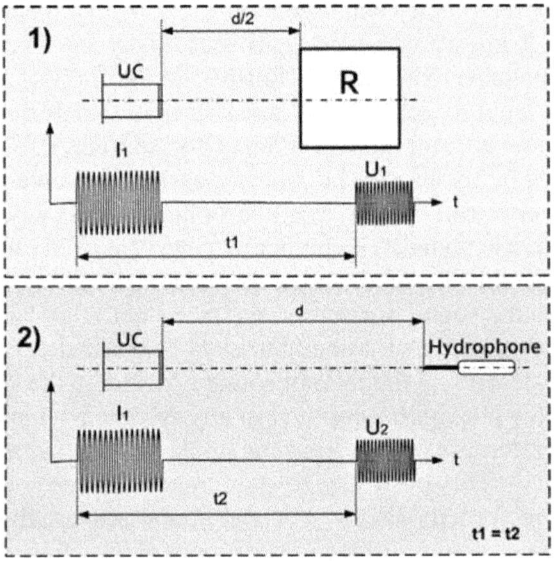


**Figure 2.9.** Determination of the hydrophone sensitivity by the method of reciprocity with two transducers: R is the reflector, UC is an auxiliary ultrasound converter; t1, t2 are the durations between the emitted and received signal at the first and the second measurement steps that are proportional to the measuring distance d.

The measuring distance d between the latter and the hydrophone is calculated for each measurement frequency by the formula:

$$d = 3 \cdot N = 3 \cdot \frac{a^2}{\lambda}, \qquad (2.9)$$

where N is the length of auxiliary converter near-field, m; λ is the sound wavelength in the water at the measurement frequency, m; a is the effective radius of the auxiliary converter at the measurement frequency,

m. The distance passed by the ultrasound beam in the first and second stages must be the same. So, the distance between the UC and the reflector in the first stage must be ½ of the calculated value of the measuring distance d, so that the length of the path of the ultrasonic beam is to the reflector and back, was equal to d. The measuring distance is set by measuring the delay time of the received signal relative to the emitted one.

The sensitivity of the hydrophone is determined by the formula:

$$M = \frac{U_2}{\sqrt{U_1 k_{U1} I_1}} \cdot \sqrt{\frac{2\pi a^2 G_1}{\rho c r}} \cdot \frac{e^{\alpha d}}{G_2}, \qquad (2.10)$$

where U1 is the output voltage of the auxiliary converter in the receiving mode, V; U2 is the voltage at the output of the hydrophone, V; I1 is the current passing through the radiating auxiliary converter, A; kU1 is a factor for voltage U1 that reduced it to the equivalent voltage of open circuit (idle voltage); G1 is the adjusting factor that takes into account diffraction losses while defining the sensitivity of the hydrophone in the acoustic field of the auxiliary converter by the method of self-reciprocity; α is the attenuation coefficient of ultrasonic amplitude in pure degassed water (depends on water temperature and signal frequency), 1/m; f is the frequency, Hz; d is the measured distance (length of the acoustic path between the UC and the hydrophone), m; r is the reflection coefficient for the reflector-water boundary equal to 0.9681 [30]; a is the radius of the auxiliary converter, m; ρ is the density of water, kg/m$^3$; c is the speed of sound in water, m/s.

Measurements of electrical signals U1, U2, I1 are carried out in pulse mode using a digital oscilloscope equipped with a current probe. The minimum number of periods of a sine wave in a pulse is calculated by:

$$n = \frac{D}{\lambda} \qquad (2.11)$$

Here D is the diameter of the auxiliary converter, m. Measurements should be made if the acoustic axes of the emitter and the hydrophone coincide revealing the maximum voltage signal at the hydrophone output. The voltage at the UC output generated by the reflected signal must be measured by setting the UC radiating surface parallel to the surface of the reflector.

- **Design of the Standard of Unit of Ultrasonic Pressure in an Aqueous Medium**

To implement the chosen method of reproducing the unit of ultrasonic pressure, a functional diagram of the standard presented in Fig. 2.10 was developed.

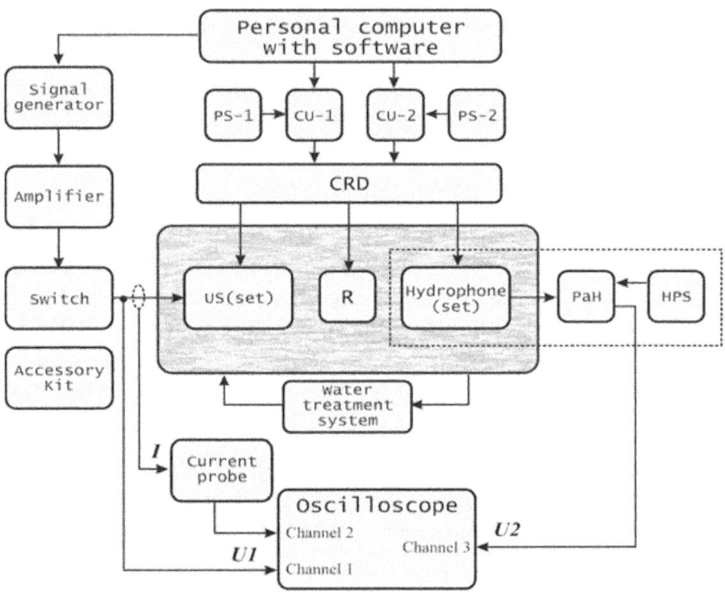


**Figure 2.10.** Scheme of the standard: CRD is the coordinate-rotary device of the positioning of hydrophones; PS1, PS2 are the power supplies; CU1 and CU2 are the control units; HPS is the hydrophone power supply; PaH is the pre-amplifier of the hydrophone; UT is the ultrasonic transducer; R is the reflector.

The standard yields: coordinate rotary-positioning device of hydrophones CRD, which provides the locating the hydrophone, auxiliary ultrasonic transducer and reflector at specified distances from each other in the measuring shell and the mutual orientation of the transducer and hydrophone to align the acoustic axes during measurements; emitting tract, which consists of a signal generator, power amplifier, and auxiliary ultrasonic transducer; a measuring block that yields a hydrophone with a preamplifier and a digital oscilloscope with the current probe; a reflector for calibration of the auxiliary converter by self-reciprocity method; a PC with special software for managing CRD and calculating G2 and G1 diffraction adjustments.

A general view of the State primary standard of the ultrasonic pressure unit in the aqueous medium is shown in Fig. 2.11.

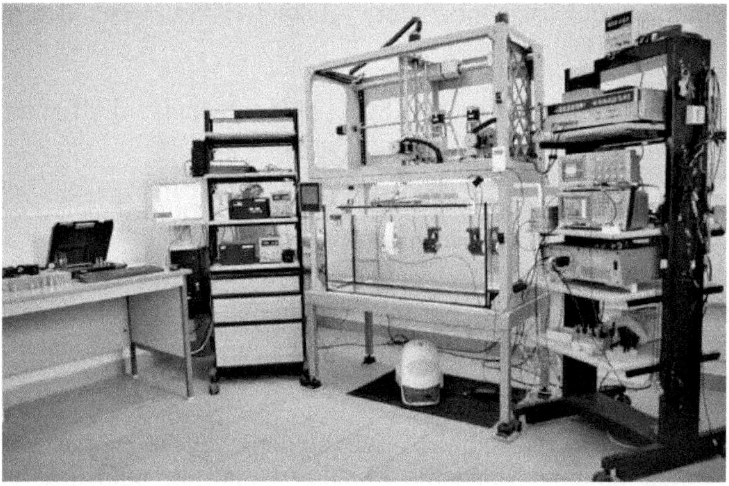

**Figure 2.11.** State primary standard for the unit of ultrasonic pressure in the aqueous medium.

Specially prepared water must be used to ensure the measurement of ultrasonic pressure in the aqueous medium. It should have the following parameters: the oxygen content of the water should not exceed 4 mg/l, the electric conductivity of the water should be at the level lower than 500 μS/m. Therefore, a water treatment system has been created that allows obtaining water with the specified parameters through its purification and evacuation. The mentioned system is included in the Standard.

- **Metrological Characteristics of the Standard**

Experimental studies of the Standard were conducted following the developed methodology. The sensitivity of the needle hydrophones included in the standard was determined during the study. Electrical parameters were directly measured, in particular the output voltage U1 of the auxiliary converter in the receiving mode, the output voltage U2 of the hydrophone, and the current I1 passing through the auxiliary converter in the radiation mode. Specially developed software was applied for computing the adjustments G1 and G2 on diffraction loss as well as the ultrasonic damping factor. Other values included in the hydrophone sensitivity calculation formula were taken from reference books. During the studies, environmental and measurement conditions, including the water temperature and conductivity, oxygen content in it, were monitored. The uncertainty characteristics were determined according to [34].

The study established that the State primary standard of ultrasonic pressure unit has the following characteristics:
- Operating frequency range from 0.5 MHz to 10 MHz;
- The range of pressure values in which the unit of ultrasonic pressure is reproduced is from 10 kPa to 100 kPa;
- Extended uncertainty U does not exceed 18.0 % (coverage factor k = 2 and confidence probability P = 0.95), including standard uncertainty of type A: $u_A$ = 5.8%; of type B: $u_B$ = 6.6 %; total standard uncertainty: $u_C$ = 8.8 %.

Comparison of the metrological characteristics of the created standard with similar national standards of other countries, published on the website of the International Bureau of Weights and Weights [35] (Table 2.1), testifies that the Ukrainian state primary standard of ultrasonic pressure in the aquatic medium corresponds to the level of other countries national standards.

In 2015-2017, the State primary standard of the unit of ultrasonic pressure in the aquatic medium has been was created, which have made it possible to ensure in Ukraine the unity and traceability of measurements of the ultrasonic pressure in the aquatic medium to SI units; its metrological characteristics have been investigated.

**Table 2.1.** Comparison of metrological characteristics of the Ukrainian state standard with similar national standards of other countries.

| **Country** | **National metrological institute** | **Frequency range, MHz** | **Extended uncertainty, %** $U$ ($k$ = 2) |
|---|---|---|---|
| Germany | PTB | 0.5 … 15.0 | 10 |
| UK | NPL | 1.0 … 20.0 | 11 |
| RF | VNIIFTRI | 0.5 … 10.0 | 20 |
| Ukraine | SE R&DI “System” | 0.5…10.0 | 18 |

### 2.2.2. One-Inch Laboratory Microphone as the Carrier-Standard Measuring Instrument. Key Regional Comparison COOMET.AUV.A-K5

Complex in performance and approach, ultrasonic standards are still classified as classic. They have not yet reached the contemporary revolution of quantization in standards' design. Although the digitization, improvement, and automation are still continuous, this testimony is the aforementioned issue of the previous point devoted to the Standard of the ultrasonic unit in the aquatic medium.

Therefore, a CCAUV of the International Bureau of Weights (BIPM) pays great attention to providing unity and traceability in the field of acoustic measurements in the world. The manifestation of this is the conduct of five international key comparisons since 2000 and three subsequent comparisons provided for in the CCAUV Strategic Action Plan for 2017-2027 [36].

The most recent international comparison, 2010-2014, was conducted under the auspices of CCAUV.A-K5 [37]. It concerned the pressure calibration of one-inch laboratory reference microphones in the frequency range from 2 Hz to 10 kHz. According to the obtained results, a new key comparison reference value (KCRV), which is, in fact, the international virtual standard of the unit of sound pressure in the air medium, has been obtained.

Thus, for the National Metrology Institute of Ukraine in the field of acoustics, there was a need to participate in the following key comparison to join the new KCRV value and to update the previously declared measurement and calibration capabilities of Ukraine in this field (SMS lines in KCDB BIPM database).

This prompted bilateral key regional comparison with the relevant Institute of Poland (GUM), which participated in the CCAUV.A-K5 comparison, which could perform the function of attaching the results to the KCRV value and assessing the equivalence of the SE R&DI "System" results to other CCAUV.A-K 5participants. This new comparison was registered with CCAUV BIPM under the name COOMET.AUV.A-K.5 [38].

- **Validation as a Path to International Recognition of Ukraine's Acoustic Measurement and Calibration Capabilities**

To achieve the above goal, a metrological program has been described and implemented. Within COOMET.AUV.A-K.5, National Metrology Institute of Ukraine has considered a one-inch standard laboratory microphone of LS1p type through verification (jointly with the GUM linking laboratory, Poland) for implementation to the international virtual standard of the unit of sound pressure in the KCRV air environment.

Before comparison was developed, approved, and registered CCAUV technical protocol [39], the requirements of which generally meet the requirements of the similar verification protocol CCAUV.A-K5. According to this protocol, each participant must calibrate one 1-inch standard laboratory microphone of LS1p type with a primary method of reciprocity as well as determine the level and phase of its pressure sensitivity for a three-octave frequency range of 2 Hz to 10 kHz. GUM Acoustic Laboratory, Poland was selected as the pilot laboratory of this study.

The calibration of the one-inch standard laboratory microphone of LS1p type, which acted as the carrier-standard in this comparison, was carried out in the acoustic laboratory of SE "System" at the State primary standard of the unit of sound pressure in the air medium DETU 10-01-11 (Ukrainian) following the requirements of [39]. The following operations were performed during calibration: a) the depth of the frontal volume of the microphone of the carrier standard is determined as also the complex electrical transmission impedance of the microphones pairs participated in the measurements; b) the resonant frequency of the microphones is defined; c) the frontal and equivalent microphone volumes are defined; d) the level and phase of the pressure sensitivity of the microphone of the carrier-standard are calculated; e) measurement uncertainty was accessed during calibration of reference microphones.

It is established that the dominant components in the total measurement uncertainty while calibration of 1-inch reference laboratory microphones of LS1p type by the primary method of reciprocity on the "DETU" 10-01-11 standard is uncertainties for determining the value of the reference capacity, frontal and equivalent microphone volumes, cross interferences, and the proper electrical noise of the measurement channels of the standard, as well as the permissible repeatability of measurements.

Since the calibration capabilities of an aforesaid type of microphone have been confirmed during the mentioned comparison, the uncertainty of type A that is usually determined by statistical analysis of the results of repeated observations [40], in our case was adopted the acceptable repeatability of the measurement results, which is a statistical characteristic of the calibration these microphones with help of the State standard.

- **Results of Determining the Sensitivity of the Carrier-Standard by the Members of the Comparison**

The calibration results of both laboratories participating in the comparison are presented in Table 2.2.

The procedure for attaching the results of the comparison of COOMET.AUV.A-K5 to the results obtained during the procedure was based on the methodology given in [41-42]. Namely, the correction Δ was calculated as the difference between the results obtained by GUM in CCAUV.A-K5 and COOMET.AUV.A-K5 comparisons. It was used to convert the results of SE R&DI "System" into COOMET.AUV.A-K5 data by directly comparing them with the results of CCAUV.A-K5. The values of this amendment and the corresponding uncertainty were computed:

$$\Delta = x_{PL} - y_{PL}, \tag{2.12}$$

Here $\Delta$ is an amendment, which is calculated for each frequency as the difference between the results obtained during GUM CCAUV.A-K5 and COOMET.AUV.A-K5; $x_{PL}$, $y_{PL}$ are the GUM results obtained during CCAUV.A-K5 and COOMET.AUV.A-K5, respectively; $u(x_{PL})$, $u(y_{PL})$ are standard uncertainties stated by GUM during CCAUV.A-K5 and COOMET.AUV.A-K5, respectively; $U(x_{PL})$ is an extended uncertainty with coverage factor k = 2, stated by GUM; $u_A(x_{PL})$ is the standard uncertainty of A-type, stated by GUM; $u(\Delta)$ is the standard uncertainty of amendment.

**Table 2.2.** The results of the calibration of the carrier-standard with the respective stated expanded uncertainty of measurement represented by the comparison participants.

| Frequency, Hz | GUM, Poland | | | | SE R&DI "System", Ukraine | | | |
|---|---|---|---|---|---|---|---|---|
| | Sensitivity | Expanded uncertainty of the level of sensitivity measurement (k=2) | Phase of sensitivity | Expanded uncertainty of phase of sensitivity measurement (k=2) | Phase of sensitivity | Expanded uncertainty of the level of sensitivity measurement (k=2) | Phase of sensitivity | Expanded uncertainty of the level of sensitivity measurement (k=2) |
| | dB rel. 1 V / Pa | | degree | | dB rel. 1 V / Pa | | degree | |
| 1.995 | -26.18 | 0.27 | 176.08 | 2.8 | -26.348 | 0.09 | 176.4 | 2.5 |
| 2.512 | -26.47 | 0.27 | 176.68 | 2.8 | -26.465 | 0.085 | 176.5 | 2.2 |
| 3.162 | -26.50 | 0.27 | 176.39 | 2.8 | -26.579 | 008 | 176.6 | 1.9 |
| 3.981 | -26.61 | 0.13 | 176.55 | 2.2 | -26.677 | 0.075 | 176.7 | 1.6 |
| 5.012 | -26.71 | 0.13 | 176.71 | 2.2 | -26.768 | 0.07 | 177.1 | 1.3 |
| 6.310 | -26.81 | 0.13 | 176.85 | 2.2 | -26.846 | 0.065 | 177.4 | 1.0 |
| 7.943 | -26.88 | 0.12 | 177.10 | 1.9 | -26.912 | 0.06 | 177.6 | 0.7 |
| 10.000 | -26.95 | 0.12 | 177.38 | 1.9 | -26.966 | 0.055 | 177.8 | 0.6 |
| 12.589 | -27.00 | 0.12 | 177.66 | 1.9 | -27.012 | 0.05 | 178.0 | 0.5 |
| 15.849 | -27.05 | 0.12 | 177.89 | 1.7 | -27.050 | 0.045 | 178.2 | 0.4 |
| 19.953 | -27.07 | 0.12 | 178.15 | 1.5 | -27.078 | 0.04 | 178.4 | 0.4 |
| 25.119 | -27.10 | 0.12 | 178.31 | 1.5 | -27.104 | 0.035 | 178.5 | 0.4 |
| 31.623 | -27.13 | 0.03 | 178.46 | 1.5 | -27.126 | 003 | 178.6 | 0.4 |
| 39.811 | -27.15 | 0.03 | 178.55 | 1.5 | -27.144 | 0.03 | 178.7 | 0.4 |
| 50.119 | -27.17 | 0.03 | 178.62 | 1.5 | -27.161 | 0.03 | 178.8 | 0.4 |
| 63.096 | -27.19 | 0.03 | 178.68 | 1.1 | -27.174 | 0.03 | 178.8 | 0.4 |
| 79.433 | -27.20 | 0.03 | 178.67 | 1.1 | -27.186 | 0.03 | 178.8 | 0.4 |
| 100.00 | -27.21 | 0.03 | 178.62 | 1.1 | -27.197 | 0.03 | 178.7 | 0.4 |
| 125.89 | -27.23 | 0.03 | 178.54 | 0.9 | -27.207 | 0.03 | 178.6 | 0.4 |
| 158.49 | -27.23 | 0.03 | 178.41 | 0.9 | -27.214 | 0.03 | 178.5 | 0.4 |
| 199.53 | -27.24 | 0.03 | 178.20 | 0.9 | -27.220 | 0.03 | 178.2 | 0.4 |
| 251.19 | -27.25 | 0.03 | 177.91 | 0.9 | -27.225 | 0.03 | 177.9 | 0.4 |
| 316.23 | -27.25 | 0.03 | 177.55 | 0.9 | -27.229 | 0.03 | 177.6 | 0.4 |
| 398.11 | -27.26 | 0.03 | 177.03 | 0.9 | -27.231 | 0.03 | 177.1 | 0.4 |
| 501.19 | -27.26 | 0.03 | 176.43 | 0.7 | -27.229 | 0.03 | 176.4 | 0.4 |
| 630.96 | -27.25 | 0.03 | 175.64 | 0.7 | -27.224 | 0.03 | 175.5 | 0.4 |
| 794.33 | -27.24 | 0.03 | 174.66 | 0.7 | -27.213 | 0.03 | 174.5 | 0.4 |
| 1000.0 | -27.22 | 0.03 | 173.23 | 0.7 | -27.193 | 0.03 | 173.1 | 0.4 |
| 1258.9 | -27.19 | 0.03 | 171.50 | 1.1 | -27.159 | 0.03 | 171.3 | 0.4 |
| 1584.9 | -27.14 | 0.03 | 169.27 | 1.1 | -27.107 | 0.03 | 169.0 | 0.4 |
| 1995.3 | -27.05 | 0.03 | 166.35 | 1.1 | -27.023 | 0.03 | 166.1 | 0.4 |
| 2511.9 | -26.92 | 0.03 | 162.44 | 0.9 | -26.892 | 0.04 | 162.1 | 0.4 |
| 3162.3 | -26.72 | 0.04 | 157.17 | 1.0 | -26.698 | 0.05 | 156.8 | 0.4 |
| 3981.1 | -26.44 | 0.05 | 149.62 | 1.2 | -26.427 | 0.06 | 149.2 | 0.4 |
| 5011.9 | -26.13 | 0.06 | 138.25 | 1.0 | -26.125 | 0.07 | 137.9 | 0.7 |
| 6309.6 | -26.04 | 0.06 | 120.87 | 1.0 | -26.051 | 0.08 | 120.8 | 1.0 |
| 7943.3 | -26.99 | 0.06 | 96.21 | 1.2 | -27.000 | 0.09 | 96.5 | 1.3 |
| 10000 | -30.00 | 0.11 | 69.88 | 1.4 | -29.992 | 0.1 | 70.7 | 1.6 |

Unilateral DoE was defined as the difference between the transformed results SE R&DI "System", $y_{UA} + \Delta$, and KCRV, calculated at each frequency as for module of the microphone sensitivity and its phase. The corresponding expanded uncertainty (k = 2) was computed:

$$u^2(\Delta) = u^2(x_{PL}) + u^2(y_{PL}) - 2 \cdot \mathrm{cov}(x_{PL}, y_{PL}), \quad (2.13)$$

$$u^2(\Delta) = 2 \cdot u_A^2(x_{PL})), \text{ for the assumption that } u(x_{PL}) = u(y_{PL}) = \frac{U(x_{PL})}{2}, \quad (2.14)$$

$$d_{UA} = y_{UA} + \Delta - x_{ref}, \quad (2.15)$$

$$u^2(d_{UA}) = u^2(y_{UA}) + u^2(\Delta) + u^2(x_{ref}) - 2 \cdot \mathrm{cov}(\Delta, x_{ref}), \quad (2.16)$$

$$u^2(d_{UA}) = u^2(y_{UA}) + u^2(\Delta) + u^2(x_{ref}) - \frac{u^2(\Delta) \cdot u^2(x_{ref})}{u^2(x_{PL})}, \quad (2.17)$$

$$U(d_{UA}) = 2 \cdot u(d_{UA}), \quad (2.18)$$

Here $x_{ref}$ is the KCRV, defined during CCAUV.A-K5; $u^2(x_{ref})$ is the standard uncertainty of KCRV; $y_{UA}$ are the results of SE R&DI "System" obtained within comparison COOMET.AUV.A-K5; $d_{UA}$ are the DoE of the results of SE R&DI "System" regarding KCRV; $u(y_{UA})$ is the standard uncertainty stated by SE R&DI "System"; $u(d_{UA})$ is the standard uncertainty of DoE results of SE R&DI "System" regarding KCRV; $U(d_{UA})$ is the expanded (k=2) uncertainty of DoE results of SE R&DI "System".

The results of the DoE calculation of the readouts of SE R&DI "System" regarding KCRV together with the expanded uncertainty are presented in Fig. 2.12 and Fig. 2.13.

Basing on the high-mentioned results, it is established that the obtained DoEs confirm the compliance of the results of SE R&DI "System", both for determining the level of pressure sensitivity of the reference microphones and for their phase with the KCRV CCAUV.A-K5 values over the entire frequency range from 2 Hz to 10 kHz within the stated measurement uncertainty. However, while international expertise results in these comparisons there were several comments, the main of which are the following. First, the measurement uncertainty stated by the pilot laboratory (GUM) at some frequencies, both for the sensitivity level

(Fig. 2.13) and for the phase, significantly exceeds the uncertainty stated by us. So, we have modified the received results to the presented below (Fig. 2.14; Fig. 2.15) limiting only the uncertainty declared by the SE R&DI "System".

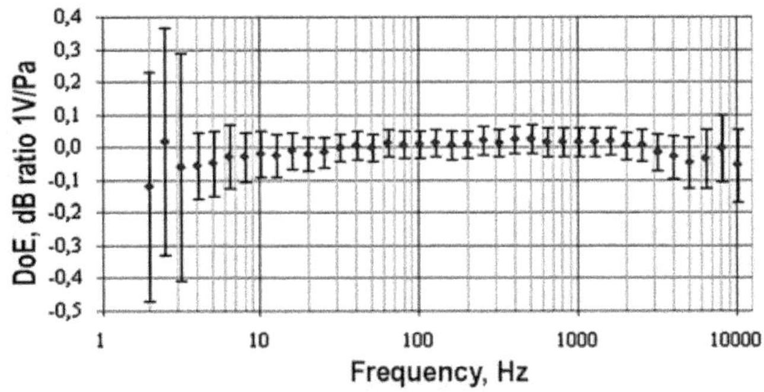


**Figure 2.12.** DoE of the measurement results of the sensitivity level of SE R&DI "System" regarding KCRV CCAUV.A-K5.

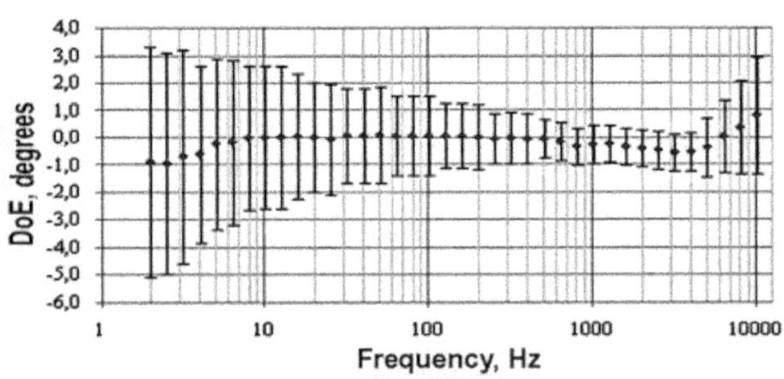


**Figure 2.13.** DoE of the measurement results of the sensitivity phase of SE R&DI "System" regarding KCRV CCAUV.A-K5.

Resultantly, there was no possibility to conclude unequivocally on the equivalence of the results of the SE System to KCRV CCAUV.A-K5 at the indicated frequencies. Therefore, on the recommendation of the experts, when submitting new SMS capabilities of Ukraine for approval in CCAUV, the uncertainty of measurements at these frequencies was increased. Also, the steady trend observed for the DoE's results of determining the sensitivity level at $f < 100$ Hz, indicates the need for further consideration. Second, the significant comment of international experts was the next one. DoE of the sensitivity level determined in the vicinity of 400 Hz, on the one hand, was quite close to the expanded measurement uncertainty of SE R&DI "System" and, on the other hand, it was the inherent rather chaotic character. In our opinion, the last can be the consequence of the quite significant value of measurement type A uncertainty in the pilot laboratory or a rather large difference in the measurement uncertainties claimed by participating laboratories that have been noted in [43]. Applying the procedure of reduction of the random component (approximation with a polynomial of the 5$^{th}$ order) to the results of the pilot laboratory, we have obtained the corrected results (Fig. 2.16) as of the "smoothed" equivalence of our previous results for the microphone sensitivity level. They are designated as DoE" received based on DoE'.

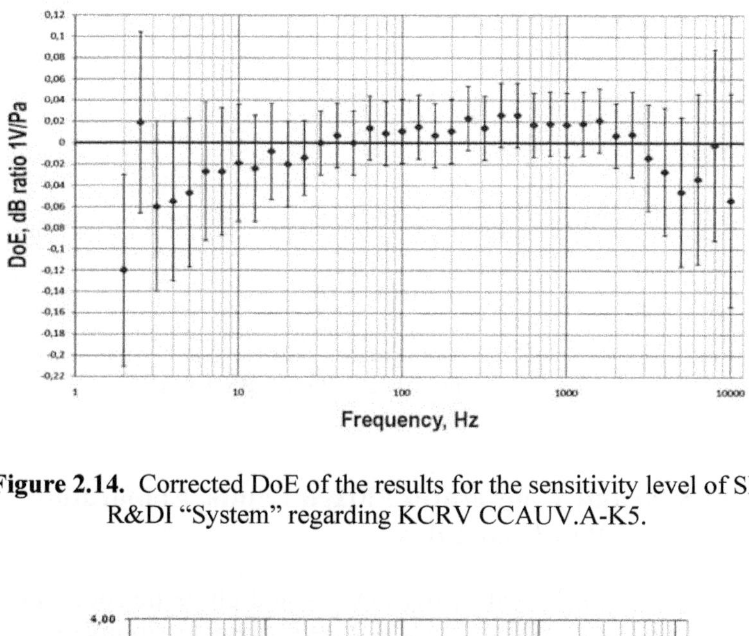


**Figure 2.14.** Corrected DoE of the results for the sensitivity level of SE R&DI "System" regarding KCRV CCAUV.A-K5.

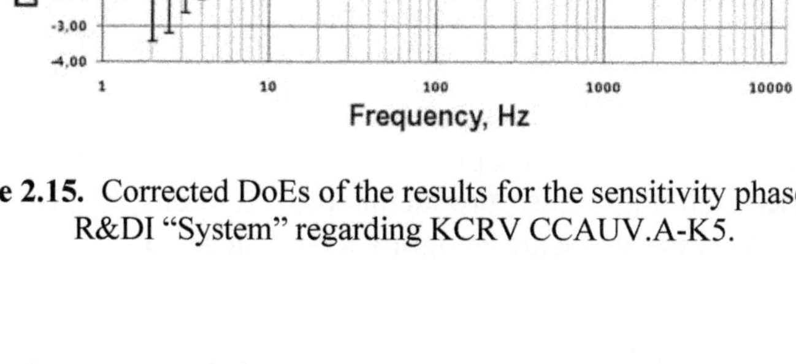


**Figure 2.15.** Corrected DoEs of the results for the sensitivity phase of SE R&DI "System" regarding KCRV CCAUV.A-K5.

So, after fulfilled approximation, the DoE "values of SE R&DI “System” ceased to be chaotic; the largest deviation from KCRV in considered frequency range did not exceed 0.021 dB (at the extended uncertainty of measurements 0.03 dB). Thus, it can be argued that the results of defining sensitivity levels in the vicinity of 400 Hz are equivalent to KCRV CCAUV.A-K5. Total, today, the results of these comparisons joined 26 countries.

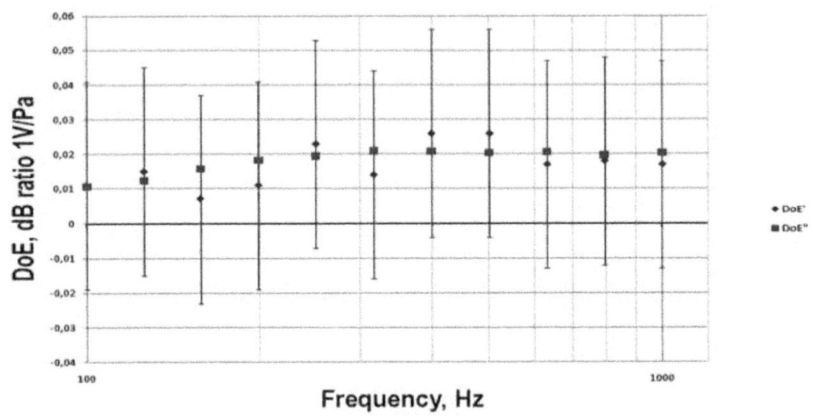


**Figure 2.16.** DoE’ and DoE’’ comparison of the results for the sensitivity level of SE R&DI “System” regarding KCRV CCAUV.A-K5.

## 2.3. ITS-90 and Temperature Fixed Points. Elaboration of Fixed-Point Assembly

The success of condensed-matter physics and phase transition physics enables us to consider that nowadays there is a real possibility to increase considerably precision, reproducibility avid coherence of monometallic fixed-point temperatures of the valid temperature scale. Besides the solution of the problem of thermodynamical temperature reproducibility precision increasing [44], the same matter of topicality is the solution to the problem of temperature scale completion with new FPTs [45]. In this regard, successful searches are focused on eutectic phase transitions.

Preference of alloy isothermal melting phenomenon usage is in the fact that such alloys are technologically suitable for reproducing of FPT being better than numerous pure metals [46]. It opens the possibility to complete a temperature scale as tightly as it is necessary.

### 2.3.1. Thermophysical Basis of Eutectic Alloys for Fived-Point Temperatures

- **Eutectic Temperature Fixed Points Problems**

It is established that the base sources of the instability of temperature reproduction and non-uniqueness of ITS-90 metal fixed points are the ambiguity of impurity concentration and nature definition as well as lack of appropriate unified technologies of crystallization phase transitions realization individual for each WS. Based on the entropy-cluster analysis, as well as radiographic and viscosimetric explorations it is shown, that eutectic melts preserve micro nonhomogeneous structure within a certain temperature interval after melting [47]. It is theoretically finished and experimentally confirmed, that due to their correlation in cluster distribution the structure of eutectic melts is sensitive to the imposing of physical fields. The methodology of estimation of the values of power influence on the melt of WS to form a fine-dispersed, even-grained, energetically-sated structure which provides the most isothermal plateau of melting is developed. The control of the formation of an optimum level of uniformity and dispersiveness of the crystalline structure of WS during its crystallization is carried out with the use of an adapted method of AE.

"ITS-90 fixed points: the reserves of thermodynamic temperature reproduction enhancing concern the valid temperature scale FPT support. Due to ITS-90 principles (regulations) and due to the evidence of numerous publications, the system analysis of the level and sources of FPT poor reproducibility is realized based on PhT heat usage: melt -- crystal and crystal-melt of pure metals. It is settled that the main factors which cause progressive in time deviation from fixed PhT temperature are: the difference of concrete thermometric characteristics as well as the difference of pure metal PhT realization methods. The first factor stipulated by the difference of national WS purity determination methods and the change of its chemical content in the exploitation process. Existing methods of metal purity determination are imperfect since they are mainly fitted only for satellite impurity control and do not take into account impurities which are being enlisted on the different stages of metallurgical processes and also at FPT fulfilling and exploitation. The $2^{nd}$ factor of the FPT method utilizing the output treatment upon the WS thermodynamic state is utterly necessary.

- **Melting / Crystallization Processes of Working Substance and Their Management**

The mechanism of mutual components' penetration at the initial stage of eutectic contact melting could be presented as substitution or penetration phase creation (a mixed mechanism is a probable one). Meanwhile, the atoms of the component in the matrix of component B are occasionally distributed on available for them knots or grating interknots. While the matrix content increasing of a component that dissolved with reaching some critical concentration distinct matrix areas are found to be "blocked" each from other by "partitions" – formed with the component that dissolves, as a result of the rupture at interatomic relations (connections) A-A (or B-B).

Specific influence on the crystallization process is made by a magnetic field. Near crystallization, clusters become commensurable with magnetic domains, and therefore, the change of magnetic perception in the process of crystal origination (conceiving) could form an essential driving force across field lines. Another mechanism of magnetic field influence is related to the braking of turbulent fluxes across the border of phase division that substantially decreases temperature fluctuations at the crystallization front. It is experimentally investigated the influence of a magnetic field on the crystallization process of melts based on Bi and Sn since at crystallization Bi change magnetic perception in 12 times, and Sn passes from a diamagnetic state to a paramagnetic one. The structure created by crystallization in a magnetic field is close to this one that is being realized under conditions of equilibrium crystallization. So, the magnetic field accelerates structural change but does not provide a necessary fine-dispersive energy saturated equal-grained structure for Bi-Sn alloys. Incomparably the most satisfactory results were reached after the melts VT. The latter was being conducted in the framework of the elaboration of technologies of FPT cell operation; there has been an increase of inner WS energy under the effect of mechanical oscillations means the raise of thermodynamic melt metastability, consequently, its crystallization normalizes and accelerates [48].

### 2.3.2. Realization of Eutectic Melting Basis Temperature Fixed Points

At the construction of eTfP cells as teflon and steel were approbated as construction materials. The high corrosion steadiness in the liquid-metal environment is inherent to the chrome-nickel steel of an austenite and ferrite class, that is why firstly the interaction of main components (Fe, Ni, Cr) and melted Gallium – the major element of low-temperature eutectics – was being studied.

In consequence of numerous explorations, the method of acoustic emission monitoring of the degree of colloidal solution equilibrium in a pre-crystallization state and correction of deviations from an equilibrium state with the help of VT were acknowledged. A degree of thermodynamic equilibrium was determined by calorimetric exploration. As well the power determination of surface tension depending on VT parameters was conducted by the methods of moistening force equilibrium. The investigations of acoustic emission ($K_p$ – parameter was chosen as a main correlation parameter of AE – radiation – see Fig. 2.17) were conducted with the double purpose: a) Determination of correlation of AE-parameters and eutectic structure formation ones; b) Determination of the duration of the residual influence of an external energetic field on a melt [47].

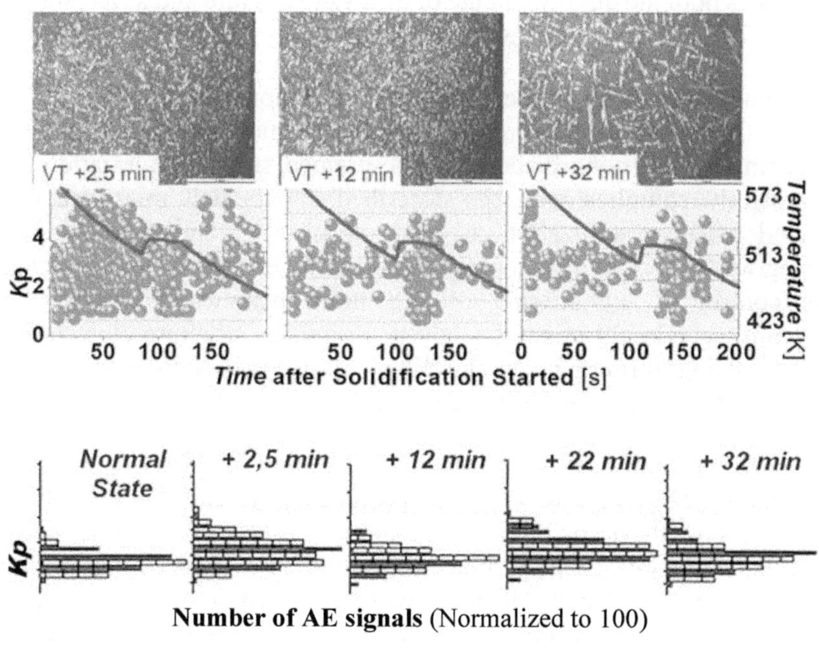


**Figure 2.17.** Time dependence of Cu-Sn eutectic melt cooling Thermograms and AE-parameters (AE process catastrophism factor $K_p$).

At the application of external nonmonotonous energetic influence on a melt, it could be possible to create the conditions which promote the shredding of structural components in the crystallization process that in turn will increase the stability of PhT temperature. Two effects provide isothermal eutectic melting in the best way that itself complies with the major requirement of FPT.

The physics basis of the AE method's usage of PhT processes is the point that they are performed non-monotonously but at the expense of the alternating changes of state clusters that form researched material. In this case, we interpret "clusters" as those material volume parts that differ from others by their energetic state. Cluster structures are presented both in solid and melted metal. Since the researched phase transition is being accompanied by substance density altering; the phase transition of a material block is being accompanied by the creation of a pressing (shock) wave, front of which is moving towards the sample boundary. This part of the wave that lies within the range 0.1÷1.5 MHz is being considered as a single AE act. At the arising of sedimentation processes (particularly, of gravitation origin) – the change of chemical content stipulates both the quality change of a cluster structure and as a result the change of AE-radiation parameters.

The main peculiarity is the complete correlation of the augmentation curve (nE and nS) during the whole duration that is persuasive confirmation of mono dispersive fine-grained structure formation and completely complies with the described above melt microstructure, gained after crystallization as a result of VT. The correlation of histogram $K_p$ form with a degree of granularity enables us to accomplish the control of the altering of WS grain parameters without cell dismantling. The 30-minutes recovery of a grain structure close to one that is being formed in an unrefined alloy enables to gain a more dispersive homogenous structure immediately after VT.

To develop non-destroying methods of the control of structural heterogeneity of eutectic WS (without cell dismantling) – the method that employs the transformation of the AE spectrum being generated during alloy melting under the condition of its content deviation from eutectic was approbated. The latter was performed on the system Ga-Sn versatile researched without the wide temperature range. The control of cooling mode was being conducted with the help of a type K thermocouple located at the wall of a cell. To arrange the correlation of AE and state diagram – dependencies gained at the output of a measured installation as the two-dimensional expressions $K_p = f(nS)$ were converted into a three-dimensional form $K_p = f(nS, Cb)$, where Cb is the concentration of the 2nd eutectic component. It is determined that at melting satisfactory correlation among nS of these values Kp and nS for maximal histogram value (max $K_p$) are taking place. Therefore, for the researched process max $K_p = f(Cb)$ was chosen as the selector curve (Fig. 2.18).

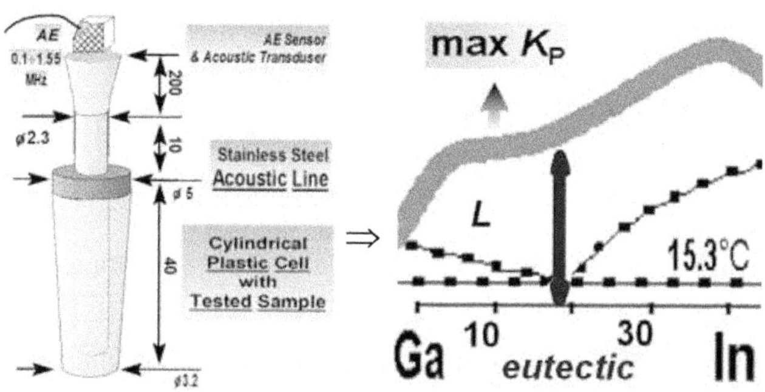


**Figure 2.18.** Eutectic materials phase-transition AE-registration. Equipment & Correlation between a liquidus curve (*L*) of a phase-state diagram and the maximal AE catastrophism factor $K_p$.

Particularly, the received dependence indicates the existence of two r1 location approximating lines. During crystallization conducting in the fixed-point cell, it was performed checking the stability of AE parameters generated by the AE unit. The periodical checking of changes of the registered signal gives information necessary for decision-making as to taking measures to enhance the homogeneity by external means of energy impact.

Within the current investigation, we have developed the mobile TfP based on the In-Ga-Sn eutectic, which could enable the conducting of the metrological checking of thermocouples and resistance thermometers of small diameter (< 3 mm). So, the spectra of recommended eutectics have revealed. Seven different alloy samples were synthesized. The research was conducted by the method of differential scaling calorimetry. Before the selection of loadings for microcalorimetric exploration accomplishing, the melts were thoroughly homogenized utilizing vibration treatment with frequency 25 Hz at a duration of 2 min. The analysis of gained results was conducted by comparison of heat flow maximum parameters "Heat flow endocep" (Fig. 2.19, enumeration of samples according to the Fig. 2.20).

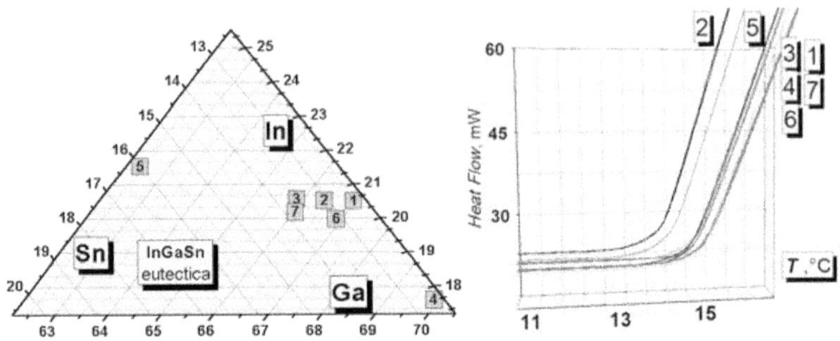


**Figure 2.19.** Heat Flow registered at the initial melting stage.

Simultaneous examination of the beginning of the phase transition of the studied samples allows us to distinguish the samples "2" and "5" and to question the samples "4" and "6".

In the Fig. 2.20 mentioned maximum parameter dependences on the concentration, sample content is presented. The analysis of PhT conventional beginning "Onset" position change; peak maximum location, and heat quantity assigned to the mass of the tested sample enable us to presume previously the possibility of "1" sample content's usage as a fixed-point one.

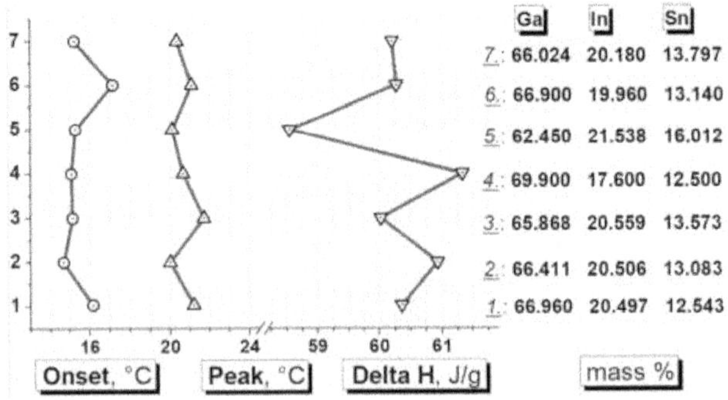


**Figure 2.20.** Heat Flow Maxima parameters received using DSC Perkin-Elmer for their Pyris™ Software analysis.

The calorimetric research of a simpler binary Ga-Sn system was previously conducted. The dependences of calorimetric curves while melting and crystallization were explored. The DSC results (Fig. 2.21) indicate the presence and importance of the correlation between the WS overheating and the parameters of heat flow while melting. Non-monotonicity of extremum that points out the furriness of the PhT structure could be explained by the presence of Ga in the alloy, which (Ga) being a half-metal changes the nature of interatomic connections and therefore is characterized by complicated transition dynamics. Overheating influence on the temperature of PhT beginning is absent. We can explain it by the mechanism of contact melting inherent in eutectics. Instead, the influence of overheating on the form of the extremum peak right part (that, particularly, correlates with fixed-point plateau stability) is quite considerable.

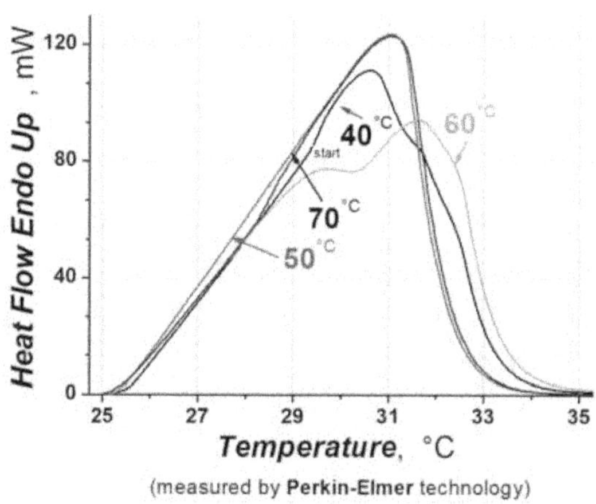


**Figure 2.21.** In-Ga-Sn melting DSC results.

Received results indicate the necessity of distinct melt thermoprocessing mode regimentation before melt crystallization, i.e. before preparing to the next cycle of operation that must be reflected in a technological instruction to the recommended temperature fixed point with eutectic alloy as working substance. A thermoprocessing mode is individual for every substance that once more proves the importance of the structural state in maintaining a stable alloy temperature.

The mobile unit of temperature fixed point with the usage of the set of WS operates within the temperature range with the upper limit 200°C defined by construction material, which is the Teflon. For temperature measuring in a fixed-point cell, the 100-Ω platinum thermometer was employed. It was verified previously at the Ga fixed point and the triple water point that provides the possibility to conduct measurement with the accuracy up to 0.0002 K. The registration of resistance change (by an H-conducting scheme) was made with help of Keithley 2700.

Melting plateau parameters are the next: $\left(PhT_{st} \times PhT_{Time} PhT_{Temp}\right) =$ $= (0.008°C \times 6hours 10.58°C)$. The temperature fixed point melting plateau with the eutectic InGaSn as a WS crystallized without previous energetic treatment and with previous melt vibration treatment is demonstrated in the Fig. 2.22.

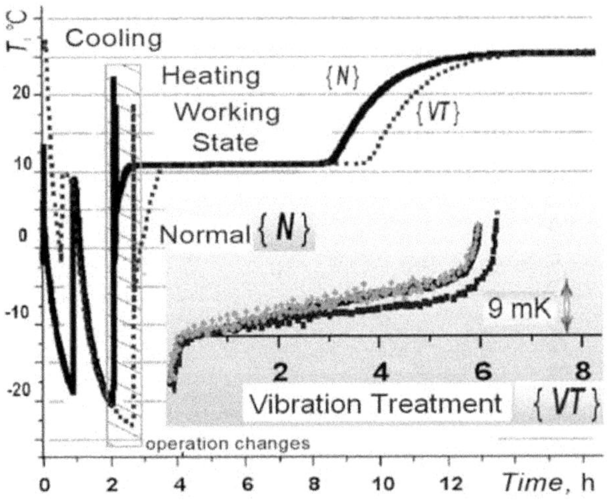


**Figure 2.22.** Thermogram and the fixed melting plateau dependency on energetic melt treatment.

The value of melting temperature 10.57 °C is adopted as the averaged result of 10 measurements. It is necessary to mark that according to deviation from plateau horizontality as a differential parameter – an energetically treated WS keeps the melting plateau isothermicity for 6 hours duration within less than 0.008 °C. Due to accuracy confining by the cycles of temperature measurements set while the thermometric bridge CA-300 and Pt100 thermometer application, the permanence in PhT realization of a melting plateau in time is adopted within limits: ±0.004 °C.

For the secondary temperature fixed points, the determined reproducibility is acceptable. Such reproducibility is achieved for following accepted conditions, which are the working substance contamination, the recommended previous overheating of considered unit, the certain speed of crystallization, and melting are compulsory for the eutectic fixed-point work cycle performance. These conditions are required to conduct for every fixed point. Another important issue seems to be creating mobile temperature fixed-points. The possibilities of considered study solving maybe perform on the eutectic In-Sn-Bi [47].

Conducted calorimetric research has demonstrated the additional structural changes nearby the melting phase transition (Fig. 2.23). The fulfilled study enables to conclude: there exists the experience of usage of PhT eutectic alloys of complicated chemical content as fixed-point working substances aiming the supplementation of the unit "FPT cell + verifying thermotransducer" by an additional device that would enable to single out in time and temperature (Fig. 2.24) the closely-placed effects with the purpose of their segregation. Such a function could be performed by the block of AE registration. It is recommended as a recorder of eutectic WS deviation from homogeneity on the cycle of crystallization of FPT. The necessity in such installations is especially caused by sensors' calibrations that can be performed at the mines, in energetics, and space.

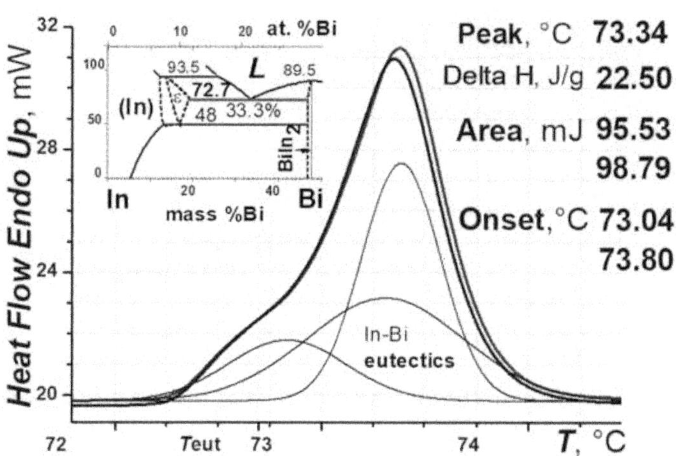


**Figure 2.23.** Test sample (In-Bi eutectic) Heat Flow registered during heating (2 passes) in the vicinity of the declared eutectic melting phase transition.

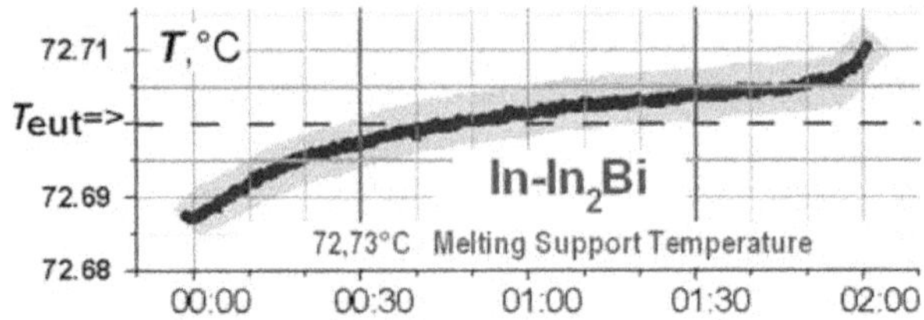


**Figure 2.24.** Received In-Bi eutectic melting plateau.

It was developed and experienced the mobile thermoelectric stabilizer (Fig. 2.25) of the boost temperature providing the FPT plateau duration. To achieve the uniform heat passing from the bottom and walls of FPT cell 1, it was installed in the densely fitted hollow in a thermo conducting block with a lid 2. A buttery 3 with four thermoelements and a resistance thermometer were put in the layer 4 of a metallic block 5 touching to the cell wall. It has provided the necessary stability of the boost temperature, that's, the temperature of overheating over the melting temperature of eutectics. The thermoelectric buttery has covered by a copper block and the radiators 6 with regulated compulsive cooling on thermoelements.

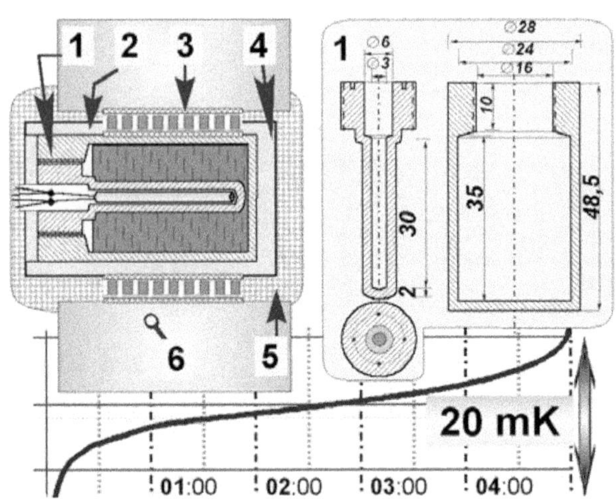


**Figure 2.25.** Section of a testing mobile cell with In-Ga-Sn eutectic melting plateau (0.02°C×5 hours 10.58°C), $T_{boost}$=11.5°C

As it was marked, the micro sedimentation processes are the considerable factor that influences on the long-term FPT cell operation stability. Their impact is checked during the crystallization of eutectic by the block of AE control. While VT, an exciting signal was given on a strain gauge installed at the cell. It produces vibrations within melt volume that stimulate "sound wind" mixing the WS. For mobile cells based on In-Ga-Sn, the duration is equal to 4 min. After VT fulfillment and registration of relevant AE-signals corresponding to the crystallization of dispersive and equally-grained WS structure, the thermometer calibration has to be conducted. The proposed cell was approbated with testing FPT cell that used the eutectic alloys In-Ga-Sn and In-$In_2Bi$ as WS.

## 2.4. Conclusions

1. Advance in technologies is impossible without temperature gauging that demands the continuous development of experimental techniques, namely in the creation of standards. Expanding the set of quantum standards of SI units towards the study of major pillars of temperature standard based on physical fundamental constants becomes possible as a result of emerged opportunities of unique electronic devices, in particular, resistance standard (based on the inverse of conductance quantum) and voltage standard (based on Josephson junctions array) combined also with cesium frequency standard.

2. Researching the foundations of the temperature standard we have derived the value of Quantum Unit of Temperature as the value of temperature jump at single electron-phonon dissipation per unit time via $h / k_B$; it is equal to $3.199\,493\,42 \cdot 10^{-11}$ K with relative standard uncertainty $59.2 \cdot 10^{-8}$. The studied quantum standard may be recommended for application as "intrinsic standard" that does not need permanently recurring measurements against the realization of the SI unit to validate its accuracy. Possibility of implementation of Quantum Temperature Standard by involving the quantum electrical resistance standard and quantum voltage standard needs the consideration of the I-T converting element as a unique electronic device which has to be significantly workload while operating. To achieve the declared exactness by Quantum Temperature Standard we have to ensure sufficient reliability of the I-T converting element by the elastic strain engineering.

3. Classic in performance ultrasonic standards have not yet reached the contemporary quantization level in world standards' design. The digitization, improvement, and automation are still continuous and

National Metrology Institute of Ukraine (jointly with the GUM linking laboratory, Poland) within COOMET.AUV.A-K.5 has verified a one-inch standard laboratory microphone of LS1p type for implementation to the international virtual standard of the unit of sound pressure in the KCRV air environment. Besides, the State's primary standard of the unit of ultrasonic pressure in the aquatic medium has been was created in Ukraine, 2015-2017, which makes it possible to ensure the unity and traceability of measurements of the ultrasonic pressure in the aquatic medium to SI units. Due to the aforementioned, the possibilities of the Consultative Committee for Acoustics, Ultrasound, and Vibration (Strategy 2017 to 2027) are strengthening continuously.

4. Eutectic alloys technologically suitable for the reproducing of ITS fixed-temperature points enable not only to widen the area in direction of higher temperatures but to complete the available temperature scale especially at the temperatures below the points of Cu, Ag, and Au. It is proved that the basic source of poor reproducibility and unity of valid ITS-80 fixed-temperature points is polysemantic of impurity concentration and nature determination. The presence of faint ordered nanosized clusters in a melt is a crucial factor of impact on crystallization. It is shown that TfP unity and reproducibility increasing requires the regulation of the structural thermodynamical state of fixed-point melt and the development of unified technology individual for every working substance. Mobile fixed-temperature point cells based on In-Ga-Sn and Sn-InzBi eutectic alloys as working substances are developed.

## References


[1]. CCU draft recommendation on changes to the definitions of ... - IUPAP. Final version, Recom. U1, 2005 (www.iupap.org/wp-content/uploads/2013/.../IU1_2002.pdf)

[2]. B. Jeckelmann, B. Jeanneret, The quantum Hall effect as an electrical resistance standard, *Rep. Prog. Phys.*, Vol. 64, 2001, pp. 1603–1655.

[3]. Ed. R. Prange, S. Girvin, The quantum Hall effect, *Springer- Verlag*, New York, 1990.

[4]. P. Joyez, D. Vion, M. Götz, M. Devoret and D. Esteve,The Josephson effect in nanoscale tunnel junctions, *Journ. of Superconductivity*, Vol. 12, Issue 6, 1999, pp. 757-766.

[5]. Ian Mills & others, The New SI: Units and fundamental constants, *BIMP*, January 2011.

[6]. The NIST References on Constants, Units, and Uncertainty, 2018 CODATA Recommended Values of the Fundamental Physical Constants (https://physics. nist.gov/cgi-bin/cuu/Value?rk)

[7]. A Practical Josephson Voltage Standard at One Volt (https://nvlpubs.nist.gov/nistpubs/ sp958-lide/315-318.pdf)

[8]. J. Kohlmann, R. Behr, and T. Funck, Josephson voltage standards, *Measur. Sc. and Techn.,* Vol. 14, Issue 8, 2003 (https://iopscience.iop.org/ article/10.1088/0957-0233/14/8/ 305/pdf).

[9]. M. de Podesta, The definition of the Kelvin in the new SI: its rationale, implementation and implications, in *Proceedings of the. 13th Int. Symp. on Temperature and Temperature Measurements in Industry and Science (*TEMPMEKO-2016), Zakopane, Poland, 2016, p. 9.

[10]. TEMPMEKO' 19. Conference Technical Summary (https://eu.flukecal.com/blog/tempmeko-2019-temperature-metrology-conference).

[11]. BIMP (https://www.bipm.org/en/measurement-units/).

[12]. P. Hohenberg, and B. Schraiman, Chaotic behaviour of an extended system, *Physica D.,* Vol. 37, 1989, pp. 109-115.

[13]. S. Yatsyshyn, B. Stadnyk, Ya. Lutsyk, Research in nanothermometry. Part 1. Temperature of micro- and nanosized objects, *Sensors and Transducers*, Vol. 140, Issue 5, May 2012, pp. 1-7.

[14]. A. Giesbers, G. Rietveld, E. Houtzager et al., Quantum resistance metrology in graphene, *Appl. Phys. Letters*, Vol. 93, 2008, pp. 222109-1 – 222109-3.

[15]. A Practical Josephson Voltage Standard at One Volt (http://www.lee.eng.uerj.br/downloads/graduacao/medidas_eletricas/ JosephsonJunction.pdf).

[16]. R. Sahoo, and R. Mishra, Simulations of carbon nanotube field effect transistors, *Int. Journ. of El. Eng. Research,* Vol. 1, Issue 2, 2009, pp. 117-125.

[17]. Ia. Mills, T. Quinn, P. Mohr, B. Taylor, and E. Williams, The new SI: units and fundamental constants, *Royal Society Discussing Meeting*, Jan. 2011.

[18]. T. Bubela, M. Mykyychuk, P. Malachivskyy, Y. Pokhodylo, O. Basalkevych, A Study of Uncertainty of Expert Measurement Results in the Quality Management System, *Eastern-European Journal of Enterprise Technologies*, Issue 3 (3-81), 2016, pp. 4-9.

[19]. L. Pitre, L. Risegari, F. Sparasci, M. D. Plimmer, M. E. Himbert, and P. G. Albo, Determination of the Boltzmann constant from the speed of sound in helium gas at the triple point of water, *Metrologia, Focus on the Boltzmann Constant,* Vol. 52, Issue 5, 2015.

[20]. C. Daussy, M. Guinet, A. Amy-Klein, K. Djerroud, et al, First direct determination of the Boltzmann constant by an optical method, *Physical Review Letters,* 98, 25, 2007, p. 250801.

[21]. K. Novoselov et al., Room-temperature quantum Hall effect in graphene, *Science*, Vol. 315, 2007, p. 1379.

[22]. O. Luste, R. Kuz', Computer control of Eddy thermoelectric current, *Journ. of Thermoelectricity*, Issue 2, 2004, pp. 11-19.
[23]. Ju Li, Z. Shan, E. Ma, Elastic strain engineering for unprecedented materials properties, *MRS Bull.,* Vol. 39, Issue 2, 2014, pp. 108-114.
[24]. A. Shpak, A. Nechytaylo, L. Kunytska, D. Leonov, L. Dziuba, M. Barabash, Strains in amorphous and nanocrystallic materials, *Nanosystems, Nanomaterials, Nanotechnologies*, Vol. 6, Issue 1, 2008, pp. 217-236, (in Ukrainian).
[25]. L. Anatychuk, William Thomson (Lord Kelvin) and thermoelectricity, in *Kelvin, Thermodynamics and the Natural World,* edited by: M.W. Collins, R.C. Dougal, C. Koenig, I.S. Ruddock, Chapter 13, *WITPress,* 2015, pp. 337-362.
[26]. Thermoelectrics Handbook: Macro to Nano, Ed. D. M. Rowe, *CRC Press,* 2018.
[27]. T. Zalutskaya, I. Likhnovsky, Ya. Lutsyk, An investigation of the acoustic properties of materials for ultrasonic thermometry, *Measurement Techniques,* Vol. 55, 2012, pp. 676–680.
[28]. IEC 61157 :2007 Standard means for the reporting of the acoustic output of medical diagnostic ultrasonic equipment (https://webstore.iec.ch/publication/4621).
[29]. IEC 61689:2013 Ultrasonics - Physiotherapy systems - Field specifications and methods of measurement in the frequency range 0.5 MHz to 5 MHz (https://webstore.iec.ch/publication/5722).
[30]. IEC 62127-2:2009 Ultrasonics - Hydrophones - Part 2: Calibration for ultrasonic fields up to 40 MHz (https://infostore.saiglobal.com/en-gb/Standards/IEC-62127-2-1-2-564956 _SAIG_IEC_IEC_1288934/).
[31]. GOST 8.555-91 (IEC 866-87) Characteristics and graduation of hydrophones for operation in the frequency range from 0.5 to 15 MHz (http://docs.cntd.ru/document /1200014699).
[32]. IEC 62127-1:2007, Ultrasonics - Hydrophones - Part 1: Measurement and characterization of medical ultrasonic fields up to 40 MHz (https://webstore.iec.ch/publication/6486).
[33]. R. Bobber, Underwater Electroacoustic Measurements, 1990.
[34]. JCGM 100:2008. Evaluation of measurement data — Guide to the expression of uncertainty in measurement. BIMP, *JCGM*, 2008.
[35]. International metrology in the field of Acoustics, Ultrasound and Vibration (https://www.bipm.org/metrology/auv/).
[36]. Strategy 2017 to 2027. Consultative Committee for Acoustics, Ultrasound and Vibration (CCAUV) (https://www.bipm.org/utils/en/pdf/CCAUV-strategy-document.pdf).
[37]. Ja. Avison and R. Barham, Final report on key comparison CCAUV.A-K5: pressure calibration of laboratory standard microphones in the frequency range 2 Hz to 10 kHz (www.bipm.kcdb.org).
[38]. Registration and progress form COOMET.AUV.A-K5 (http://kcdb.bipm.org/appendixB/appbresults/COOMET.AUV.A-K5/COOMET.AUV.A-K5_registration_and_ progress_form.pdf).

[39]. COOMET.AUV.A-K5 Technical Protocol (https://kcdb.bipm.org/appendixB/appbresults/COOMET.AUV.A-K5/COOMET.AUV.A-K5_Technical_Protocol).

[40]. IEC 61094-2:2009 Measurement microphones - Part 2: Primary method for pressure calibration of laboratory standard microphones by the reciprocity technique.

[41]. I. Kharitonov, A. Chunovkina, Evaluation of regional key comparison data: two approaches for data processing, *Metrologia*, Vol. 43, 2006, pp. 470-476.

[42]. COOMET R/GM/14:2006 Recommendation, Guidelines for data evaluation of COOMET key comparison (http://www.coomet.org/).

[43]. D. Dobrowolska, A. Kosterov, Report on key comparison COOMET.AUV.A-K5: Pressure calibration of laboratory standard microphones in the frequency range 2 Hz to 10 kHz, *Metrologia, Techn. Supplement*, Vol. 53, 2016, pp. 22 (https://www.bipm.org/utils/common/pdf/final_reports/AUV/A-K5/COOMET.AUV.A-K5.pdf).

[44]. I. Palyanytsya, M. Procevyat, V. Kovaleva, V. Bryk, Z. Kolodii, New developments in industrial low-temperature thermometry, *Meas. Tech., (USA), Journ. Vol. 30:3; Engl. transl from Izmer. Tekh.,* Vol. 30, No.3, 1987, pp. 28-29.

[45]. F. Pobell, Temperature Scales and Temperature Fixed Points, in Matter and Methods at Low Temperatures, *Springer*, 1996, pp. 226-240, (https://link.springer.com/book/10.1007/978-3-662-03225-1)

[46]. Guide to the Realization of the ITS-90, Metal Fixed Points for Contact Thermometry, *Bureau Int. des Poids et Mesures, Cons. Comm. for Thermometry under the auspices of the Int. Comm. for Weights and Measures,* 1 Jan. 2018 (https://www.bipm.org/utils/common/pdf/ITS-90/Guide_ITS-90_2_4_MetalFixedPoints_2018.pdf)

[47]. G. Machin, Twelve Years of High Temperature Fixed Point Research: A Review(Invited Paper), in *Proceedings of the 18 Int. Conf. on Positron Annihil. (AIP'18),* Vol. 1552, September 2013.

[48]. S. Prokhorenko, W. Prokhorenko, S. Mudry, A. Bylica, J. Navotska, Structure of a melt, thermal and ultrasonic properties during gallium alloys crystallization, *Archives of Foundry*, Vol. 4, Issue 11, 2004, pp. 123-130.

# Chapter 3

# Nanothermodynamics and Nanometrology

**Svyatoslav Yatsyshyn, Bohdan Stadnyk, Khrystyna Melnyk, Pylyp Skoropad and Roman Baitsar**

## Definitions

- **Nanothermodynamics** is a section of thermodynamics, which deals with the limited size thermodynamic system consisting of a defined number of primary subjects (atoms or molecules) or embryos of the second phase.

- **Nanothermometry** as a branch of nanometrology implies measuring the temperature of nanoobjects with the given accuracy (error, uncertainty), repeatability, discreteness, and under the preliminary guaranteed instability of a thermometer transformation function.

## 3.1. Thermodynamics and Metrology 4.0

The basic principles of the thermodynamic method in the phenomenological sense were created in half of the last century based on classical thermodynamics and described in the thermodynamics of irreversible processes [1]. In linear thermodynamics, when the system is not too far from the equilibrium state, the thermodynamic flows and forces are related by the Onsager reciprocity relations.

The thermodynamic approach considers thermodynamic forces and flows, including in the presence of deformation. Here, due to technological factors of the manufacturing of sensitive elements of thermoelectric transducers [2] or, f.i. of multigate FETs [3], the level of stress in the substance becomes significant and it can be accounted by the mechanical degree of freedom. In the case of an elastic continuum with dislocations that is the case of nanomaterials [4], this component of the basic equation ($p\frac{dV}{dt}$) of thermodynamics can be replaced by the product of strain $\hat{\varepsilon}$ and stress tensors $\hat{\sigma}$:

$$p\frac{dV}{dt} = \hat{\sigma}\frac{d\hat{\varepsilon}}{dt}, \quad (3.1)$$

For example, phenomenological consideration of stressed thermoelectric material of thermoelectric transducer shows an increase in Gibbs energy, and in the basic equation of thermodynamics an additional thermodynamic force due to this energy appears $X_J = \nabla\left(\frac{\sigma^2}{2E_U}\right) = \frac{\sigma}{E_U \nabla_x \sigma}$, where $\nabla_x \sigma$ is the gradient of mechanical stresses along the length of a thin cylindrical thermoelectrode. The energy accumulated during deformation significantly affects the transformation function of the thermoelectric transducer by forming a mechanical impact function (Fig. 3.1).

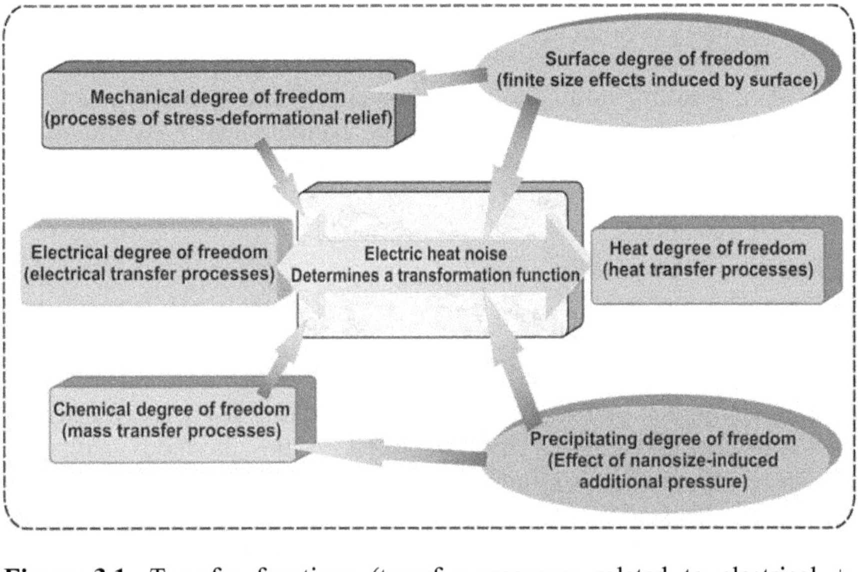


**Figure 3.1.** Transfer functions (transfer processes related to electrical + chemical and thermal degrees of freedom forming thermo-EMF) and effects (transfer processes related to a mechanical degree responsible for thermo-EMF drift).

That is how thermodynamics works in bulk materials. Nanostructured materials are the materials considered to be inherent in nanostructured properties due to the presence of nanostructured particles in the structure of bulks. The required experience, encouraged by the endeavor to cognize the nature of MI drifting which is related to the processes in a thermometric substance [5], proved to be precious since being concerned with the profound processes in the substance, namely at the micro- and nano-levels. Besides, with the development in nanotechnology, nanometrology [6] accompanied by nanothermodynamics has arisen [7].

Metrology is inseparably related to thermometry. The same concerns the nanothermometry as an integral part of nanometrology. Its development implies both a direct task, to measure temperatures and other substance thermo-properties in the nanoworld, and an opposite one, to ascertain the reasons for particular behavior of metrological characteristics, linking them with fluctuation-dissipation processes in thermometric substance during manufacturing and exploiting at a micro- and nano-level, based on high experimental experience and metrological culture due to the measurement results with the use of nanosized and nanostructured sensitive elements of thermometers.

In nanotechnology, the term "Temperature" acquires ambiguous statistical thermodynamic sense [8]. Hereby, the absence of knowledge about the temperature of the researched micro- and nano-objects does not allow to assure the reproducibility of technologies, and measurement leads simultaneously to the shift of a temperature field, appearance of a methodical error of measurement instrument and sometimes to the destruction of a nanopattern despite applying contact and/or remote method [9].

Keeping the deep insights of metrological approaches into the essence of processes within the thermosensitive substance, and mastering the achievements of nanotechnology, we have widened the spectrum of thermometric methods: assimilating Raman thermometry [10-11] of micro- and nanopatterns; conducting the research in thermometry based on solid- and liquid-phase sensitive elements at decreasing their sizes down to a nano area, and studying the role of a thermosensitive substance superficial tension gradient as a main thermodynamic force of nanothermodynamics in the forming of metrological characteristics [12] besides the already ascertained role of a mechanical micro tension gradient [13] in the forming of thermometer transformation function drift.

The branches of thermometry were singled out to study the nanostructured thermosensitive substances to improve durability or increase the temperature limits of sensors. Those are the branches of ultrasonic thermometry [14], noise thermometry [15], and resistive thermometry [16]. Owing to the means of engineering maintenance including the traditional equipment for calibrating, thoroughly conducted resource tests, and the row of structural methods for characterizing a structure and its changes during exploitation, we managed to associate the effect of nanostructuring with quite precisely controlled micro- and macro-characteristics.

To generalize thermodynamics on a scale, we should understand well the unique properties of nanosystems. One of the characteristic features of nanosystems is their high surface-to-volume ratio. As the results of surface effects becoming increasingly important with decreasing size, the Gibbs free energy relatively increases for some thermodynamic equilibrium systems. Therefore, the behavior of such nanoscopic clusters differs significantly from the usual thermodynamic limit. It is known that when the system size decreases, one has to consider the fluctuations. Therefore, these valuable hints motivate researchers to pursue the thermodynamic description at the nanometer size for the nucleation of a metastable phase.

The research of the processes of energy-transmission, based on thermodynamics, enables us to determine a methodical error component as well as cognizable part of systematic component of an instrumental error component, and thus to decrease substantially the guaranteed by the manufacturer of MI's total error of measuring the physical values in exploitation conditions.

Thermodynamically stipulated phase equilibrium is replaced due to the contribution of division surfaces or superficially predetermined mechanical tensions to the free energy of thermosensitive substance system which enables us to produce new quasi-nonequilibrium materials with high stability of calibrating characteristics for thermoelectric thermotransducers, and also to create functionally gradient thermocouples that are a bright example of structures, quasi-distributed in space [16].

## 3.2. Nanothermodynamics and Nanothermometry. Resistance Thermometers with Nanostructured Thermosensitive Substance

MG with an amorphous structure undergoes our issues as the new-gained materials with the high specific resistance ρ at the small value of its temperature coefficient α. The particular efforts are made in the endeavor to bind the electrical MG properties with the peculiarities of their manufacturing technology which could be profitable in evolving the special electrotechnical materials, spintronics [17], etc. The study of nanostructured materials confirms the existence of fields of considerable mechanical micro tensions whose influence is equal to the doping with many admixtures. In this case, we have applied nanothermodynamics to explain MG electrokinetics and other properties.

Tetradic metal amorphous alloys consisted of two transitive ferromagnetic metals (Fe, Ni) and two amorphizators (B, P) form the content of $Fe_{40}Ni_{38}P_{13}B_9$. MG $Fe_{40}Ni_{40}P_{14}B_6$ similar in composition has been studied quite in detail [18]. In general, 5 groups of patterns are gained as MG that at the moment of casting differed by (a) the alloy temperature $T_p$, (b) the rotation speed of the massive disk covered by the alloy at the moment of casting n, and accordingly by (c) the thickness of the gained foil δ (Table 3.1).

Two groups of experimental dependencies R(T) and α(T) were formed (Fig. 3.2). Measurement was conducted according to the known methods in the reference points: 4.2 K, 13.8 K, 20.4 K, 77.4 K, and 300 K. The error of electric resistance measuring made ±0.005%, i.e. the reliability of the first four signs of the measured value was assured. The processing of the gained result was made with the help of the software "Curvescript".

**Table 3.1.** Main technological MG parameters and thickness.

| Pattern group number | $T_p, K$ | $T_{\alpha}, K$ | $n, 1/min$ | $\delta, \mu m$ |
|---|---|---|---|---|
| 1 | 1373 | 653 | 1000 | 100 |
| 2 | 1373 | 653 | 3000 | 30 |
| 3 | 1273 | 653 | 3000 | 30 |
| 4 | 1523 | 653 | 3000 | 30 |
| 5 | 1100 | 653 | 2000 | 60 |

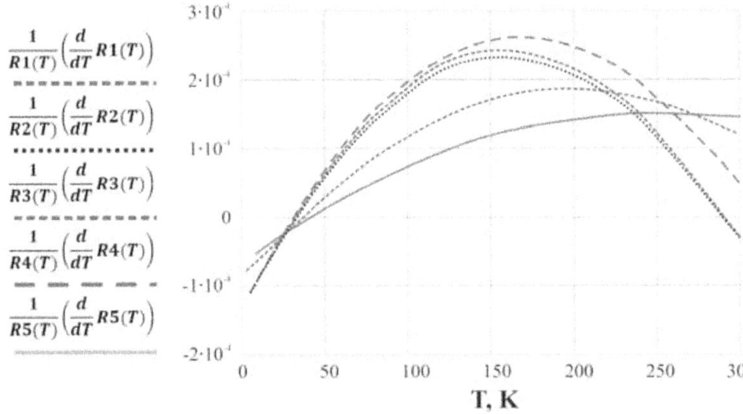


**Figure 3.2.** Temperature dependence of the temperature factor of electric resistance $\alpha$ of the studied MG.

Experimentally gained dependencies R(T) are approximated (error $\leq 10^{-3}$% at the correlation factor 0.99999998) by polynomials of the 3rd degree: $R(T) = A + BT + CT^2 + DT^3$. To compare, the description by the polynomial of the 2nd degree has given the error 1.6% at the correlation factor 0.998. The character of the temperature dependencies R(T) and ρ(T) affirms the following fact: the remaining MG resistance at the low temperatures is considerably larger as compared to their crystalline analog; the temperature coefficient of the MG resistance – on the contrary, less, and could be $\alpha > 0$, $\alpha < 0$, $\alpha = 0$; low-temperature minimums pertain to temperature dependencies of MG electric resistance, and maximums at 160 ... 260 K (Fig. 3.2) pertain to α.

According to [7, p.106], the described above the character of nanostructured material manufacturing corresponds to the MPNUR-mechanism of forming the metastable structural state of an M-phase as well as organizing the new B-region of a pseudo stable M-phase consequently of the additional substance-nanostructuring pressure effect. The existence of distension areas in MG with considerably smaller material density – defects of the n-type as 12-20 atom-sized clusters with material density, approximately 10% lower as compared to the unified characteristics -- is affirmed by our investigations [6]. Their number as well as the number of symmetrical contraction defects could be estimated at the level of 2,5% from the general volume. During 30 minutes annealing at 623 K, 10% of defects recombine. Moreover, the inner mechanical strains reach 15 MPa and decrease sevenfold while annealing. Density fluctuations caused by the non-uniform distribution of free volume in a pattern could become the sources of inner strains.

Structural relaxation attributed to the changes caused by the reformation of a local atom structure is researched in amorphous and crystalline alloys $Fe_{40}Ni_{40}P_{14}B_6$ at the temperature from 340 K (start) till 600 K (finish). It is shown that this phenomenon is related to diminishing free volume in an amorphous structure. The study following the method of nuclear quadrupole resonance spectroscopy has shown that a local redistribution is fixed around the atoms of Fe.

The results of relaxation processes study at the low temperatures 1…30 K where the changes in sound speed are referred to as the electronic states of a two-level system are of interest. The groups of atoms are remaining in two states: in one potential pit or another, isolated from the former by a potential barrier. The difference of energies between those states is 2Δ, moreover $\varepsilon^2 = \Delta_0^2 + \Delta^2$, here $\Delta_0$ is an energy of an overhead cover, 2Δ is asymmetry energy. The energy of an overhead cover is $\Delta_0 = h\omega e^{-\lambda}$, here $\lambda = \left(\frac{2mv_0}{h^2}\right)^{\frac{1}{2}} d$; where ω is the fluctuation frequency of an atom of the mass m in the potential minimum; d is the width of the system.

Two-level systems cause a considerable influence on the spreading of phonons. Phonons of the deformation field of a resilient wave modulate the energy and in case $h\omega \approx \varepsilon$, the phenomenon of absorption appears in the system. The excited state arisen could relax either while excitement intensifying or spontaneous transferring. Therefore, anomalies of sound speed in MG could be explained by resonance / relax absorption in a two-level system. The contribution of a relaxing process is getting less at temperature fall which leads to the increase in sound speed and to other similar anomalies, e. g., to the minimum of inner friction, etc.

The effects like these confirm the existence of fields of considerable mechanical micro tensions whose influence is equal to the doping by several admixtures. In this case, we should apply nanothermodynamics to explain the MG electrokinetic and other properties. The latter stipulates the introduction of two additional freedom degrees into the main equation of thermodynamics: σdM (M is the surface area), caused by the superficial tension σ and ɣdV, caused by the expenses of the specific energy ɣ for the formation of precipitations of the second phase with the volume V in the matrix of output substance.

First of all, let us ascertain the reason for the description of the researched temperature dependences of electric resistance by the four-member polynomial of the 3rd order. All other standard functions of 16 types, involved in the program software Curvescript, have proved to be incapable of describing the gained results accurately.

A. Let us consider the two-phase MG model, i.e., assume that there are precipitations of other phases in a matrix, e.g. pseudo-phases which could be represented by the mentioned above micro volumes with different density. If the specific electrical resistance of a matrix is described by the equation $\rho_1 = a + bT$, according to the rule of Mattison, and that of another similar phase is represented by the equation $\rho_2 = c + dT = a + bT + a' + b'T = \rho_1 + \Delta\rho$, and taking into account that the volume part of the second phase makes $\frac{\Delta V}{V}$ on the efficient intersection area $q = \frac{\Delta S}{S}$, we gain the equation of the third order for the specific electrical resistance of two-phase material:

$$\rho_a(\Delta\rho)^2 = (1-q)\rho_1(\Delta\rho)^2 + (\Delta\rho)^3 - q\rho_1^2, \qquad (3.2)$$

Analyzing factors of its components, we come to the expressions:

$$A = a'\left[(1-q)aa' + (a')^2 - qa^2\right], \qquad (3.3)$$

$$D = (1-q)b(b')^2 + (b')^3 - qb^2b' \qquad (3.4)$$

At the $q = \frac{\Delta S}{S}$ being equal to 0.1, the factor A is larger than 0 at $a' \geq 0.1a$ that meets realities. Simultaneously the factor D, experimentally determined to be lower than 0, is negative in the case of slight deviations from the linearity of electric resistance temperature-dependence $b' \leq 0.1$. The comparison of the factors B and C allows representing that if $B < 0$, then $C > 0$, and the factor A is proportional to the factor C having the same sign; the factors B and D stay in the analogical dependence.

The following issues are: the basis of MG is homogeneous material manufactured due to the usual technologies (avoiding its quenching and the following heat processing); $\rho = a + bT$, here the factors a and b do not depend on the manufacturing technology. Whereas the factors $a'; b'$ as well as q, characterizing the precipitations of real MG, depending on the manufacturing technology. The increase of the constant q corresponds to the decrease of factor A – the 1st member in the temperature dependence of electrical resistance. It is supported by the experimental data where A decreases from 1.9426 through 1.7518 till 1.3109 at the increment in the temperature of fusion before pouring out on the disk from 1273 K through 1373 K till 1623 K.

The explanation is that the specific volume depends on the speed of chilling, determined due to the mentioned temperature. The increment in the MG volume reaching some percent and considerably influencing the transmission processes occurs with the temperature rising. Moreover, the structures with less average density have higher values of speed transference. The first member of the expression (1) behaves itself in the same way at the decreasing of the rotation speed of the disk where quenching is taking place. So, at 3000 rpm (thickness 30 μm), it makes 1.7518; at 2000 rpm (60 μm) – 0.995, and 1000 rpm (100 μm) – 0.958.

B. Let us compare our investigations with the works made in the Saint Petersburg school of deformation physics where the decrease in the real limit of material strength concerning the theoretical values is explained by the presence of tensile defects [19] as micro- or nano- concentrators of mechanical tensions. According to the results of analyzing the Raman spectrum at 632 $cm^{-1}$, the appearance of dot radiators with the temperature in 2...5 times prevailing the average is fixed in the transparent polyethylene terephthalate in the presence of extension deformations or without them. The availability of low- and high-frequency satellites regarding the main maximum (see Fig. 3.3 with maximums at 624 and 640 $cm^{-1}$ respectively) confirms the existence of micro- and nano-areas of extension and contraction, respectively. The displacement of frequencies for 8 $cm^{-1}$ is explained by the deformation of interatomic bonds of tensile quasi defects for the value $\varepsilon = -\frac{1}{G}\frac{\Delta\nu}{\nu}$, where G is the mode Grüneisen parameter, $\Delta\nu$ is an alteration of light frequency.

While capturing the phonons by the tensile quasi defect, its local temperature is rising, considerable thermo-extension takes place that in some moment could cause the rapture of interatomic bonds and thus microcrack appearance. Using the intensity of fluctuation stripes in the Raman spectrum of combinational dispersion (Stokes and Anti-Stokes bands), the phonon value or capture by tensile quasi defects were determined. If usually, they are much less than equilibrium, determined from the formula of Bose-Einstein, then in the case of the extension effect they are on contrary much higher. The local temperature of abovementioned quasi defects (with linear size till 10 atoms) is determined as $T_d = \frac{h\nu}{k[\ln(n+1) - \ln n]}$. Regarding the dependence given in Fig. 3.3, it is obvious how essential the dot temperature increasing is on these defects, being especially important for the thermosensitive substance [20].

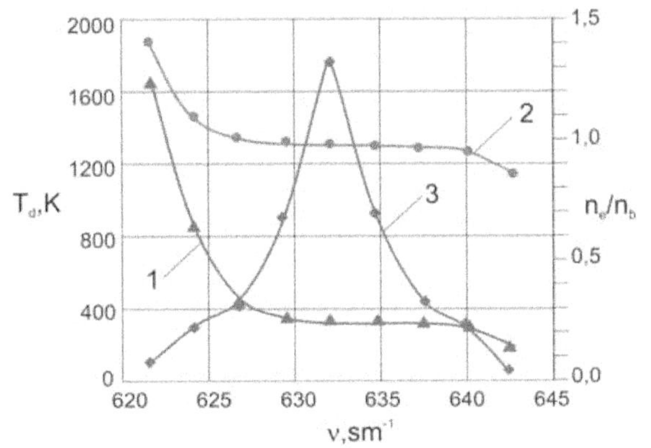


**Figure 3.3.** Dependence of local temperature (1) and phonon occupation value (2) vs the deformation of interatomic bonds in polyethylene terephthalate (3 is the shape of the relative intensity of Raman spectrum).

Micro concentrators serve as the trigger that launches the transformation of the energy applied to the material into microcracks. The appearance of the latter is indissolubly related to the emission and redistribution of energy, considerably exceeding the energy capacitance of elementary deformation acts. The indication of tensile quasi defects as physically elementary nanosized subsystems of structure enables us to stipulate the introduction of thermodynamic values i.e. the application of nanothermodynamically nonequilibrium processes to the researched substance.

Considering the mentioned defect in its closest surroundings as a locked thermodynamic system, we could show that the production of entropy within this system is determined by the sum of products of correspondent thermodynamic powers and flows. The product of the tensor gradient of mechanical tension $\nabla\overline{\sigma}$ and deformation tensor $\overline{\varepsilon}$ could be regarded as the main among them. As a result, the electromigration parameters change, leading to the increment in electric resistance as compared to the resistance of a crystal pattern. The reasons to treat the given substance as metallic glass, disappear at the removal of mechanical tension gradients nearby the defect, at least due to electromigration. The immediate crystallization is taking place, moreover, the proceeding to this process is described by the diffusion factors: the temperature exposure $T_{exp}$ and the time $t_{p.cryst}$ of maturing at this temperature without crystallization:

$$t_{p.cryst}^2 T_{\exp} = 600...800 B t_g \tag{3.5}$$

Here $t_c$ is the crystallization time, studied in the process of researching its dynamics; B is the introduced criterion called the temperature exposition: $B = 0{,}75 T_g t_{an}$; $T_g$ is the glass-forming temperature; $t_{an}$ is the temperature of isothermal annealing.

## 3.3. Nanothermodynamics and Nanothermometry. Liquid-in-Tube Nanothermometers

Due to the development of nanotechnology, significant attention is paid to improving the metrological characteristics of thermometers. An important aspect of further progress in nanotechnology is considered compliance and reproducing a given temperature of nanoproducts' manufacture and operation. It requires both the achievements of nanotechnologies toward producing nanothermometers and simultaneously of metrology - in specific issues of nanothermometry.

There are many types of liquid-in-tube thermometers, which differ since they apply the dependence of different properties of the measuring medium on temperature. These thermometers have become widespread as do not require auxiliaries and energy sources. Among them the thermometers with the mercury and non-mercury temperature-sensitive liquid (alcohol, xylene, toluene that are used for temperature measurement in the area from minus 200°C; helium, for temperature measurements up to 120°C) and the others, are distinguished.

Modern technologies, in particular of the cancer treatment, require the provision of the ability of small-sized thermometers, for example, while irradiating the operating volume with high-frequency electromagnetic waves. Under these conditions, the performance of most types of common thermometers becomes questionable. Therefore, we consider the types of thermometers that can operate in such conditions. Moreover, the thermometer should be small-sized and low-inertial. If the liquid-in-nanotube thermometer is of the appropriate size (~ 10 μm), it has the disadvantages associated with nanoscale. The first and most important of these seems to be inertia. The decreasing the intersection of the channel of the capillary made from nanotube raises the relative value of the friction force of mercury against its walls, acting on the mercury column of the thermometer during its movement along the capillary. As a result, in a narrow capillary, the mercury column moves unevenly, abruptly. When the measured temperature decreases, it can break. Therefore, in the manufacture of common mercury liquid-in-tube thermometers (macroworld) use capillaries with the channel diameter of not less than 0.1 mm. When checking and operating such thermometers with a narrow nanotube duct, measurements at sufficiently low temperatures should be avoided. Nanothermometers with capillary diameters of ~ 0.1 - 10 μm are required for modern technologies including the medicine. There arise some problems, in particular of inertia of thermometer, since the filling of the capillary by fluid and its overflowing is complicated, and its regularities are not studied yet. Let's consider peculiarities of liquid-in-tube thermometers while diminishing their sizes to nanosized.

- **Macro Thermometers**

The principle of operation of a liquid-in-tube thermometer is as follows. When thermal expansion occurs, the volume of fluid increases. The column of liquid in the capillary tube rises and it can be seen this change by the scale. The process is the opposite when the temperature drops. It takes time to increase or decrease the temperature of the fluid in the thermometer, as well as change its volume; similarly, it takes time to establish a thermal equilibrium between the measured object and the thermometer. To measure the temperature of the body, contact between the body and the thermometer must be maintained for some time. To measure the temperature of the object, it is necessary to ensure contact between it and thermometer for a while. Then, there is a clear relationship between the measured temperature and the height of the liquid column. The length of the column becomes a measure of temperature, and more precisely the upper end of the column that is the meniscus marks the value of temperature serving as an indicator of temperature.

The change in volume, depending on the temperature for any liquid, is characterized by the coefficient of volumetric thermal expansion β. The average value of β in the temperature range 0 - t is calculated by the formula:

$$\beta = \frac{V_t - V_0}{V_0 t}, \tag{3.6}$$

where $V_t$ and $V_o$ are the volumes corresponding to temperatures t and 0°C. In a liquid-in- glass-tube thermometer, as the temperature changes, the volume of not only the thermosensitive fluid but also the volume of the glass tank changes. Therefore, there is only a visible change in the volume of the liquid in the thermometer, which is equal to the difference of changes in the volumes of the liquid and the tank (and partially the capillary) of the thermometer. In this regard, when calculating the design of the thermometer apply the so-called apparent coefficient of expansion of the thermometric fluid in glass $\beta_1$. The phenomenon of thermal expansion of a fluid ΔV (volume change, m3) is described by the equation:

$$\Delta V = \beta V \Delta T, \tag{3.7}$$

where β, 1/K; ΔT is the temperature change, K. The liquid in the thermometer rises since the coefficients of volume expansion β of the liquid and glass differ significantly: $1.81 \cdot 10^{-4}$ 1/K for mercury; $10.6 \cdot 10^{-4}$ 1/K for alcohol; $9.16 \cdot 10^{-4}$ 1/K for toluene; and $0.25 \cdot 10^{-4}$ 1/K for the glass.

Due to the advances of science, biotechnology and nanotechnology are gaining in popularity. This requires the development of high-precision temperature measurement technologies based on nanosensors including nanothermometers.

- **Micro- and Nanothermometers**

The most used are the liquid-in-(carbon)nanotube thermometers. They resemble a conventional mercury-in-tube thermometer that has been reduced by a billion times. The carbon nanotube is a series of closed at one end concentric cylinders with a length of ~10 μm and a diameter of ~75 nm; this is the body of the nanothermometer. As a temperature-sensitive element, metal gallium may be used, which fills the inner space of the nanotube. When heated, the gallium column elongates; simultaneously forms an oxide layer. Due to its properties, the length of the metal column is fixed and can be stored for a long time, which is convenient in measurement. However, a major drawback of nanothermometer concerns in its miniature, which necessitates a microscope for taking readouts [21].

### 3.3.1. Development of Liquid-in-Tube Thermometers

To measure temperature, we should improve thermometers' sensitivity without losing precision. In the world of nanotechnologies, unlike in the case of macro objects, where the accuracy of research can be improved by extending the scope of the experiment, improving measurement conditions, and minimizing the influence of external factors, we face the problem of comparing expediency and significance of energy interference when using the measuring instrument, to determine quantitative characteristics of an object, as well as the problem of the reproducibility of research results obtained with the help of different instruments of the same type.

On the other hand, nanotechnologies develop the methods of influence on materials, which are aimed at creating miniature and super-miniature devices, including those in which dimensional quantization plays the key role. These are methods of applying thin and superfine films during thermal evaporation, ion beam sputtering, and plasma chemical precipitation; ion implantation in semiconductors; plasma chemical etching of structures; various methods of annealing crystal lattice defects (laser annealing, electron beam annealing). Such technologies are being developed and are widely used when manufacturing integrated circuits for microelectronics, acoustic-optic electronics, and micromechanics, as well as for the synthesis of new materials. In recent decades, these technologies have been used to create devices based on nanoscaled objects (quantum wells, wires, dots). If a technological operation still requires high temperatures, the process is not performed in inertial tempering furnaces, where there are several dozens of substrates at the same time. It is instead performed in individual treatment reactors, where a single substrate is heated much (hundreds of times) faster due to the effect of powerful optical radiation, while reactor walls remain practically cold. Typical characteristics: The substrate heating rate reaches 30 - 100 K / s., heating time is 10 - 30 s. (for comparison, in tempering furnaces, the heating rate is only 0.1 ÷ 0.3 K / s., and the heating takes about 1 hour). Therefore, it is important to develop fast-response temperature sensors, which, however, may have a significant methodic component of measurement error due to the discrepancy between their size and the size of the monitored object.

- **Micro Thermometers and Their Design**

Nanothermometers with carbon nanotubes are the most commonly used ones; they resemble a typical mercury thermometer which has been reduced in size by billions of times. A carbon nanotube is arranged in several concentric cylinders with capped ends with a length of about 10 microns and a diameter of about 75 nm – this is the sensitive element of a nanothermometer. Metallic gallium, which fills the internal space of a nanotube, is used as a thermosensitive substance (Fig. 3.4). When heated, the temperature of a stem of gallium rises, and an oxide layer is formed; due to its properties, the height of the metallic stem settles and can remain in this position for a long time, which is convenient when taking readouts [1]. However, nanothermometers are inherent in the drawbacks: repeatability of metrological characteristics and their uncertainty caused by the energy exchange between them and monitored objects.

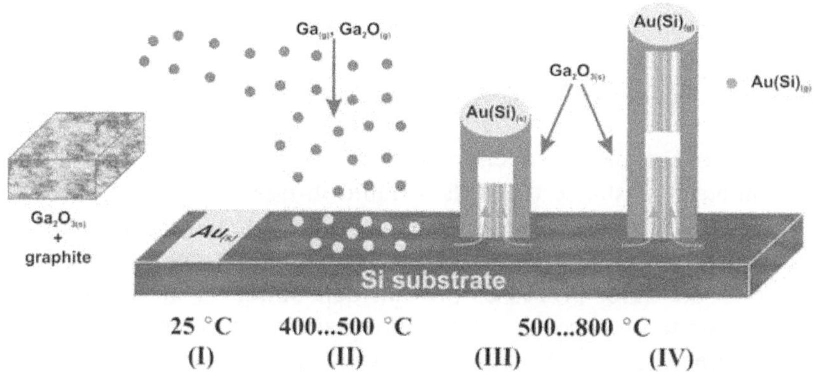


**Figure 3.4.** Scheme of micro thermometer manufacturing.

An important aspect of further progress in nanotechnologies is the establishment and reproduction of the needed temperature mode of manufacturing with the help of designed micro thermometers. This issue has not been fully addressed yet, because it required both the development of nanotechnology (in particular, manufacture of micro- and nanothermometers), as well as the development of metrology – the emergence of nanothermometry. These outlined tasks contributed to the further advancement of nanotechnologies. Production of quantum-dot diodes as highly efficient light sources can serve as an example. For their manufacture, it is necessary to maintain the temperature regime ~ 1300 °C with the permissible error not exceeding ± 2 °C, otherwise, the output of finished products will be reduced by 5 times.

- **Influence of the Dimensions of Thermometers on Their Metrological Characteristics**

The production of liquid-in-nanotube thermometers is rather trivial and consists of the superficial forces applying for the filling up the micro- and nanotubes with the thermosensitive substance. Nevertheless, metrological characteristics of the similar thermometers at their diminishing were not specially considered yet. Let's analyze the sources caused by the effect of thermodynamic forces and flows on a temperature-sensitive substance of a thermometer under the conditions of a reduction of its size into the micro- and nano area. They lead to the formation of a calibration curve and its change as a result of the effect of impact function. The latter forms an instrumental component of the thermometer's error in case of a change in the thermodynamic state of a thermosensitive substance. To assess the influence of various factors on the readouts taken using thermotransducer during different transfer processes, the Gibbs and Gibbs-Duhem equations were used.

The effect of complex transfer processes in a thermodynamically isolated thermometric substance that fills a liquid-in-tube thermometer may be presented as follows. Volume transfer processes and surface transfer processes form the calibration characteristic of this thermometer; and the former or the latter processes prevail, depending on the thermometer size, or rather on the size of thermosensitive substance. Also, other transfer processes can become responsible for the impact of function formation.

In linear thermodynamics, if a thermodynamic system of the given substance is not too far from the equilibrium state, the active thermodynamic flows J and forces X are connected by Onsager reciprocal relations. In the case of a liquid-in-tube thermometer, we deal with the forces of mechanical, superficial, and heat freedom degree. Then the system of transposition equations for the thermosensitive substance of a liquid thermometer is:

$$\begin{cases} I_m = -L_{11}\nabla V - L_{12}\nabla M - L_{13}\nabla T \\ I_n = -L_{21}\nabla V - L_{22}\nabla M - L_{23}\nabla T \\ I_T = -L_{31}\nabla V - L_{32}\nabla M - L_{33}\nabla T \end{cases}, \tag{3.8}$$

Here, $I_e; I_h; I_m$ are the transposition flows of thermometric liquid consequently of volume change, superficial tension, and temperature changes, respectively; $L_{ij}$ is the transposition coefficient. Transposition $I_m$ is used for the description of the graduating characteristic of a liquid thermometer in the macroworld. It is necessary in the case of minimizing the error caused by the influence of surface tension and the temperature gradient. The latter has been applied to a thermometer in whole and the vessel with thermosensitive substance particularly. Mainly, thermometers measure the temperature in the areas where a space temperature gradient is supposed to be negligibly small so that (3.8) is reduced to:

$$\begin{cases} I_m = -L_{11}\nabla V - L_{12}\nabla M \\ I_n = -L_{21}\nabla V - L_{22}\nabla M \end{cases}, \qquad (3.9)$$

The 1st equation (3.9) describes mechanical flow as a flow of thermosensitive liquid movement under the influence of thermodynamic forces caused by the gradient of its volume and the gradient of its surface area. The 2nd equation (3.9) is to the flow of transfer of superficial degree of freedom – the flow of reduction (increase) in the size of the stem of liquid under the influence of thermodynamic forces described above. Using these equations, we can obtain an equation for the graduating characteristic of a liquid-in-micro tube thermometer for the micro world.

In the macro world the consideration for a liquid-in-tube thermometer under the condition of neglecting superficial force, when the diameter of the thermometric tube is quite large and thermosensitive substance is practically incompressible, could derive the equation describing the interrelation of the volume of the mentioned substance and its temperature:

$$V = Vo(1+\alpha_V \Delta T) \text{ or } \Delta V = V_o \alpha_V \Delta T \qquad (3.10)$$

Here, $V_o$ is an initial liquid volume; $\Delta V$ is the thermo expanding factor. The latter indicates the extent of alteration in the initial liquid volume 1 $m^3$ at the rising of temperature for 1 K. Considering that a spherical container for liquid of the diameter D in the spheroidal lower part of a thermometer practically does not change its dimensions with temperature raise, whereas the liquid itself is expanding in volume, the growth in volume is displacing into the thermometric tube of the diameter d. It gives the possibility to bind the changes of column sizes $\Delta h$ of a thermometer and the increase in the temperature $\Delta T$ by the proportionality:

$$\Delta h = A\,\Delta T,\ \mathrm{mm} \tag{3.11}$$

Here, A is the constant. It is worth noting that the leading or diagonal member of the matrix corresponds to thermometer stem movement, whereas a member associated with the effect of surface tension – a non-diagonal member – causes problems with measuring temperature based on its projection on the scale due to the distortion of the surface of its edge. Therefore, the meniscus can be either concave or convex, which depends on whether the thermosensitive liquid is wetting tube walls. Meniscus curvature results in errors in liquid-in-tube thermometer readings in the macro world.

Micro and nanoworld. Under the condition of neglecting the influence of mechanical freedom degree and thus caused transposition processes, two interrelated factors, superficial tension, and temperature that determine the shape of a graduating characteristic of liquid-in-tube nanothermometer, could be found from equation (3.8).

Our previous study [21] has considered the Eötvös rule [22] and Rayleigh-Shield equation that determines the dependence of surface tension of any clear liquid on the temperature: the coefficient of superficial tension is a linear function of temperature. The rule was proved correct for the majority of known equations. If one has constructed the graph of dependence of a superficial tension coefficient on the temperature, he could observe a quite straight line intersecting the X-axis at the critical temperature. For example, in the case of water, this temperature makes 547 K, the coefficient of superficial tension is equal to zero.

The temperature-dependence of superficial tension could be depicted for all liquids so that the data are placed along one common curve. To perform this, we should know the molar mass, density, or molar volume of the appropriate liquid. If V is the molar volume and $T_c$ is the critical temperature for the certain liquid, then the coefficient of superficial tension $\gamma$ is defined as:

$$\sigma V^{2/3} = k\left(T_c - 6 - T\right) \tag{3.12}$$

Here, k is a constant for all liquids (Eötvös constant makes $2.1\times10^{-7}$ J/K mole$^{-2/3}$). Having an idea about the molar mass M and density ρ: $V = \frac{M}{\rho}$, the molar volume V could be determined. It is useful to transform the formula so that the units mole$^{-2/3}$ would be absent. Avogadro number could be used for that purpose:

$$\sigma = k'\left(\frac{M}{\rho N_A}\right)^{-2/3}\left(T_c - 6 - T\right) = k'\left(\frac{N_A}{V}\right)^{2/3}\left(T_c - 6 - T\right) \tag{3.13}$$

Thus, taking into account (3.12), (3.13), the equation of graduating characteristic of a liquid-in-tube micro thermometer could be determined (Fig. 3.5):

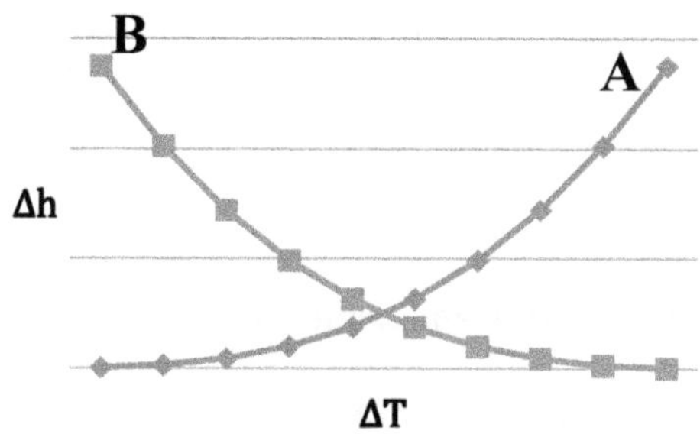


**Figure 3.5.** Graduating characteristics $\Delta h = F(\Delta T)$ of macro (A) and micro (B) thermometers.

$$\Delta h = \frac{4k'}{\gamma d}\left(\frac{N_A}{V}\right)^{2/3}\left(T_c - 6 - T\right) = C\left(T_c - 6 - T\right) = C(K - \Delta T) \tag{3.14}$$

Here, C is a constant, $K = T_c - T_0$ ; $\Delta T = T - T_0$.

The factor C of graduating characteristic of a micro thermometer with a liquid-in-tube sensitive element decreases when the tube inner diameter d grows. It is the main difference between the latter and the macro thermometer whose sensitivity falls with diameter.

Considering the equation system (3.9), we can deduce that while liquid-in-tube micro thermometer application in the area of the temperature gradient, the effect of thermocapillary flow arises. It lies in the appearance of superficial tension difference and thus, in the difference of capillary pressure in the liquid, which leads to the transposition of the liquid itself in the unevenly heated medium. This factor could become determinative in forming the additional source of a thermometer error in the nanoworld. Anyway, it is already applied in nanotechnology while manufacturing nano engines [23] whose motive force is the effect of a thermocapillary flow.

We underline that the 2nd equation (3.9) describes thermodynamic flow as a flow of thermosensitive liquid reducing/increasing under the influence of thermodynamic forces that are caused by gradients of volume and surface area. The latter becomes predominant in the microworld and determinative in the nanoworld. Thus, in a liquid-in-micro tube thermometer, we deal with the movement of liquid relatively to a certain mark due to changes in superficial force, caused by temperature changes. This movement serves as the basis for the graduating characteristic formation. The effect of a non-diagonal term (mechanical degree of freedom) leads to the formation of deviations from the graduating characteristic, that is, to an error in temperature measurement caused by a micro thermometer design.

### 3.3.2. Methodic Error of Measurement

Consideration of the liquid-in-tube thermometers envisages that the performance depends on their linear sizes and the row of thermophysical properties.

In the macro world, it is assumed by default that linear dimensions of a sensor do not exceed 1/10 of the linear dimensions of a controlled object (the ratio of volumes is 1/1000). Then the relative procedural component of an error in temperature measurement taken by a sensor of the thermal converter does not exceed 1/1000 of 0.1%. As a rule, this value is lost among some other components of measurement error. In particular, it is smaller than the instrumental component. Therefore, there are grounds not to consider and not to take into account the methodic component of measurement error when measuring temperature in the macro-world. Moreover, in the case of a contact method of measurement, information about the state of the monitored object is transferred to the sensor as a result of heat exchange between them. If there is thermodynamic equilibrium (if there is also long-term thermal contact), measurements are classified as equilibrium, and in the case of short-term contact, they are considered as non-equilibrium.

In the micro world, when measuring temperature of objects, a sensor (for example, an ultra-thin wire or a laser beam which contains temperature information) violates the thermodynamic equilibrium of an object to such an extent that there arises a significant methodic component of an error. Moreover, in the case of a contact method of measurement, information about the state of the controlled object is transferred to the sensor as a result of heat exchange between them.

Let us assume that the controlled object has a temperature somewhat lower than the thermotransducer. Then, if there is long-term thermal contact of the sensor and the controlled object, the latter is heated or cooled as a result of heat exchange $Q_H$:

$$Q_H = c_{ob} m_{ob} (T_x - T_0), \qquad (3.15)$$

where $T_0$ is the temperature of the controlled object before measurement; $T_x$ is the temperature of the monitored object, which is established due to its thermal contact with the sensor; $c_{ob}$; $m_{ob}$ are the specific heat of an object and its mass, respectively. It should be noted that in this case, the sensor will measure the resultant or weighted average temperature of the pair "monitored object - sensor", which exceeds the initial value of the temperature of the object by $\Delta T_{met} = T_x - T_0$. While measuring the cold object surface, the sensor cools down, transmitting heat to the object. So:

$$c_{ob} m_{ob} (T_x - T_0) = c_{sen} m_{sen} (T_{sen} - T_x), \qquad (3.16)$$

$$c_{ob} m_{ob} \Delta T_{met} = c_{sen} m_{sen} (T_{sen} - T_x) \qquad (3.17)$$

If we express the mass in terms of the specific mass of material w and volume V and also assume that the object and the sensor have the same form of a flat disk, we obtain:

$$c_{ob} w_{ob} V_{ob} \Delta T_{met} = c_{sen} w_{sen} V_{sen} (T_{sen} - T_x), \qquad (3.18)$$

$$c_{ob} w_{ob} D^2 H \Delta T_{met} = c_{sen} w_{sen} d^2 h (T_{sen} - T_x), \qquad (3.19)$$

where D and d are the diameters of the disks of the monitored object and the sensor, respectively; H and h are the heights of the disks of the monitored object and the sensor, respectively. After dividing the left and the right part of (3.19) by $T_x$, we obtain the thermal energy balance equation for the long-term contact of the sensor and the monitored object:

$$c_{ob} w_{ob} V_{ob} \delta T_{met} = c_{sen} w_{sen} V_{sen} \frac{Tsen - Tx}{Tx} = c_{sen} w_{sen} V_{sen} \left(\frac{Tsen}{Tx} - 1\right) \qquad (3.20)$$

Hence, we express the relative methodic error component in temperature measurement as the required value:

$$\delta T_{met} = \frac{c_{sen} w_{sen} V_{sen}}{c_{ob} w_{ob} V_{ob}} \left( \frac{T_{sen}}{T_x} - 1 \right) = \frac{c_{sen} w_{sen} d^2 h}{c_{ob} w_{ob} D^2 H} \left( \frac{T_{sen}}{T_x} - 1 \right) = \frac{c_{sen} w_{sen} sh}{c_{ob} w_{ob} SH} \left( \frac{T_{sen}}{T_x} - 1 \right), \quad (3.21)$$

where S; s are the areas of the end face of disks of the monitored object and the sensor, respectively. As we can see, if thermo-physical and mass-dimensional characteristics of a sensor and the monitored object are similar, the methodic component of measurement error depends on the ratio of their volumes or linear dimensions. In Fig. 3.6, the value of the relative methodic component of an error in the measurement of the temperature of an object taken by a sensor is shown for the ratio of their volumes — 1:1, modified, taking into account thermo-physical and mass-dimensional characteristics.

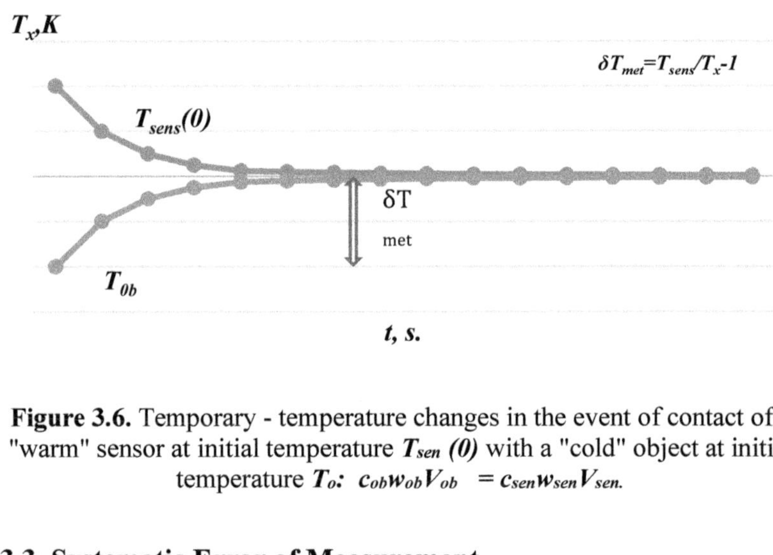


**Figure 3.6.** Temporary - temperature changes in the event of contact of a "warm" sensor at initial temperature $T_{sen}$ *(0)* with a "cold" object at initial temperature $T_o$: $c_{ob} w_{ob} V_{ob} = c_{sen} w_{sen} V_{sen}$.

### 3.3.3. Systematic Error of Measurement

The liquid-in-tube thermometers are characterized by different types of systematic error components. When reducing their linear dimensions, which is extremely necessary for modern micro- and nanotechnologies, in particular in science, industry, medicine, etc., there arise additional components of a systematic error in measuring temperature. The latter is associated with both the technology of their manufacture and the practice of their application. Deviation of the characteristics of thermometers from the standard when measuring temperature leads to variations in calibrations and is often caused by insufficient development. The spread is particularly characteristic for micro thermometers, often unique modern thermometers required, for example, in medicine [24] with intense energy irradiating the tissue by electromagnetic fields.

Thus, each measurement of specific MI is inherent in a systematic error, which in turn is the result of some independent components. The latter has been studied previously, mainly on samples of macro thermometers, which, due to the continuous miniaturization of both technologies and means of their control, require constant improvement and study.

The 1st component that arises is the component of a systematic error caused by the heat exchange between the measured object and the thermometer. This is the methodic error considered higher. Shown that it is determined by the ratio of mass-dimensional and thermophysical characteristics of the mentioned object and the thermometer.

The 2nd component of the systematic error is related to the mismatch of the thermometer calibrating and operation conditions. If the thermometer is calibrated being fully immersed to the reference mark in the measuring medium, then in practice it is positioned not as previously (part of the column may be outside the measured medium or it may be completely in the medium, which also does not meet the conditions of thermometer calibration). This leads to the above-mentioned error component since the temperatures of the protruding part of the column and the part immersed in the medium are different. Since the expansion of the immersed and protruding parts of the fluid is different, this causes arising of this component of error. Preferably, this component is essential only for macro thermometers; it is much smaller for a micro thermometer, and difficult to access for nanothermometers.

The 3rd component of the systematic error inherent in glass liquid-in-tube thermometers is considered to be the offset of the zero point of the thermometer. It is observed after the thermometer heating up to temperatures close to the higher boundary of the application. While subsequent cooling of the thermometer to 0 °C, the glass capillary does not immediately acquire the same dimensions as it had before the previous cycle of heating. Therefore, a thermometric liquid, such as mercury, the volume of which becomes equal to the initial volume, is partially located in the capillary, the intersection of which has not yet decreased to the original. As a result, the zero-point offset of the thermometer can reach 3 °C in technical thermometers with a scale of 0–600 °C.

The next, 4$^{th}$, component of the systematic error (deadlock error) is caused by micro-irregularities inside the thermometer capillary, which leads to the delay of movement of the thermometric fluid when the temperature changes, and then to its abrupt movement. This error is observed, even in the macro thermometer, namely in mercury thermometers. Here, due to the forces of surface tension, when measuring the liquid thermosensitive medium responds to the alterations in temperature not smoothly, but abruptly. The jump value is determined by the elastic properties of the reservoir and the capillary diameter. Thermometers with a channel diameter of less than 50 μm are inherent in a significant "dead stroke". To reduce this error component, it is enough to shake the thermometer before measurement.

The next, 5$^{th}$, component of the systematic error is close to the previous one, since the capillary channel in a liquid thermometer may have a distinct form - for example, a smooth capillary narrowing due to the overheating of the glass microtube while it is manufacturing. There is no component of the error caused by the "deadlock", but there is precisely this component of the displacement of thermometer characteristics due to the raise of the capillary hydraulic resistance at the considered spot.

To eliminate these last two error constituents (4$^{th}$ and 5$^{th}$), capillary tubes should be especially carefully sealed, after which it needs heating the spot and adjacent sections, then cooling and grinding the capillary duct from the inside. These constituents are powering while minimizing the capillaries' dimensions.

The 6$^{th}$ component among the similar components of the systematic error occurs when the interpolation method is used to determine the corrections at the intermediate points of the thermometer scale. This is not considered since anybody can introduce his component. In such a manner it arises the uncertainty of determination.

- **The Component Caused by the "Dead-Stroke" of the Thermometric Fluid**

Let's take a closer look at the last two components of a systematic error in temperature measurement with liquid-in-tube thermometers since they have not been studied methodologically, especially concerning linear thermometer dimensions. This component is related to the properties of thermometric liquids, their fluidity and movement in the capillary by the action of surface tension and/or external pressure, wetting of the capillary walls, etc. The measurement is based on the visual observation of changes in the volume of thermometric fluid. The distance between the marks on the scale is directly proportional to the capacity of the thermometer tank and the difference of the average coefficients of thermal expansion of the thermometric fluid and the glass from which the thermometer tank is made, and inversely proportional to the square of the diameter of the capillary channel.

To evaluate the component of a systematic error caused by a "dead stroke", let's consider the equation of motion of a thermometric fluid in the capillary of a micro thermometer. Write the equation of equilibrium of forces acting on the fluid flow in the channel, taking into account the direction of forces action:

$$F_p - F_{fr} - G\sin\alpha - F_1 = 0 \qquad (3.22)$$

Here $\alpha$ is an angle of inclination of the capillary to the horizon. The pressure force is defined as:

$$F_{fr} = (\rho_1 - \rho_2)S \qquad (3.23)$$

Here $\rho_1$; $\rho_2$ is the pressure values, respectively, in the different sections of the capillary; S is the cross-sectional area of the capillary. The friction force is:

$$F_{fr} = \tau S_{surf} = \tau\pi dl = \tau\frac{Sl}{R_h}, \qquad (3.24)$$

where $\tau$ is the friction stress, $S_{surf} = \pi dl$ is an area of the inner surface of the capillary, $R_h = S/\pi d$ is the hydraulic radius of the capillary. The projection of liquid weight is equal to:

$$G\sin\alpha = mg\frac{z_2 - z_1}{l} = (\rho Sl)g\frac{z_2 - z_1}{l} = \rho gS(z_2 - z_1), \qquad (3.25)$$

where m is the mass of the liquid in the capillary between the intersections; $z_2, z_1$ are the locations of the appropriate intersections; $\rho$ is the liquid density; g is the gravity acceleration:

$$F_1 = ma = (\rho Sl)a, \qquad (3.26)$$

Here a is the acceleration of fluid motion. With the help of the given equations (3.23) and (3.26) it can be derived:

$$(p_1 - p_2)S - \tau \frac{Sl}{R_h} + \rho g S(z_1 - z_2) - (\rho S l)a = 0 \qquad (3.27)$$

Dividing all the terms of this equation in ρgS, we obtain:

$$\frac{p_1 - p_2}{\rho g} - \frac{\tau}{\rho g} * \frac{1}{R_h} + (z_1 - z_2) - \frac{a}{g} l = 0 \qquad (3.28)$$

We rewrite the equation as follows:

$$\left(z_1 + \frac{p_1}{\rho g}\right) - \left(z_2 + \frac{p_2}{\rho g}\right) = \frac{\tau}{\rho g} * \frac{1}{R_h} + \frac{a}{g} l \qquad (3.29)$$

This equation describes the unsteady motion of a fluid with constant acceleration in the capillary of a permanent cross-section. Taking into account that the difference of velocity head is zero, the left part in (3.29) is the difference of the total hydraulic heads of the fluid; then the right part is the sum of the pressure-loss accounting the friction along the length of the capillary pressure $h_{fr} = \frac{\tau}{\rho g} * \frac{1}{R_h}$ and pressure-loss, which overcomes the action of the inertia force $h_{in} = \frac{a}{g}$. Their sum can be as positive as in the case of accelerated motion, and negative if the motion is equally hampered since the inertial pressure-loss is continuously substituted with a minus sign. The friction pressure-loss along the length of the capillary can be calculated by the Darcy-Weisbach equation [25]:

$$h_{fr} = f_D \frac{\rho}{d} \frac{\upsilon^2}{2}, \qquad (3.30)$$

from which the content of the hydraulic resistance factor $f_D$ or Darcy friction factor becomes clear (It represents the head loss of the flowing fluid due to friction. As the latter increases, the mentioned factor decreases). To determine the energetic causes of movement and stoppage of thermometric fluid in the capillary of a micro thermometer, let's consider below the energetics of the fluid's displacement with the processes that occur. The compression energy of the fluid is equal to the operation of the pressure force $F_x$ in the cross-section of the capillary S by a distance Δx, which corresponds to the change in the volume of fluid ΔV due to the pressure raise:

$$E_{com} = F_x \Delta x = \frac{1}{2} \Delta p S \Delta x = \frac{\Delta p \Delta V}{2} \quad (3.31)$$

The change in the volume of the fluid can be determined by the modulus of the fluid elasticity K:

$$\Delta V = V \frac{\Delta p}{K} \quad (3.32)$$

Then

$$E_w = V \frac{\Delta p^2}{2K} \quad (3.33)$$

The deformation energy of the walls is equal to the product of the force $F_r$ (on the inner surface of the capillary) by the distance $\Delta h$, which corresponds to the real displacement of the capillary wall:

$$E_{def} = F_r \Delta h = \frac{1}{2} \Delta p 2\pi r^2 l \frac{\Delta h}{h} = \Delta p V \frac{\Delta h}{h} \quad (3.34)$$

According to Hooke's Law, the relative radial deformation of the capillary wall $\frac{\Delta h}{h}$, caused by the increase in pressure by $\Delta p$, is equal to the ratio of normal stresses $\sigma$ in the capillary wall to Young's modulus E of that wall:

$$\frac{\Delta h}{h} = \frac{\sigma}{E} \quad (3.35)$$

On the other hand, the stresses in the capillary wall resulting from the increase in pressure by the value of $\Delta p$ can be determined by the formula:

$$\sigma = \frac{\Delta p d}{2\delta} \quad (3.36)$$

Here d is the capillary diameter, $\delta$ is the wall thickness of capillary. Taking into account (3.35) - (3.36), the dependence on the energy of the capillary wall deformation becomes:

$$E_{def} = V \frac{\Delta p^2 d}{2E\delta} \quad (3.37)$$

In result:

$$\rho V \frac{v^2}{2} = V \frac{\Delta p^2}{2K} + V \frac{\Delta p^2 d}{2E\delta} \tag{3.38}$$

Multiplying this expression by the density of the fluid, we can determine the value of the pressure increase Δp required to move the thermometric fluid in the capillary of the micro thermometer; that is, we specify the conditions for eliminating the component of the systematic error caused by "deadlock":

$$\Delta p = \rho\, v \left(\frac{\rho}{K} + \frac{\rho d}{\delta E}\right)^{-1/2} \tag{3.39}$$

- **Component Caused by the Imperfections of the Inner Diameter of the Capillary**

In capillaries of radius r, the level of wetting by the liquid to the height h is described by the Jurin's Law [26]:

$$h = \frac{2\sigma \cos\theta}{r\rho g} \tag{3.40}$$

Here θ is an edge angle. As the radius of the capillary channel changes by Δr, the lifting height of the liquid Δh will change to:

$$\Delta h = \frac{2\sigma \cos\theta}{\Delta r \rho g} \tag{3.41}$$

Let's analyze this expression. For example, as the diameter of the capillary decreases, the liquid rises higher by Δh. We estimate below the change in micro thermometer readouts in the presence of similar deviations within the capillary radius (Table 3.2).

**Table 3.2.** Effect of changing the radius of the inner surface of the capillary $\Delta r$ (at the edge angle of wetting $\theta = 0°$ and capillary's radius $r = 500$ μм) on the raising of the fluid column $\Delta h$ in it.

| Thermometric fluid | Parameters | | | | | |
|---|---|---|---|---|---|---|
| | $\Delta r_1$, μm | $\Delta h_1$, μm | $\Delta r_2$, μm | $\Delta h_2$, μm | $\Delta r_3$, μm | $\Delta h_3$, μm |
| Purified water | 0.1 | 14.709 | 0.5 | 2.942 | 1.0 | 1.471 |
| Alcohol | 0.1 | 5.608 | 0.5 | 1.122 | 1.0 | 0.561 |
| Mercury | 0.1 | 7.076 | 0.5 | 1.415 | 1.0 | 0.708 |

Capillary pressure depends on the shape of the meniscus. In the case of a concave (convex) surface, it is smaller (larger) than above the flat surface by the value $\Delta P_M$:

$$\Delta P_M = \frac{\rho}{\rho_1 - \rho} P_{rm} \tag{3.42}$$

Here ρ is the density of the saturated vapor, $\rho_l$ is the density of the liquid, $P_{rm}$ is the additional pressure associated with the curvature of the surface. When moving the fluid from the wider part of the capillary channel to the narrower one, the capillary is distorted. So, the concept of the determined hydraulic capillary radius (Fig. 3.7) is introduced:

$$R_{r1} = S/\pi d \tag{3.43}$$

When passing the thermosensitive fluid of the narrow section of the capillary, the hydraulic resistance alters to:

$$R_{r2} = S/(\pi d + \Delta d) \tag{3.44}$$

The fluid then extends to the expanded section of the capillary duct with another hydraulic radius:

$$R_{r3} = S/(\pi d - \Delta d) \tag{3.45}$$

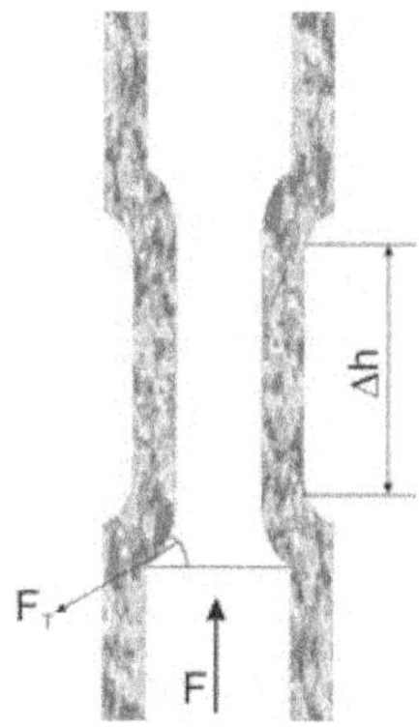


**Figure 3.7.** Deviation of the size of the capillary of the liquid-in-tube thermometer, due to fluctuations in its production parameters (tensile strength, temperature change, the chemical composition of the material).

To pass the fluid throughout narrowing of the capillary channel, that is, to lift the column of fluid must overcome such factors as the force of friction $F_{fr}$, the projection of weight P. Setting the condition of the fluid stoppage in the channel, we obtain accordingly necessary for this change (increase or decrease) alteration of the inner capillary radius on $\Delta R_h$:

$$\Delta R_h = -\frac{S}{\Delta d} \tag{3.46}$$

When the capillary duct narrows linearly, the thermosensitive fluid stops when Δd becomes negative. In the case of nonlinear constriction, we obtain $h_{ideal}$. With long-sized constriction, thermometric fluid rises higher at the capillary duct. When the liquid is raised to almost the limit of lift, the capillary duct narrowing acts as a resistance gain and the column stops. In such a way appears an error component caused by the deviation of the inner diameter of the capillary duct:

$$\Delta h = h_{ideal} - h_{R_{r3}} \tag{3.47}$$

It is estimated for thermometers with capillaries of different diameters (Fig. 3.8), and for some kinds of thermometric liquids (water, mercury, alcohol).

After the transition from absolute to relative errors the following data are obtained (Table 3.3).

**Table 3.3.** Component of relative error of liquid-in-tube thermometer, caused by deviation of capillary inner diameter.

| Thermometric fluid | Parameters | | | | | |
|---|---|---|---|---|---|---|
| | $d_1$, µm | $\delta h_1$, % | $d_2$, µm | $\delta h_2$, % | $d_3$, µm | $\delta h_3$, % |
| Water | 1000 | 0.125 | 100 | 0.499 | 10 | 0.787 |
| Mercury | 1000 | 0.141 | 100 | 0.496 | 10 | 0.705 |
| Alcohol | 1000 | 0.112 | 100 | 0.235 | 10 | 0.561 |

Thus, the value of the considered component of the systematic error caused by the deviation of the inner diameter of the capillary due to technological deviations in its manufacture is estimated at up to 1%.

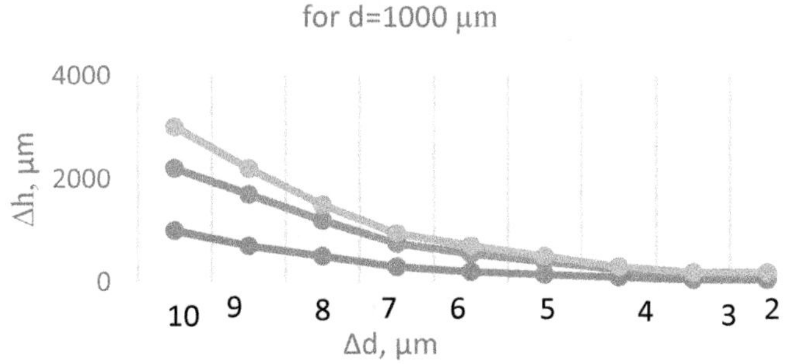


(a)

μm

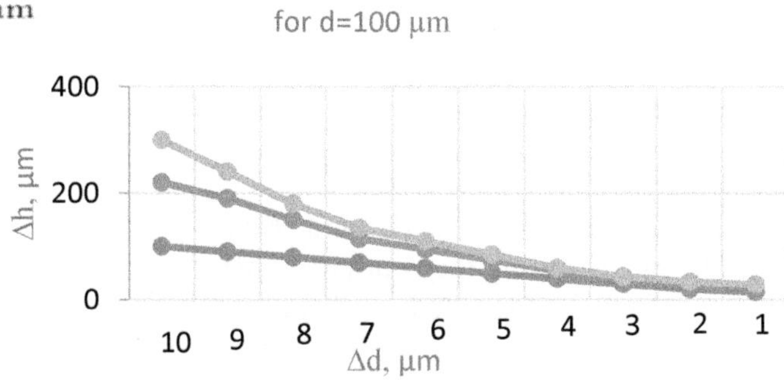


(b)

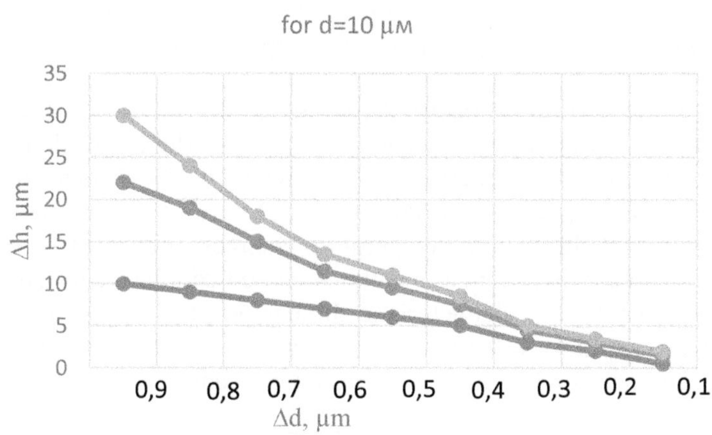


(c)

**Figure 3.8.** The 5th component of systematic error of liquid-in-tube thermometer with different thermometric liquids (water - grey, mercury - red, alcohol - blue) caused by deviations in capillary inner diameter at its size: a) 1000 μm; b) 100 μm; c) 10 μm.

### 3.3.4. The Inertia of Liquid-in-Tube Micro- and Nanothermometers

In reality, the movement of a liquid in a capillary of a thermometer, which determines the readout of the temperature value, due to the change in the thermodynamic conditions during the measurement, is not so rapid to satisfy the metrology. The experience of studying the spontaneous penetration of liquids in porous micro and nanochannels of natural structures is limited mainly by works that capture the significance of the influence of initial conditions.

The features of filling capillaries of different internal diameters at different angles of inclination are considered below. It enables us to work out the technology of manufacturing and application of these thermometers, and thus to predict the duration of setting the micro thermometers' indications not limited by thermal inertia.

For this purpose, the Washburn's and Bosanquet's equations are considered. The rate of filling the capillary is affected by fluid viscosity, surface tension, length of the filled capillary, and its diameter. Since the viscosity of the liquid increases slightly with pressure and decreases exponentially with temperature, it is inversely proportional to the rate of liquid penetration or the rate of moving meniscus. We have analyzed the impact of these factors on the micro thermometers' performance that promotes their efficient production and application.

If for a liquid-in-tube macro thermometer the thermodynamic force caused by the volumetric expansion of a sensitive fluid is the determining one, then for the micro thermometer we deal with the displacement of a fluid column regarding a certain mark due to a change in the surface tension of that fluid as the temperature changes. Important for macro-, micro- and nanothermometers is the so-called capillary constant a, as a value describing the characteristic dimension of the system: $L < a$ in which the capillary effects (capillary suction, the capillary motion of fluid, capillary condensation, etc.) become essential:

$$a = \sqrt{\frac{2\sigma}{(\rho_1 - \rho_2)g}} \tag{3.48}$$

For example, for water at 293 K, the capillary constant is equal to 3.8 mm. Note that:

$$a^2 = hd/2 \tag{3.49}$$

Here h and d are the height and the diameter of the liquid column in the capillary tube. To determine the dominance of certain forces in the design of a considered thermometer while reducing its size to the micro and further to the nano dimensions, it is advisable to use the introduced dimensionless criterion that is the Bond number $B_o$:

$$B_0 = \rho g d^2/\sigma \tag{3.50}$$

When $B_o < 1$, the capillary forces dominate; at $B_o > 1$, the gravitational forces prevail. The fluid penetration into the horizontal capillaries is characterized by a more substantial duration. An asymptotic solution determines the maximum duration of readouts' stabilization of a micro thermometer. It is described by the Washburn's equation, which is a limiting case of sufficiently long exposure durations. The simplified Washburn's equation deduced to fluid movement in a capillary tube at the absence of gravitation is quite accurate when the capillary force is still much greater than the gravitational force. It looks like this:

$$L^2 = \frac{\gamma D t \cos(\varphi)}{4\eta} \tag{3.51}$$

Here t is the time; L is the distance; $\gamma$ is the surface tension; $\eta$ is the dynamic viscosity, $\varphi$ is an angle between the liquid and the solid. When deriving the equation, the inertia of the fluid is ignored as insignificant, giving rather a considerable velocity for small values of t. However, the constant current Poiseuille conditions do not take into account the effects of mass transfer, which lead to the acceleration of the transfer flow at points of changing the inner geometry of the capillary. An improved Washburn's equation is obtained, which takes into account the inertia of fluid [27] - the Bosanquet's relation, that is, the differential equation of the 2nd order in time derivatives. It describes the movement of a liquid in a capillary tube or in a porous material that can be regarded as a set of capillaries. Using the notation x for the distance passed by the fluid, and assuming that the motion is completely caused by the surface tension forces, without applying pressure to the ends of the capillary, the Bosanquet's equation was derived as:

$$\frac{d}{dt}\left(\pi r^2 \rho x \frac{dx}{dt}\right) + 8\pi\eta x \frac{dx}{dt} = 2\pi r \gamma \cos\theta \tag{3.52}$$

The next equation studies the initial motion of a fluid with a relatively short (close to 0) duration of the experiment:

$$x^2(t) - x^2(0) = \frac{2b}{a}\left[t - \frac{1}{a}(1 - e^{-ct})\right] \qquad (3.53)$$

Here $c = \frac{8\eta}{\rho r^2}$, $b = \frac{2\gamma \cos\theta}{\rho r}$. Under this condition, the initial position of the meniscus is determined, which, as can be seen from the equation, is proportional to time. Eventually, the equation decomposes to the known Washburn's form, which reproduces the dependence of the meniscus shift on time and viscosity [28]. The latter increases slightly with pressure and decreases exponentially with temperature.

Thus, over a period determined by the inertia of the thermometer the latter mounted in a medium with of another temperature should set its readout, significantly changing the height of meniscus in the capillary. Therefore, the above parameters rather characterize the ability to perform micro- and nanothermometer of structural materials and the certain design (capillary selection, selection of thermosensitive substance, their mutual wetting and penetration of the mentioned fluid into the capillary), the movement (horizontal, inclined, vertical) of the latter depending on the position of the thermometer in space, the correct choice of the ratio of the length and diameter of the capillary. That is, the liquid filling (capillary constant a) has to be matched with optimization of the design provided the needed Bond number.

According to Jurin's law, the liquid height in the capillary is proportional to the coefficient of surface tension of the liquid and inversely proportional to the radius of the capillary and the density of the liquid. The values of height depending on thermosensitive liquid for different diameters of capillaries are shown in Table 3.4.

$$h = \frac{2\sigma}{\rho g r} \qquad (3.54)$$

The capillary filling rate is determined by the surface tension, the viscosity of the fluid, the radius of the capillary, and the length of the filled part of the capillary z, and is described by the equation:

$$\frac{dz}{dt} = r\sigma \cos\theta / 4\mu z \quad (3.55)$$

Examples of filling capillaries of different diameters with water, alcohol, mercury in their horizontal arrangement are given in Table 3.5.

**Table 3.4.** Height of capillary filling by water, alcohol, and mercury at the radiuses of capillaries 0.1; 0.5 and 1 μm.

| **The radius of capillary r, μm** | **Height of column, μm** | | |
|---|---|---|---|
| | Water, $h_1$ | Alcohol, $h_2$ | Mercury, $h_3$ |
| 0.1 | 2940 | 1122 | 1410 |
| 0.5 | 2210 | 841 | 1058 |
| 1.0 | 1470 | 561 | 705 |

**Table 3.5.** Capillary filling rate dz/dt by water, alcohol, and mercury at $0^0$ depending on radius r.

| **Capillary filling rate by water, alcohol, and mercury** | | | | | | | | |
|---|---|---|---|---|---|---|---|---|
| $r_1$, μm | z, μm | dz/dt, μm/s | $r_2$, μm | z, μm | dz/dt, μm/s | $r_3$, μm | z, μm | dz/dt, μm/s |
| Water | | | | | | | | |
| | 15 | 13.42 | | 15 | 67.11 | | 15 | 134.23 |
| 0.1 | 20 | 10.07 | 0.5 | 20 | 50.34 | 1 | 20 | 100.67 |
| | 40 | 5.034 | | 40 | 25.17 | | 40 | 50.34 |
| Alcohol | | | | | | | | |
| | 15 | 3.41 | | 15 | 17.07 | | 15 | 34.14 |
| 0.1 | 20 | 2.56 | 0.5 | 20 | 12.80 | 1 | 20 | 25.61 |
| | 40 | 1.28 | | 40 | 6.40 | | 40 | 12.80 |
| Mercury | | | | | | | | |
| | 15 | 5.13 | | 15 | 25.67 | | 15 | 51.33 |
| 0.1 | 20 | 3.85 | 0.5 | 20 | 19.25 | 1 | 20 | 38.50 |
| | 40 | 1.93 | | 40 | 9.62 | | 40 | 19.25 |

As can be seen, the fill rates of capillaries (diameter 1.0 μm) with water in the horizontal position are 50… 134 μm/s, with alcohol - 13… 34 μm/s, and with mercury - 19… 51 μm/s. This means that a liquid-in-tube micro thermometer of 100 μm length provided almost filled the capillary (diameter 1.0 μm) while altering its readings by 10 μm, is characterized by a setting time of 0.20… 0.07 s. for water,

0.80… 0.30 s. for alcohol and 0.5… 0.02 s. for mercury. When the diameter of the capillary is reduced by 10-fold, i.e. up to 0.1 μm, the time of setting the thermometer readouts of the same length increases by 10 times as the filling rate decreases by an order of magnitude. For mercury, it can reach 5.0 s., which may not meet the requirements.

Thus, the filling rate of the thermometer's capillary does not fully characterize its inertia. By correct selection of the capillary diameter, it is possible to achieve the compatibility between the influence of thermophysical characteristics and the influence of the characteristics caused by the movement of the thermosensitive fluid due to the change of the surface tension with the temperature, on the micro thermometer inertia.

### 3.3.5. Liquid-in-Nanotube Thermometers in Tumours Treatment

Magnetic fluid hyperthermia is the cancer treatment method based on the heating of oncologic injured tissues using MNPs. Their controllable size allows bond and interaction with biological entities once they are coated with biodegradable and biocompatible molecules such as dextran, polyvinyl alcohol, and phospholipids. These MNPs can also be propelled through blood vessels in the external magnetic field gradient. Nowadays are widely researching composites based on hydrogels and MNPs. They are applied in practice since their physical properties are similar to those of living tissue [29]. Heating MNPs together with hydrogels by altering the magnetic field enables combining magnetic hyperthermia and chemotherapy. It increases the efficiency of treatment by 1.5-2.5 times [30]. But heating to the target temperatures causes cellular inactivation in a dose-dependent manner. Cancerous tissue does not possess an advanced vascular system; therefore, it does not cool itself as effectively as healthy tissue. When cancer cells are heated higher than 42 °C they begin to die, while healthy cells are unharmed till 48 °C [31]. Based on numerous studies it is widely considered that the maximum safe temperature is 43 °C [32-33].

Temperature studies in micro volumes of human tissue become the contemporary challenge since they have to be exactly performed conveniently in-situ.

- **Temperature and Heat Transfer in Biological System**

To apply successfully the hyperthermia, it becomes necessary for the provision of the desired temperature regime in the entire volume to be treated. Temperature measurement during hyperthermia realization is mostly invasive and provides results only at certain points. Therefore, and for safety reasons the temperature field is previously examined by modeling only for body parts in which is planned to create hyperthermia. This helps to prevent the formation of dangerous dots of overheated tissue. For modeling of thermal processes in the body, the different equations of heat transfer in biological systems are enjoyed. Often researchers apply Pennes' bioheat transfer equation [24]. Due to relative simplicity, the Pennes' model is quite spread since it takes into account the heat transfer between tissue and blood.

Thus, the MNPs key parameter in terms of heating by the alternating magnetic field is the released specific heat. Defining this option and thermal conditions, it may be estimated the temperature field in the body part being treated. Unfortunately, the gap arises between studies of the heating of the material during its engineering testing and research on objects. Thus, the need to measure temperature directly in the field of irradiation occurs. The magnetic losses for MNP heating beyond human tissue can be determined by the calorimetric method. To exclude the impact of the magnetic field on temperature, an optical fiber thermometer was applied [34].

While radiating the tissue by the alternating magnetic field the very process of heating even of introduced MNPs complicates and becomes nonstationary. Another nonstationary process overlaps with it; that is the process of heat removal from particles dependent not only on temperature difference and MNPs concentration but on the concentration of moving particles of blood, lymph, and more. Therefore, preference should be given to direct measuring the temperature of tissue irradiated micro volumes, and it can be realized utilizing micro and nanosensors [35].

Currently, the established method for measuring the temperature and its gradient at the nanoscale are absent. Below we consider some possible methods of temperature measurement of biological objects, taking into account their noninvasiveness, possibility of measurement with minimal methodological error, and so on. Important is the last point. Our studies have demonstrated that minimum methodological error in case of monitoring small objects is guaranteed only when the ratio of volumetric and thermal characteristics of the nanosensor and the object is $10^{-3}$. That is, to control the objects of linear dimensions $10^{-3}$ m, the sensor's linear size should not exceed $10^{-4}$ m. Only the few temperature sensors meet such requirements. Exactly they are considered below.

For instance, liquid crystal films that change colors depending on temperature operate at microsized thickness [36]. Fluorescence polarization anisotropy of green fluorescent protein can fast and accurately measure the temperature not harming the cell. It bases on the next. When the temperature increases, the Brownian motion of the fluorophores is accelerated. Hence, the molecules rotate more, and more re-emitted photons lose the memory of the incident light polarization. Consequently, the temperature jump causes decreasing a degree in fluorescence polarization [35]. Unfortunately, bioluminescent methods refer to qualitative methods of thermometry. They are described by metrological characteristics, in particular, repeatability, reproducibility, reliability, and so on.

- **Contact Nanothermometry**

Sensors with liquid- and solid-phase sensitive elements are already well-known in thermometry. Their construction implies the availability of a narrow tube with a movable thermosensitive substance inside. Principles of sensor functioning root in expanding the substance volume with temperature rise. Consequently, getting the larger volume, substance moves along the tube marked by a temperature scale. It was designed the nanothermometer of µm-length based on the beta-$Ga_2O_3$ nanotube of diameter 70-100 nm filled with Au/Si alloy [36].

For liquid-state macro thermometer exists the possibility of binding the changes of column length $\Delta h$ and the increase of temperature $\Delta T$. For example, when thermo expanding coefficient $\alpha_V = 10^{-4}$ 1/K and diameter of the spherical container for liquid D=10d (d is the diameter of the thermometric tube), the result is developed into the equation of graduating characteristic: $\Delta h = 0.067\, d\Delta T$. Hereby, the sensitivity of a sensor is reducing with the decrease in the diameter d. A similar equation is deduced also for micro- and nanosensors filled with solid-phase sensitive substances.

Taking into account behavior of liquid in a capillary tube, the equation of graduating characteristic of a nanothermometer with a liquid-phase sensitive element was defined as $\Delta h = C\,(T_c - 6 - T)$, K. Here, $T_c$ depends on thermosensitive material, constant C of graduating characteristics comprises the size of the inner diameter of thermometric tube d. The sensitivity of a nanosensor with a liquid-phase sensitive element rises relatively to a decrease in tube diameter, whereas the sensitivity of a nanosensor with a solid-phase sensitive element falls.

- **Contactless Nanothermometry**

In nanothermometry most universal is Raman method, apt for the direct temperature measurement of micro- and nanoobjects within the range 100 nm – 100 μm from cryogen till mid-high temperatures. It does not demand calibration. Determining the temperature by it is based on the dependencies of a) ratio of intensities of Stokes and Anti-Stokes components of scattered radiation, and b) the frequency shift of the object proper radiation. Progress in nanotechnologies permits certain nanoobjects to serve as standard patterns. Classist of nanotechnology [37] deems that CNTs represent exactly such patterns. Their Raman spectrum is simple and well reproducible and assures good reproducibility of received results. To decrease different components of error, He-Ne laser of continuous effect and small power 1.2 mW as well as MS3501i spectrometer were used. It is necessary to underline that we need 50 mg of CNTs to cover the square 1 $mm^2$ chosen for research. Moreover, a thin plastic film package does not distort the structure and intensity of light in the studied waveband as its excitation lines are shifted to the UV spectrum.

The 2$^{nd}$ method bases on the temperature determination due to the frequency shift of object radiation. Our studies envisage that shift is reducing within 15 ... 250 °C from 1585.6 $cm^{-1}$ at 15 °C to 1576.1 $cm^{-1}$ at 250°C. The temperature coefficient of frequency change is -0.041 $cm^{-1}$/°C. Gained result at 287.27 K was estimated with the expanded error of 0.58% and combined standard uncertainty 0.3% at the credence level 0.95, and expanded coefficient 1.96. The method was elaborated; just several modes for improving the accuracy and reducing the measurement uncertainty twice were studied. They included the modes of defining shift by a) maximal value of the peak intensity; b) averaging the mean integral value of the anti-Stokes component based on perfecting the hardware and software.

## 3.4. Frequency Strain Gauge Development

It has been concluded by experts that MEMS and nanotechnology are two different labels for what is essentially a technology encompassing highly miniaturized things that cannot be seen with the human eye. Note that a similar broad definition exists in the integrated circuits domain which is frequently referred to as microelectronics technology even though state-of-the-art IT technologies typically have devices with dimensions of tens of nanometers. Whether or not MEMS and nanotechnology are one at the same, it is unquestioned that there are overwhelming mutual dependencies between these two technologies that will only increase in time. Perhaps what is most important are the common benefits afforded by these technologies, including increased information capabilities; miniaturization of systems; new materials resulting from new science at miniature dimensional scales; and increased functionality and autonomy for systems [38]. In the current point, we discuss below the abilities of a similar issue on MEMS on one hand and nanosensor on the other hand aiming further development of the last. Special attention is devoted to implementing the frequency type of MI since it features the main advantage, namely the possibility within the MI to move from measurand to frequency. An exactness of measuring the latter is the best currently.

New severer requirements for measuring transducers, together with significantly increased information flow generate a need for digital computing techniques and new designing opportunities for sensors, the output signal of which is integrated with an input of computing complexes.

A strain gauge is a sensor whose resistance varies with applied force; It converts force, pressure, tension, weight, etc., into a change in electrical resistance which can then be measured. When external forces are applied to a stationary object, stress and strain are the results. Stress is defined as the object's internal resisting forces, and strain is defined as the displacement and deformation that occur. The strain gauge is one of the most important sensors of the electrical measurement technique applied to the measurement of mechanical quantities. As their name indicates, they are used for the measurement of strain. As a technical term "strain" consists of tensile and compressive strain [39].

Comparing with other types of frequency transducers the resonant one has demonstrated a row of advantages [40-41]. Resonant sensors provide, in addition to their excellent stability, resolution, and accuracy, a signal output in frequency form that is immune to intensity fluctuations, and allows the devices to be easily connected with digital systems, which is required for its effective employment as a measurement device. The structure should only respond to changes in the load to be measured [42].

We will discuss some of these aspects in more detail in the following point but focus here on semiconducting FSC as a studied object. Since most device designs for sensor systems are inherently difficult to comprehend, we will first give a design example for the FSC application.

For instance, nanomaterial-based cantilever sensors [43] mainly are resonant sensors that take advantage of the small mass. CNTs can be used as a mechanical resonator, which allows detecting the molecules on the nanometer scale. A nanotube resonates at a particular frequency due to the size and shape of the tube. If the mass is loaded onto this tube, the resonant frequency will change, due to the orders of magnitude in the size difference between CNTs and other MEMS. A radio-frequency electric field is applied to the CNT to forces it into resonance. The field emission current from the nanotube to the counter electrode is measured to monitor the nanotube resonance. In this case, gold atoms evaporated from a resistive heater attach to the nanotube. A quartz crystal membrane sensor is used to stabilize the evaporation rate.

Often, the tube actuators (single tubes and tube arrays with different dimensions) are fabricated to optimize the field-induced strain in the axial (tube height) direction. Tube actuators with helical electrodes were also fabricated to determine the possibility of enhancing field-induced displacement along with the tube height. Such tubes with helical electrodes were made of a piezoelectric material and silver–palladium electrode [44].

Conventional strain gauges are made of wire, metal foil, or semiconductor unit that is mounted to the surface of the object. As the underlying object is strained, the strain is transferred to the strain gauge material and a change in electrical resistance is exhibited proportional to the deformation (i.e. strain) of the strain gauge substance. Some semiconductor strain gauges are made of a piezoresistive material such as silicon. As the piezoresistive material is stressed, the fundamental materials properties of the material change causing the resistance to change. This change in electrical resistance is measured to provide an indication of the stress in the piezoresistive material and therefore also an indication of the strain in the underlying object. MEMS devices are extremely small machines fabricated using integrated circuit techniques or the like. The small size of MEMS devices makes possible the mass-production of high-speed, low-power, high sensitivity, and high-reliability mechanisms that could not be realized on a larger scale. It would thus be desirable to provide a MEMS device that can be mounted onto an underlying object that is operable to provide an indication of strain experienced at the object [45].

Further development of electromechanical resonators for vibration frequency transducers is based on the semiconductor FSCs with piezoresistive effect [46]. The performance of the structure and related thoroughness of their physical and mechanical properties enable considering the FSCs as a mechanical medium, strength, and elastic properties of which are best suited for the vibration system with maximum possible Q-value, stability, and reliability. As a rule, the choice of a string substance determines the automatic selection of the vibration excitation system. Certain advantages are provided by the electrostatic method for excitation of FSC vibrations. It provides extreme convenience to design vibration transducers based on the string type of the strain gauge method.

### 3.4.1. Model and Principle of Sensor Design

For this purpose, the Si and Si-Ge single-crystals are studied below as they enable better implementation of advantages of frequency transducers with mechanical resonators (Fig. 3.9).

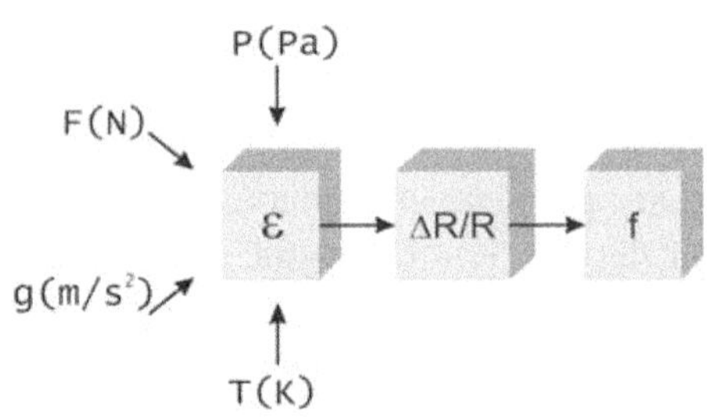


**Figure 3.9.** Strain gauge principle of converting vibrations of FSC in the frequency.

First, a direct transformation of mechanical vibrations into electrical vibrations inside of a vibrating crystal precludes the use of special transforming devices that simplifies the design of the MEMS, enhances its reliability and metrological quality. Second, elevated elastic properties of filamentous crystals, absence of their plastic deformation, and hysteresis enable achieving higher sensitivity and reproducibility of results. Since FSCs are capable of retaining their elastic properties almost up to the ultimate strength boundary, the deformation measurement range is essentially extended. Third, the fixation of crystal ends in the separate points enables the application of short and thin filaments that ensures high sensitivity.

While modeling the operation of a frequency vibration strain gauge, the FSC in the form of a thin elastic string with rigidly fixed ends (Fig. 3.10) was considered. Such approximation is justified by the rigidity of FSC's fixing utilizing glass-ceramic cement. The energy losses in mounting points were neglected.

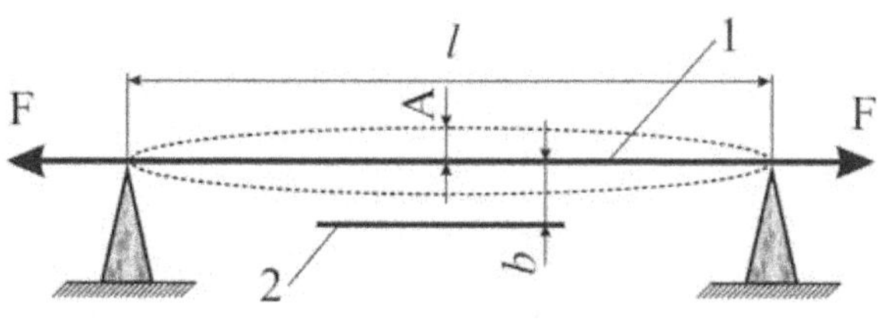


**Figure 3.10.** Scheme of the operation of MEMS with FSC electrostatic excitement.

An alternating voltage is generated between the crystal 1 and the exciting electrode 2 that results in arising of the electrostatic interaction force:

$$P = \frac{c^2 U^2}{4\pi\varepsilon_0 b^2} \cdot \sin^2 \omega t = p_0 \left(1 - \cos 2\omega t\right) \tag{3.56}$$

Here b is the distance between FSC and exciting electrode; c is the system FSC-exciting electrode capacity; $c_o$ is the absolute dielectric constant. This force is a gravity force, that is determined by the permanent component $P_o$ and the variable harmonic component $P_0 \cos 2\,\omega t$. Under the action of the harmonic exciting force, the sine harmonic vibrations are established in the crystal. The transverse vibration equation of the crystal can be written as follows:

$$y(x,t) = \frac{4P}{n\pi\mu\omega_0} \sin \omega_0 t \cdot \sin \frac{n\pi x}{l} \tag{3.57}$$

Here $\omega_0$ is the basic frequency of FSC vibrations (resonance condition); n is the harmonic number; vibration amplitude; and a maximum deviation of the point x at the moment t:

$$A = \frac{4P}{n\pi\mu\omega_0}, \quad y_m(t) = \frac{4P}{n\pi\mu\omega_0}\sin\omega_0 t \tag{3.58}$$

The operating principle of the vibration frequency strain gauge is based on a deformation of the FSC while vibrating, i.e. its length changes periodically. The total length of the crystal l at any moment of time t can be written as the sine segment $A(t)\sin\pi x/l$ between the points $x = 0, x = l_0$, for which the displacement is equal to y=o, where $A(t) = A_0 \sin\omega_0 t$. Considering that the length l of the curve $y = y(x,t)$ is defined by:

$$l = \int_0^{l_0} \sqrt{1 + \left(\frac{\partial y}{\partial x}\right)^2}\, \partial x \tag{3.59}$$

we determine:

$$l = l_0\left[1 + \frac{\pi^2 A^2(t)}{4l^2}\right]. \tag{3.60}$$

Hence, the relative elongation of the string (deformation) is as follows:

$$\frac{\Delta l}{l} = \varepsilon = \frac{\pi^2 A_0^2}{8l^2}(1 - \cos 2\omega_0 t) \tag{3.61}$$

Thus, the transverse vibrations of the initially deformed crystal produce due to the strains of the crystal substance the harmonic component with the frequency equal to the doubled frequency of FSC's vibration and the amplitude proportional to the squared ratio $A_0/l$. The relative change of the crystal resistance is also proportional to the deformation with the factor K:

$$\Delta R / R_0 = K\varepsilon \tag{3.62}$$

The variable deformation ε causes a change of the crystal resistance in the result of the mentioned effect:

$$\Delta\tilde{R} = -\frac{\pi^2 K A_0^2}{8l^2} R_0 \cos 2\omega_0 t \tag{3.63}$$

If the string is fed with direct current $I_o$, the resistance vibrations result in voltage vibrations:

$$\tilde{U} = -I_0 R_0 \frac{\pi^2 K A_0^2}{8l^2} \cos 2\omega_0 t \qquad (3.64)$$

The frequency of this voltage can be defined using the known approximation formula for an ideal string:

$$f = \frac{1}{2l}\sqrt{\frac{\sigma}{\rho}}, \qquad (3.65)$$

where l is the string length; σ is the mechanical stress inside the string (tensile force); ρ is the substance density. It is easily seen that at $\sigma_{tension} \approx 10^9 \mathrm{N/m^2}$ for the FSC with a minimum diameter of ~5 μm the maximum value of its frequency per unit length of the crystal is provided. The small value of semiconductor substance density (for Si $\rho = 2.33 \cdot 10^3$ kg/m$^3$) enables keeping vibrations active with minimum values of excitation energy.

The block scheme of frequency strain gauge yields the next units (Fig. 3.11). Single-crystal as an elastic element, which senses a measurement value and transforms a deformation into the frequency. Practically absence of the hysteresis (~$10^{-9}$ at 300 $^0$C in Si) provides the replicability of measurements. In a considered issue the vibrating and transducing functions are combined in a single unit, which is a semiconductor FSC. Vibrating element – an electromechanical resonator, which changes the frequency of its natural vibrations depending on the level of crystal deformation. Electrode for the supply of exciting alternating voltage. In our case, the surface of the piezoelectric element adjacent to the crystal is used as an exciting element. The device for registration of signal frequency is a digital frequency meter with the class of exactness $\geq 10^{-4}$.

The operation of the proposed strain gauge does not significantly depend on the temperature dependencies of sensitivity factor K and the value of resistance since the parameter to be measured is the vibration frequency. The latter is defined by the crystal geometry, mechanical properties, and applied force. The defined sensitivity factor K of such strain gauge is $10^9$ Hz/relative deformation.

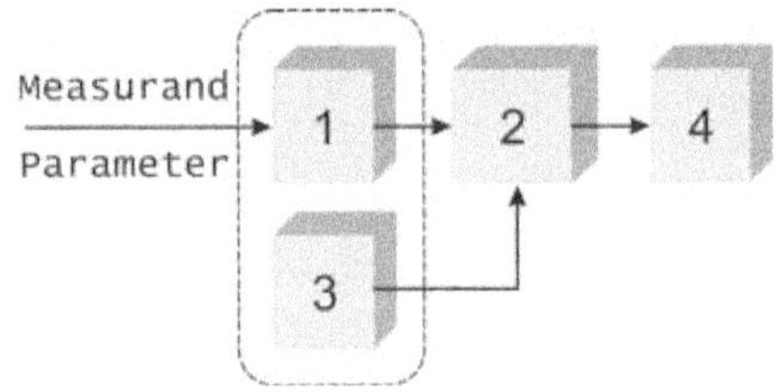


**Figure 3.11.** Block diagram of frequency strain gauge: 1 - Single-crystal; 2 - Vibrating element – an electromechanical resonator; 3 – Electrode; 4 - Registration device.

The techniques developed for research of the FSCs under cyclic loads made it possible to study the specifics of their properties and structure, to raise the measurement accuracy of the output signal parameters. We have investigated vibration parameters while loading filamentous crystals of different geometry and using magnetoelectric and electrostatic modes of excitation (Fig. 3.12).

We have studied the optimal conditions of excitation and the influence of the FSC's ends fixing method on the vibration mode. It was established that the fixing of the FSCs employing glass cement with appropriate mechanical and temperature properties similar to the same of FSCs, ensures a durable connection. It has been determined since the better technique for excitation of mechanical vibrations in a crystal is an electrostatic one. In modeling the resonator operation in the function of a transducer, we studied the dependence of the FSC's natural frequency of vibration depends on the tensile force applied to it. The obtained results envisage that the special feature of such resonator is a broadband of vibration frequency that changes under deformation for a sufficient level of the input signal.

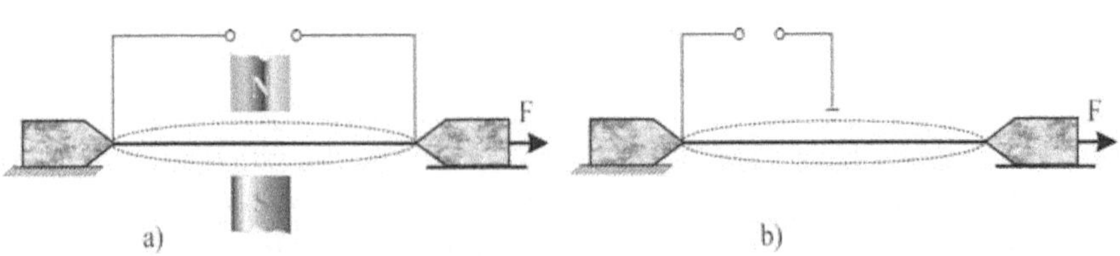


**Figure 3.12.** Excitation methods of FSC vibrations: a – Magnetoelectric; b – Eelectrostatic.

The output signal after bending and stretching of the crystal while vibrating was estimated. We have established the relationship between the parameters of this signal and geometrical + elastic parameters of the FSCs, their axial stress, Q-value of the vibration system, amplitude, frequency, and phase of mechanical vibrations.

It has been determined that the FSC relative resistance changes:

$$\frac{\Delta R''}{R''_0} = \frac{K^2}{4}\left(\frac{r}{l}\right)^2 + \left(\frac{A}{l}\right)^2 \int_0^l \left(\frac{d^2U_0}{d\xi^2}\right)^2 d\xi \qquad (3.66)$$

is caused by its bending deformation under vibration and approximates to $10^{-4}$, variable integration $\xi = x/l$, $U_o$ is the function of FSC point maximum deviation from the equilibrium position, A is the amplitude of maximum deviation; l, r is the length and radius of the crystal.

The relative change of crystal resistance due to its bending and stretching under vibrations is proportional to the relative deformation and the squared amplitude of vibrations, being in two orders bigger:

$$\frac{\Delta R''}{R''_0} = \frac{K}{2} + \frac{A^2}{l^2} \int_0^l \left(\frac{d^2U_0}{d\xi^2}\right)^2 d\xi \approx 10^{-2}. \qquad (3.67)$$

The most acceptable physical model for resonator calculation based on the FSC is an elastic homogeneous rod with fixed ends. Considering the boundary conditions, the equation of natural transverse vibrations of such rod is as follows:

$$1 - chk \cdot \cos h + \frac{v^2}{2\sqrt{\lambda}} shk \cdot \sin h = 0\,. \qquad (3.68)$$

For an unstrained rod, where $v^2 = 0$, the equation is simplified to:

$$1 - chk \cdot \cos h = 0. \qquad (3.69)$$

The frequencies of the 1st harmonics are defined with a consistent approximation. Received results of computing are provided on the Fig. 3.13 as the dependencies between frequency and mechanical stress σ that affects the unit of the crystal cross-section area.

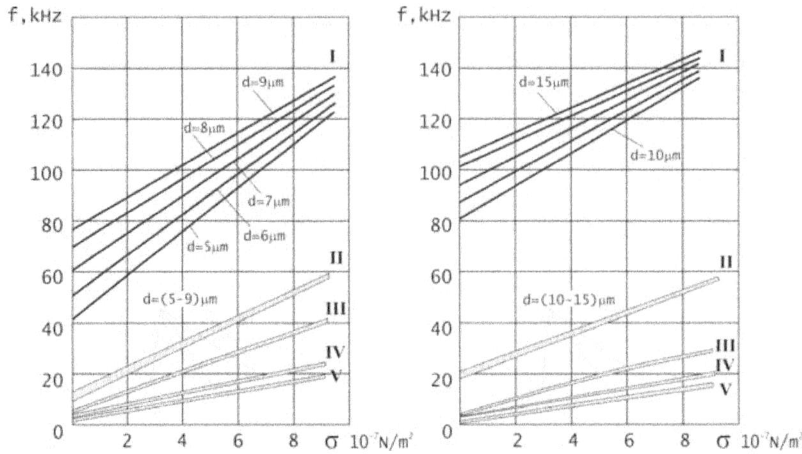


**Figure 3.13.** The dependence of the frequency *f* on the stress $\sigma$ for FSCs of different lengths and diameters: I – *l*=1 mm; II – *l*=2 mm; III – *l*=3 mm; IV – *l*=4 mm; V– *l*=5 mm.

While altering σ the changes in frequency become more significant for the shorter and thinner FSCs. This is since while FSC diminishing the vibrating element starts to resemble an ideal string. Its bending moment resistance decreases proportional to diameter, and the ratio of strain to the linear density for permanent σ does not dependent on diameter. The established linear dependence of the natural frequency of FSC vibrations on the diameter and inversely proportional to the exponential dependence on the length, confirm the correctness of the considered approach.

## 3.5. Conclusions

1. Since the sizes play a crucial role in ascertaining the properties of nanomaterials and new phenomena are observed in nanopatterns, based on nanothermodynamics and nanometrology we have considered the next peculiarities:

- The decreasing to zero the temperature coefficient of electric resistance in MGs caused by the precipitating the 2$^{nd}$ phase; it seems to be important not only for exact resistance thermometers production but for understanding the nature of the effects occurring in strained-engineered MOSFET [3, p.115];
- The action of surface tension force (the main component of the

state equation of nanothermodynamics) provides establishing the graduating characteristics of liquid-in-tube thermometers while decreasing sizes to a nano area.

2. Analysis of the studied temperature dependences of electric resistance of $Fe_{40}Ni_{38}P_{13}B_9$ metallic glass enables the evaluation of its structure as a structure of heterogeneous two-phase material. The researched MG favorably differs by the null value of the electric resistance temperature factor and therefore could be employed for manufacturing the high-precision resistance thermometers and strain gauges.

3. The development of nanotechnologies is impossible without measuring the temperature of micro- and nanosized objects, which requires further development of liquid-in-tube thermometers, whose thermometric characteristics vary depending on linear dimensions. The force of the surface tension is decisive in the formation of the graduating characteristic of the micro- and nanothermometers. This characteristic is inversely proportional to the temperature, as the surface tension force is falling while the temperature raising and at a certain high temperature becomes equal to zero. The latter determines the upper limit of a liquid-in-nanotube thermometer application.

4. An important feature is the ratio of thermo-physical and mass-dimensional characteristics of a sensitive element of thermometer concerning the measured object. It determines the methodic error component of temperature measurements. For nanotechnology, the issue of nanothermometers becomes crucial since minimizing the methodic error to a level below 1% means that each linear dimension of the thermometer should be 1 order smaller than the size of the measured object.

5. Results of the study of liquid-in-tube thermometers characteristics while diminishing their sizes, f.i. the diameter of the capillary, demonstrate:

- The thermal inertia of the thermometer decreases significantly due to its intense heat exchange with the environment, it can drop to 0.001 s.
- While the diameter of the capillary decreases, its inertial characteristics can no longer be fully described by the thermophysical phenomenon alone. The hydrodynamic phenomenon estimated by the filling rate of the capillary and the movement of the thermosensitive fluid in it directly characterizes the time of setting the readouts of the

micro thermometer. This time is powering while diminishing the sizes. So, it may be in 1-2 orders of magnitude higher than similar value caused by the heat exchange.

- By correct selection of the capillary diameter in creating micro thermometer, we can achieve conjugation of thermophysical characteristics impact, which determines the 1st constituent of thermal inertia, and the impact of the thermosensitive fluid movement with temperature caused by the surface tension changes (the 2nd constituent of thermal inertia).
- Challenge in medicine is to develop current and/or create new treatment methods, including measurement techniques and standards, which rely on nanoscaled materials. Within range 270-330 K for Raman method, as one of the few direct methods of thermometry, the carbon nanotubes are treated to be standards or universal calibration artifacts for measuring the temperature of micro- and nanosized objects. So, CNT nanosensors are recommended to apply for temperature measurements at tumors' magnetic fluid hyperthermia treatment.

6. The research of FSCs confirms their promising application as a sensitive element of the frequency strain gauge, metrological parameters of which is determined by the density of the sensitive substance, by FSC length and cross-section as well as by an elasticity modulus. The measurement range of the frequency strain gauge becomes wider since the FSC retains the elasticity modulus almost at the highest permissible mechanical loads. The tensile deformation of FSC determines its oscillating frequency and sensitivity.

## References

[1]. S. R. De Groot, Thermodynamics of irreversible processes (https://pdfs.semanticscholar.org/cd9f/be1dfb317dd729ff8e4be108411dead1eb56.pdf).

[2]. F. Bernhard, Handbuch der technischen temperaturmessung, *Springer Vieweg*, 2014.

[3]. C. K. Maiti, T. K. Maiti, Strain-engineered MOSFETs, *CRS Press, Taylor & Francis Group*, 2013.

[4]. Handbook of nanomaterials properties, Coord. editor B. Bhushan, Editors D. Luo, S. Schricker, W. Sigmund, S. Zauscher, *Springer*, 2014.

[5]. S. Yatsyshyn, B. Stadnyk, Z. Kolodiy, Development of Noise Measurements. Part 1. Fluctuations and Thermodynamics, Proper Noise

and Thermometry, *Sensors and Transducers*, Vol. 150, Issue 3, March 2013, pp.59-65.

[6]. R. Leach, R. Boyd, T. Burke, The European nanometrology landscape, *IOP Publishing, Nanotechnology*, Vol. 22, 2011.

[7]. H. Hofmann, Advanced nanomaterials, Course support, *Powder Technology Laboratory, IMX, EPFL*, Version 1, Sept 2009.

[8]. B. Stadnyk, S. Yatsyshyn, Ya. Lutsyk, Research in Nanothermometry. Part 1: Temperature of micro- and nano-sized objects, *Sensors and Transducers*, Vol. 140, Issue 5, May 2012, pp. 1-7.

[9]. B. Stadnyk, S. Yatsyshyn, O. Kozak, Research in Nanothermometry. Part 2: Methodical error problem of contact thermometry, *Sensors and Transducers*, Vol. 140, Issue 5, May 2012, pp. 8-14.

[10]. B. Stadnyk, S. Yatsyshyn, O. Seheda, Metrology of Temperature Transducer based on Raman Effect, *Sensors and Transducers,* Vol. 117, Issue 6, June 2010, pp. 78-84.

[11]. B. Stadnyk, S. Yatsyshyn, O. Seheda, Research in Nanothermometry. Part 6: Metrology of Raman thermometer with universal calibration artifacts, *Sensors and Transducers,* Vol. 142, Issue 7, July 2012, pp. 1- 9.

[12]. S. Yatsyshyn, B. Stadnyk, Ya. Lutsyk, Research in Nanothermometry. Part 3: Characteristics of the thermometers with liquid- and solid-phase sensitive elements, *Sensors and Transducers,* Vol. 140, Issue 5, May 2012, pp. 15-23.

[13]. T. Dominiuk, S. Yatsyshyn, About thermodynamic consideration the impact of mechanical stresses and strains on the thermo-EMF, *Measuring Equipment and Metrology,* Issue 59, 2002, pp. 66-69 (in Ukrainian).

[14]. S. Yatsyshyn, B. Stadnyk, Ya. Lutsyk, Research in Nanothermometry. Part 7: Ultrasonic thermometer with nanostructured thermometric materials, *Sensors and Transducers,* Vol. 143, Issue 8, August 2012, pp. 1- 9.

[15]. S. Yatsyshyn, B. Stadnyk, Z. Kolodiy, Research in Nanothermometry. Part 5: Noise thermometry and nature of substance, *Sensors and Transducers,* Vol. 141, Issue 6, June 2012, pp. 8-16.

[16]. B. Stadnyk, S. Yatsyshyn, P. Skoropad, Research in Nanothermometry. Part 4: Amorphous alloys of thermoresistive thermometry, *Sensors and Transducers,* Vol. 141, Issue 6, June 2012, pp.1-7.

[17]. S. Das Sarma, Spintronics, *American Scientist,* Vol. 89, November-December 2001, pp. 516-523.

[18]. B. Stadnyk, S. Yatsyshyn, P. Skoropad, Analiza efektywnosci stosowania w termometrii funkcjonalnie-gradientowych czujnikow, *Pomiary, Automatyka, Kontrola,* Issue 12, 2006, pp. 42–45 (in Polish).

[19]. W. Petrov, Thermal fluctuations as a generator of initiating cracks, in *Proceedings of the Conference on Physics of Strength and Ductility*, Nauka, Leningrad, 1986, pp. 11-17 (in Russian).

[20]. V. Vettegren, Spectroscopic study of the disruptive density fluctuations, in *Proceedings of the Conference on Physics of Strength and Ductility*, Nauka, Leningrad, 1986, pp. 17-27 (in Russian).

[21]. S. Yatsyshyn, B. Stadnyk, R. Samchenko, CNT nanosensors in the tumors' treatment, *Internat. Journ. Biosensors & Bioelectronics*, Vol. 2, Issue 6, 2017, pp.188-189.
[22]. L. Eötvös, Ueber den Zusammenhang der Oberflächenspannung der Flüssigkeiten mit ihrem Molecularvolumen (PDF), *Annalen der Physik,* Vol. 27, No.3, 1886, pp.448–459.
[23]. Powering nanotechnology devices with novel surface energy generators, *Nanowerk Nanotechnology*, *Spotlight*, 2010.
[24]. V. Zablotskii, O. Lunov, P. Gomez, Magnetic heating by tunable arrays of nanoparticles in cancer therapy, *Acta Phys. Pol.*, Vol. 115, 1, 2009, pp.413-417.
[25]. Standard handbook of petroleum and natural gas engineering, Editors W. Lyons, G. Plisga, M. Lorenz, 2016 (https://www.sciencedirect.com/book/9780123838469/standard-handbook-of-petroleum-and-natural-gas-engineering).
[26]. D. Quéré, F. Brochard-Wyart, P.-G. Gennes, Capillarity and Gravity, in Capillarity and Wetting Phenomena, *Springer,* New York, 2004, pp. 33–67.
[27]. Edward W. Washburn, The dynamics of capillary flow, *Physical Review*, 1921, Vol.17, Issue 3, p. 273.
[28]. Micromeritics, A new method of interpolation and smooth curve fitting based on local procedures, *Journ. Ass. Comp. Machinery*, Vol. 17, Issue 4, 1970, pp. 589-602.
[29]. S. N. Tabatabaei, J. Lapointe, S. Marte, Magnetic nanoparticles encapsulated in hydrogel as hyperthermic actuators for microrobots designed to operate in the vascular network, in *Proceedings of the IEEE/RSJ International Conference on Intelligent Robots and Systems*, St. Louis, MO, 2009, pp. 546-551.
[30]. A. Jordan, R. Scholz, P. Wust, H. Fähling, R. Felix, Magnetic fluid hyperthermia: cancer treatment with AC magnetic field induced excitation of biocompatible superparamagnetic nanoparticles, *J. of Magnetism and Magn. Mat*, Vol. 201, 1–3, 1999, pp. 413–419.
[31]. L. Blue, Fluid dynamics and heating of magnetic nanoparticles in simulated blood vessels, 2009.
[32]. Sh. Maenosono, S. Saita, Theoretical assessment of FePt nanoparticles as heating elements for magnetic hyperthermia, *IEEE Transactions on Magnetics*, Vol. 42, Issue 6, 2006, pp. 1638-1642.
[33]. K. Khanafer, K. Vafai, Synthesis of mathematical models representing bioheat transfer, *Adv. in Num. Heat Transfer*, 2009 (books.google.com).
[34]. S. Yatsyshyn, B. Stadnyk, Ya. Lutsyk, L. Buniak, Handbook of Thermometry and Nanothermometry, *IFSA Publishing*, Barcelona, 2015.
[35]. A precise nanothermometer for intracellular temperature mapping, *Nanowerk Spotlight,* Post, March 19, 2012.
[36]. N. Gong, M. Lu, C. Wang, Y. Chen, L. Chen, Au(Si)-filled b-$Ga_2O_3$ nanotubes as wide range high temperature nanothermometers, *Appl. Phys. Let.*, Vol. 92, Issue 7, *Nanosc. Sc. and Des*., 2008, 3, 073101.

[37]. M. Dresselhaus, G. Dresselhaus, R. Saitoc, A. Joriod, Raman spectroscopy of carbon nanotubes, *Physics Reports,* 53, 2004.
[38]. What is MEMS Technology? (https://www.mems-exchange.org/MEMS/what-is.html).
[39]. Strain Gauge, Introduction to Strain Gauge, Omega, Prospekt (https://www.omega.co.uk /prodinfo/StrainGauges.html).
[40]. C. J. Welham, J. Greenwood, M. Bertioli, A lateral resonant pressure sensor fabricated via fusion bonding, wafer thinning and reactive-lon-etching, in *Proc. of the 12th Eur. Conf. on Solid-State Transducers and the 9th UK Conf. on Sensors and their Appl., Southampton, UK, 13-16 Sept. 1998, EUROSENSORS XII, Ed. By N.M. White*, Inst. Physics Publ. Bristol and Philadelphia, Sensors Series, 1, 1998, pp.353-356.
[41]. M. Haueis, J. Dual, C. Cavalloni, M. Gnielka, R. Buser, Packaged bulk micromachined resonant force sensor for high temperature applications, *SPIE - Design, Test, Integration and Packaging of MEMS/MOEMS*, Paris, May 2000, 4019, pp. 379-388.
[42]. R. Luttge, in *Nano- and Microfabrication for Industrial and Biomedical Applications*, 2016 (https://www.narcis.nl/publication/RecordID/).
[43]. J. Zhang, K. Hoshino, Chapter 6 - Mechanical Transducers: Cantilevers, Acoustic Wave Sensors, and Thermal Sensors, *Molecular Sensors and Nanodevices,* 2014, pp. 321-414 (https://www.sciencedirect.com/science/article/pii/B9781455776313000065).
[44]. A. Safari, M. Allahverdi, Electroceramics: Rapid Prototyping, in *Encyclopedia of Materials: Science and Technology,* 2001, pp. 2510-2518 (https://www.sciencedirect.com/science/article/pii/B008043152600454X)
[45]. V. Varadan, K. Vinoy, S. Gopalakrishnan, Smart Material Systems and MEMS: Design and Development Methodologies, *John Wiley & Sons*, 2006.
[46]. R. Baitsar, V. Voronin, E. Krasnogonov, N. Bogdanova, Operation of monocrystalline silicon resonator in a measuring circuit, *Sensors and Actuators*, A 30, 1992, pp.175-178.

# Chapter 4

# Metrology 4.0 and Standardization for Agricultural Cyber-Physical Systems

**Tetiana Bubela, Tetiana Fedyshyn, Andriy Midyk, Ivan Pytel, Roman Ivakh, Yevhen Pokhodylo, Yuriy Khoma, Vasyl Yatsuk, Viktor Semerak and Olha Lysa**

## 4.1. Cyber-Physical System as a Tool of Monitoring, Control, and Management

Modern advanced technologies in the agricultural sector are based on the use of cyber-physical systems to ensure the possibility of data collection and network monitoring and control [1]. Modern agriculture faces enormous challenges to create a sustainable future in different regions of the globe. Examples of such global challenges are population growth, urbanization, climate change. Global economic problems must be addressed in such a way that the potential of agricultural sectors does not jeopardize the satisfaction of world food needs. Various factors are connected with the concept and needs of innovations in agriculture, among which it is expedient to single out the following: environment, biodiversity, and health care. The link between agriculture and the environment is a source of challenges and technological optimization. Mankind benefits from intensive agricultural production at the cost of losing the natural state of ecosystems.

Thus, excessive application of fertilizers can lead to the threat of environmental pollution, while insufficient amounts can lead to soil degradation and loss of fertility [2-5]. In these circumstances, agricultural production systems need to focus more on the effective conservation and management of biodiversity and ecosystem services to address the dual challenge of environmental sustainability and food security through the introduction of such a robust mean as the CPS.

The development of automation and GPS-assisted control of vehicles in the agricultural sector established the concept of precision agronomy and precision farming, as well as automation in the production chain. However, such systems are inherent in the significant disadvantages associated with flexibility, rationing, reliability, real-time data presentation.

- **Precision Farming as a Guarantee of Creating Smart-Technologies in Agriculture**

Advances in recent decades in the field of automation and remote sensing have introduced the concepts of precision agronomy and precision farming, as well as automation in the food production chain. Precision farming and automation have already established paradigms to increase productivity, quality, and improve working conditions by reducing manual labor. All these factors play an important role in ensuring the sustainable development of the agricultural sector. Many modern farmers already use high-tech solutions, such as digitally controlled agricultural equipment, GPS-enabled vehicles, and unmanned aerial vehicles for monitoring and forecasting. There are partially and fully automatic devices for most aspects of agriculture: from grafting to sowing and planting, from harvesting to sorting, packaging, and boxing, and for livestock. Although approaches to precision farming can be effective and useful for farmers, after proper training, they are usually calibrated only for a specific task, without providing a holistic view of agricultural processes.

It is worth noting that agriculture can be viewed as a dynamic system in which for each set of input data required to receive the result or final/intermediate products: however, different conditions can change the results (e.g., climatic conditions, soil quality, pests). External factors that positively or negatively affect the agricultural system, it is difficult to predict or control. For this reason, predictive models cannot always guarantee the expected results. Therefore, it is necessary to create cyber-physical systems with multiple functionalities. Farmers should know the state of their crops and collecting monitoring data in real-time are the best solution to achieve innovation and sustainable performance. Consequently, the combination of modern measuring technology and instrumentation, robotics is the basis for the concepts and methodologies of intelligent agriculture.

### 4.1.1. CPSs' Classification. Types of Risks While Developing and Operating

Analyzing the known CPS, their metrological and software, it can be concluded that with the continued growth of their number are expanding their scope, and the software and metrological services are evolving to support the work of existing CPS and their shelf life when designing a new one. Therefore, the requirements that are behind the creation of CPS, is the security, privacy, reliability, stability, warranty, relative to common interconnected devices and infrastructures, dynamism, compatibility (possibility to place various computational models), support various modes of communication, problem-solving difficulties (problems of sensing and feedback control for the certain CPS's architecture), synchronization, interaction with the operational environment, the ability to cooperate to enable the combination of several objectives.

Intending to create appropriate regulatory, technical, and metrological support of CPS, it is reasonable to classify according to different criteria. In particular, depending on the use of CPS can be divided (Fig. 4.1) [6]:

**Figure 4.1.** CPS classification by purpose.

It should be noted that in Europe one of the main factors in the use of CPSs in agriculture is the optimization of operational costs [7]. CPSs can be used and are used in almost any field of agriculture: in crop production, animal husbandry, during the processing of raw materials, transportation, storage, sales, etc. In this regard, the classification of CPSs in the field of agro-industry is of great theoretical and practical importance (Table 4.1).

**Table 4.1.** Classification of agricultural CPS.

| Criteria for classification of CPSs | Classification of CPSs |
|---|---|
| by industry | - in animal husbandry;<br>- in crop production;<br>- in auxiliary production |
| by types of work performed | - sowing of crops;<br>- spraying of plants with pesticides and fertilizers;<br>- removal, weeding;<br>- control of laboratory germination of crops;<br>- harvesting fodder crops;<br>- fruit picking;<br>- care for vineyards and gardens trees;<br>- transportation of seedlings in greenhouses;<br>- watering plants in greenhouses;<br>- mechanized soil preparation works performed by unmanned (autonomous) transport<br>- monitoring of agricultural lands;<br>- sorting of agricultural products;<br>- packaging of agricultural products |
| by the nature of the movement | - stationary man-in-loop CPSs;<br>- mobile CPS;<br>- man-out-loop CPSs |
| by type of management | - man-in-loop CPS;<br>- man-out-loop CPS |

It should be noted that the control system of modern industrial CPS is based on the repetition of programmed movements in fixed zones, while agricultural CPS applies the control systems that operate in volatile natural and climatic conditions.

In this case, it:

- Provides operation with living organisms that are the plants and animals, with unsorted and disordered objects (different varieties of plants, shrubs, fruit trees, etc.);
- Uses tools and other equipment designed for human work;
- Provides safety for people and animals working nearby.

From a technical point of view, several risks can be identified during the construction and operation of the CPS, taking into account which is important in particular for the agricultural sector (Fig. 4.2):

1) Data diversity. Data diversity is a serious problem that can negatively affect the effectiveness of interactions and the development of communication protocols. Systems must be able to support a large number of different applications and devices.

2) Reliability. Cyber-physical systems can be used in critical areas such as health, infrastructure, transportation, agribusiness, and many others, the main requirements of which are reliability and security, as actuators affect the environment and human health. Also, CPSs must be able to continue to operate in unpredictable conditions and adapt to any influences.

3) Data management. You need to store and analyze large amounts of data from different network devices, process them, and output the results in real-time. Data can be managed using deferred or operational current processing, depending on the purpose of the system. When using real-time streams, information can change frequently and processing should be based on adaptive and constant queries.

**Figure 4.2.** Risks of creation and functioning of CPSs.

4) Confidentiality. The challenge is to maintain a balance between maintaining privacy and protecting personal data and the availability of

data to provide better service. Because CPSs manage large amounts of data that contain confidential information, there are serious problems in ensuring their confidentiality.

5) Security. CPSs must ensure the security of communications, as all actions are coordinated between devices in real-time. They expand the scope and extent of interaction between physical and computer systems, which complicates the task of security. Traditional infrastructures are not enough to solve this problem, so new solutions need to be found. Both current data and stored data collected for future use must be protected. Besides, CPSs use a variety of applications and wireless communications, which often complicates security.

6) Real-time. CPSs control large amounts of data received from sensors, the processing of which must be efficient and timely because physical processes continue regardless of the results of calculations. To meet this requirement, CPSs must have a correspondingly high bandwidth because untimely actions can lead to information distortion.

- **Regulatory Support for the CPS Creation and Operation**

Ensuring an effective state policy in the field of creation and functioning of the CPS in Ukraine requires a detailed study of the current state and analysis of trends in the PFS, their types in the world in various sectors of the economy, features of management mechanisms in leading countries and especially in the EU. In Ukraine, the adoption of domestic and international standards as national with the various requirements and technical characteristics necessary for the functioning of the CPS, the certification of their functional compatibility, providing conditions for professional training of qualified personnel, and improve their skills for managing CPS. In connection with the rapid development of information technology, the expansion of services in cyberspace, and the expansion of the scope of cyber-physical systems, there is a need to develop a unified common definition of cyber-physical systems. Therefore, it is advisable to consider in more detail the features of regulatory support for the functioning of the CPS in the leading countries of the world.

The IEC, the ISO, and the Standards Association have created some standards for CPS [8], which can be systematized by their levels of operation, namely the level of intelligent connection that is responsible for obtaining data from physical objects. The most common technique is Automatic Identification and Data Capture, i.e. the use of automatic identification and data collection. ISO / IEC 19762: 2016 provides terms and definitions for AIDC. The ISO / IEC 15459 series defines unique identification for registration procedures, general rules, individual transport units, individual products and packages of goods, individual goods to be returned, transport goods and groups. The CPS needs to use sensors to automatically collect data from production systems. ISO / IEC / IEEE 21450: 2010 defines the basic functions required to control the system and control smart sensors. The ISO / IEC / IEEE 21451 series describes the Network Capable Application Processor and the information model and communication protocols, Transducer Electronic Data Sheet for smart sensors. Standard sensor control methods are important. Thus, the IEC 61131 series defines the main functional characteristics of programmable control systems. IEC 61499 defines a common model for distributed control systems based on IEC 61131, IEC 61131, and IEC 61499 provides guidelines for establishing reliable, interchangeable communication with control systems.

The level of data conversion into information includes data processing in terms of the level of intelligent connection and information analysis (Fig. 4.3). Thus, the standards IEC 61804-3 - IEC 61804-6 are used to describe the characteristics of devices. The IEC 61360 series provides the basis for a clear and unambiguous definition of the characteristic properties (types of data elements) of all elements of electrical systems from basic components to components and complete systems. Also, the IEC 62714 series provides a communication format called Automated Markup Language. The above standards ensure the availability of unified data. IEC / ISO13236: 1998 states that a high-quality system applies to the IT environment. As data security is an important issue, ISO 27000 guides best practices in information and risk management and control to achieve security. IEC 62443 series (ISA99) is used to ensure the safety of industrial automation and control systems and provides comprehensive protection and safety.

The level of cyber computing. Communication is the most important element considered at the level of CPSs and computing. CPS data and information exchange requires several relevant standards, including wired and wireless standards. ISO / IEC 8802 provides a set of international standards that describe local area networks. There are several standards for wired communication. The IEC 61158 and IEC 61784 series are standards for bus types and profiles, including common industrial protocols, PROFIBUS and PROFINET, P-Net, WorldFIP, INTERBUS, SwiftNet, CC-Link, HART, VNET / IP, TCnet, EtherCAT,

Ethernet POWERLINK Ethernet for Plant Automation (EPA), Modbus, SERCOS, Rapi Net, SafetyNet p and MECHATROLINK.

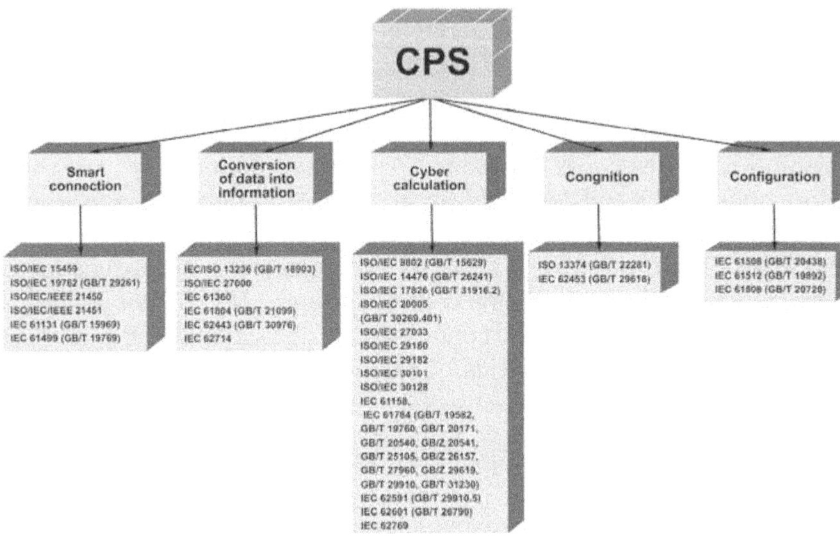


**Figure 4.3.** CPS international standard structure.

These protocols contain real-time distributed control in the CPSs with wireless communication. IEC 62591 (Wireless HARTTM) and IEC 62601 (WIA-PA) should be used for industrial wireless communication and industrial measurements, monitoring, and control. The ISO / IEC 14476 series improves the transport communication protocol to ensure good quality of service. A quality industrial network should use the above standards to connect the sensor and machine network. ISO / IEC 20005, ISO / IEC 29180, ISO / IEC 29182, ISO / IEC 30101 and ISO / IEC 30128 are used to create intelligent, reliable and secure sensor networks. Besides, ISO / IEC 17826 specifies an interface for accessing cloud storage and managing data stored internally. ISO / IEC 27033 standards ensure network reliability. IEC 62769 is used to integrate devices using communication technologies.

The level of cognition involves monitoring and decision-making. Series ISO 13374 provides the basic requirements for open software specifications that allow machines to control data processing and communication. IEC 62453 regulates the process of integration of all devices, regardless of manufacturer.

A configuration level includes standards of common control of CPS. In particular, IEC 61512 determines the control models, terminology, and data model. IEC 62264, used to integrate the enterprise management system, promote uniformity and consistency of the interface. The standard reduces the risk, cost, and errors associated with implementing these interfaces. IEC 61508 increases the safety and reliability of the life cycle and control of the industrial process.

Cybersecurity seems to be an integral necessary component of the operation of cyber-systems. Ukraine also pays attention to the development of regulatory documents on security [9-10]. Nevertheless, the Strategy of cybersecurity does not contain the concept of cyber-systems, although they are increasingly developing in the world. Thus, there is a need for a clear formulation of the concept of CPS and its main universal features and proper regulation of their operation. Based on this, it is proposed to introduce the concept of the CPS in the Ukrainian and other countries' normative documents on metrology with the definition given in Chapter 1.

It can be concluded that both the German Framework Document on the CPS and the US Framework Document draw attention to the need for appropriate certification of the CPS, which will confirm the requirements of reliability, security, stability, the confidentiality. The need for certification also is emphasized in the EU framework document Cyber-Physical European Roadmap & Strategy Research Agenda and Recommendations for Action, according to which modern methods and tools for certification are not adequate for CPS certification at present due to their non-compliance with modern CPS requirements. Another issue highlighted in the German CPS Framework Document and the relevant US Framework Document is the adoption and application of uniform CPS standards. To give a universal full definition of CPS, it is necessary to analyze aspects of the creation of CPS. This has done thoroughly in the US Framework for CPSs Release May 2016 CPSs Public Working Group.

Today, numerous efforts of various international organizations in the field of creating standards for CPSs, including ISO, ITU, Industrial Internet Consortium, IoT-A, etc., have not yet ensured full compatibility of these standards for heterogeneous CPS. Therefore, the issue of CPS standardization is open and it needs to be addressed both at the international and national levels.

### 4.1.2. Cyber-Physical System of Grain Production Control. Design Principles

Today CPSs are actively introduced in agriculture. The most important area is the cereal production. Modernization of such agricultural systems can be carried out by introducing innovative technologies based on the CPSs [11]. Manufacturers and investors who are interested in implementing such systems on their farms can become potential users of such CPS. The latter could solve the tasks, namely: starting from production control, ending with monitoring the process of selling the finished products. The architecture of the CPS may contain several levels of processing of the obtained data, each of which performs its function in the operation. Different authors demonstrate dissimilar approaches to the CPS design [11-14], but each one is developed taking into account its specific application. Therefore, the main task is to create an apparatus for making management decisions in a particular branch based on a universal intelligent system [15] (Fig. 4.4). Below we consider a CPS for cereal production.

| |
|---|
| 1.Physical world |
| 2. Means of interaction with the physical world |
| 3. Means of information acquisition and information delivery |
| 4. Means of information processing |
| 5. Decision-making means |
| 6. Means of personal service |

**Figure 4.4.** Elements of the universal structure of the CPS.

Thus, exists the need to study grain production technologies from the standpoint of creating a CPS. Indicators and the type of information can be structured. The values of the indicators can be established and the management decisions of the CPS can be made.

To adapt the CPS general structure [16] (Fig. 4.5) to managing the production of cereals, this task can be divided into subtasks, such as preparation of agricultural soil for sowing; cultivation; certification of the gained products and more. For each level of CPS, the structural elements have to be subjected but the unification of requirements for them are provided both at the level of the object of study (soil, water, air, etc.) and the level of the finished product (f. e. the grain). At each of these levels, the man-in-loop CPS has to make the managing decisions to optimize the process.

A. Determining the place of cereals in grain rotation

B. Cultivation

C. Fertilizer application

D. Preparation of seed

E. Sowing

F. Crop care

G. Harvesting

J. Quality control of harvest

**Figure 4.5.** The main stages of technology for cereals production.

Therefore, for the element № 6 of the CPS structure (Fig. 4.4), the subject's users of such a system must be defined. Farms or regulatory authorities can become such entities. At this level, the row of problem classes Z = {z1, z2,..., zi,..., zn} must be set. The CPS system is focused on the research, technological, and management issues described by the parameters of soil, air, crops, harvest as the main objects. Decision-making tools № 5 (Fig. 4.4) are designed to form optimal executive actions. For example, at stage A (Fig. 4.5) it is necessary to decide on the place of sowing, depending on the type of precursors and moisture content in the soil (Fig. 4.6) (Table 4.2).

**Table 4.2.** Structuring of the type of information for the work of the CPS in stage A.

| Indicator | Type of information |
|---|---|
| Nomenclature of predecessors | Reference information |
| Soil moisture content, % | Measuring information |

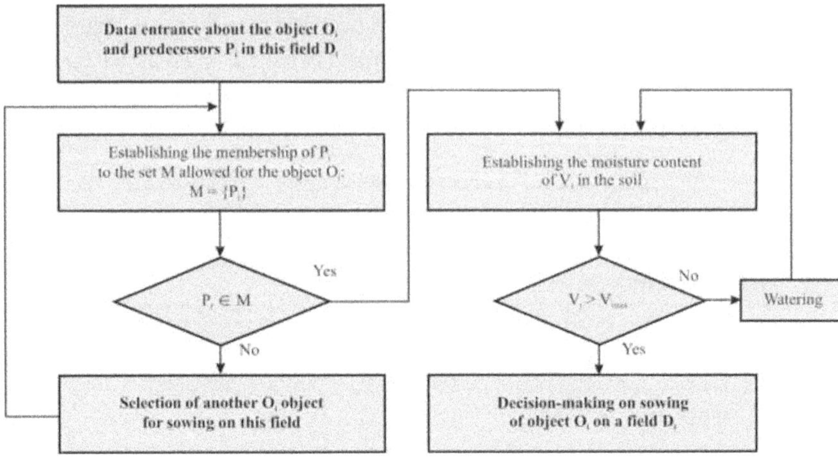


**Figure 4.6.** Algorithm for decision making at stage A (Fig. 4.5) for cereals production technology.

At stage B (Fig. 4.5), namely, at the stage of tillage, decisions on the procedure for this process will depend on the following indicators: type of soil zone, type of precursors, type of weeding, moisture content, time of harvesting the predecessor (Table 4.3). At stage B (Fig. 4.5), namely, at the stage of fertilizer application, soil analysis is an important process. The analysis was carried out to determine the content of such important indicators as nitrogen, phosphorus, and potassium (Table 4.4). At this stage, samples should be taken correctly to adequately assess the distribution of fertilizers in the field. To this end, a method of controlled soil sampling is proposed. It is based on the use of admittance sounding of the soil (establishing the dependence of the admittance on the physicochemical parameters of the soil).

During its implementation, it is possible to monitor the territorial distribution of mineral salts according to the indicators of admittance (complex conductivity) (Fig. 4.7). The study of the sensitivity of the reactive component of admittance B for different soil types to the same amount of applied pesticide was performed. According to its results, it was determined that the sensitivity for loamy soil is 39 μSm, for sandy soil - 42 μSm, for black soil - 30 μSm (Fig. 4.8). It is concluded that it is possible to evaluate the type of electrolytes (which are pesticides) by the reactive conductivity component.

**Table 4.3.** Structuring the type of information for the work of the CPS at stage B.

| Indicator | Type of information |
|---|---|
| Nomenclature of predecessors | Reference information |
| Type of soil zone | Reference information |
| Type of weeding | Video information |

**Table 4.4.** Structuring the type of information for the CPS operation at stage C.

| Indicator | Type of information |
|---|---|
| Soil nitrogen, phosphorus, potassium content, mg/100g | Measuring information |
| Soil acidity | Measuring information |
| Fertilizer application rate D, kg/ha | Estimated, reference, measuring information |
| The territorial distribution of mineral salts | Measuring information |

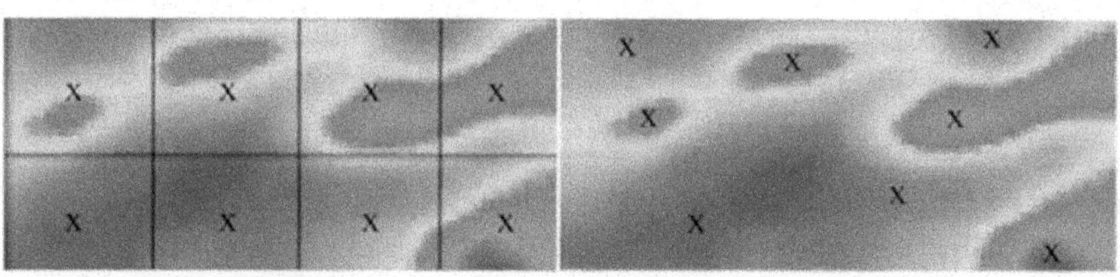

*a)* *b)*

**Figure 4.7.** Soil admittance map: a) Traditional method of sampling and fertilization; b) Controlled method of sampling and application of fertilizers taking into account the territorial distribution of mineral salts.

Next, the rate of fertilizer application for the planned yield should be calculated. For this purpose, the formula of the work [16] is used, the implementation of which requires reference and measurement information. Other levels of the CPS should provide it:

$$D = \frac{U_{pl} \cdot b - G_3 \cdot KG_3 - D_o \cdot K_o \cdot C_0}{K_m}, \quad (4.1)$$

where D is the rate of nitrogen, phosphorus or potassium, kg/ha, $U_{pl}$ is planned yield, 100 kg/ha, b is the removal of the battery per 1 quintal of grain, $G_3$ is soil reserves of the available form of the battery, $KG_3$ is the coefficient of absorption of the battery from the soil reserves, $D_0$ is the dose of organic fertilizers, t/ha, $C_0$ is nutrient content in 1 ton of organic fertilizers, kg/t, $K_0$ is the utilization factor of the nutrient from organic fertilizers, $K_m$ is the utilization factor of the nutrient from mineral fertilizers.

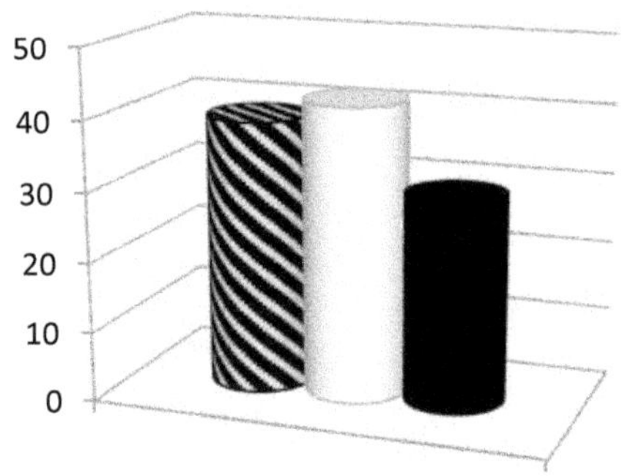


**Figure 4.8.** Sensitivity of the admittance reactive component B, μSm, for different soil types to the same amount of pesticide: the loamy soil - 39 μSm, the sandy soil - 42 μSm, the black soil - 30 μSm.

At stage D (Fig. 4.5) of seed preparation, the decision on the readiness of seed for sowing should be the task of the CPS. The normative indicators of laboratory germination, purity, growth force, the weight of 1000 grains (Table 4.5), determine this process. Control of the grain sorting fraction process is important at this stage. Also, the seeds should be treated with pesticides. The methods of their introduction are also regulated and need to be monitored by the CPS, as they are carried out automatically.

**Table 4.5.** Structuring the type of information for the work of the CPS at stage E.

| Indicator | Type of information |
|---|---|
| Laboratory germination | Estimated information |
| Purity | Estimated information |
| Mass of 1000 grains | Measuring information |

Stage D (Fig. 4.5) of sowing requires control and decision-making CPS on indicators: timing and depth of sowing, which depends on moisture content, temperature and soil density, as well as the mass rate of sowing (Table 4.6), which is calculated according to the expression [17]:

$$M = \frac{h \cdot a \cdot 100}{p \cdot c} \tag{4.2}$$

Here h is the number of millions of pure seeds sown per 1 ha in the area, million/ha; a is the weight of 1000 seeds, g; p is the seed purity, %; c is laboratory germination, %.

**Table 4.6.** Structuring the type of information for the work of the CPS at stage F.

| Indicator | Type of information |
|---|---|
| Soil moisture content, % | Measuring information |
| Soil density | Measuring information |
| Temperature | Measuring information |
| Seeding rate | Estimated, reference, measuring information |

Care of crops at stage E (Fig. 4.5) is to monitor soil moisture to decide on watering; by monitoring the condition of plants (video surveillance, expert measurements, or the use of sensors to determine the level of chlorophyll fluorescence in plants) to disinfect them from diseases and pests by treatment with appropriate pesticides. The number and types of these pesticides are normalized (Table 4.7).

Harvesting at stage E (Fig. 4.5) requires the CPS to decide on the moment of harvesting, which depends on the moisture content of the grains and the height of the crop. The latter is determined by the height of the stem. The parameters required for the operation of the CPS at this stage are set by a set of measurements (Table 4.8).

**Table 4.7.** Structuring the type of information for the CPS operation at stage G.

| Indicator | Type of information |
|---|---|
| Soil moisture content, % | Measuring information |
| Condition of plants | Video and measuring information |
| Type and rate of pesticides, kg/ha | Estimated information |

**Table 4.8.** Structuring the type of information for the work of the CPS at stage J.

| Indicator | Type of information |
|---|---|
| Moisture content in the grain, % | Measuring information |
| Stem height, m | Measuring information |
| A set of indicators of finished grain | Measuring and estimated information |

At the final stage, the control of normalized quality indicators of grown grain is carried out. According to the results of control, the grain should be assigned to the appropriate group - feed or food.

For element № 4 (Fig. 4.4) of the CPS structure of obtained measurement data (OMD) processing, comparison of the measurement information with the maximum allowable values P, modeling M, correlation analysis of the spacecraft, coding CD, data transmission PD, etc.) and marker data type identification, for example, OD = {OMD, P, M, KA, KD, PD,...} (Table 4.9) are characteristic.

For means of acquisition and delivering information № 3 (Fig. 4.4) of CPS - the correct choice of characteristic parameters from the relevant measuring arrays and databases, is inherent. Within the CPS operation, they can be structured based on content, for example, soil temperature T, humidity B, indicators of the ecological and toxicological condition of soil ET, agrochemical indicators ACh, as well as quantitatively, for example, the volume of OD and frequency of BH, etc. : B = {[ T, B, ET, ACh,...], [OD, BH,...]}. Establishing the boundary values of object or process parameters, errors, and uncertainty of measurements and calculations are crucial in the choice of measurement methods and application of data processing algorithms.

Means of interaction № 2 with the physical world №1 (Fig. 4.4) is a network of sensors of different purposes and levels (smart-sensor), measuring information which is an important source for the functioning of other levels of CPS [18-24]. Today promising systems for interaction and information acquisition can be considered WSNs, the main advantage of which is the ability to monitor the state of plants or environmental parameters in large areas in real-time. Based on the analysis of the architecture of WSN nodes, it was concluded that for monitoring and management of cereal production it will be appropriate to use WSNs with stellar and cluster architecture of CPS (Fig. 4.9).

**Table 4.9.** Structuring the type of information processing for the CPS operation.

| Task | Type of information processing |
|---|---|
| Production control, the certification procedure | ➢ Processing of data arrays PDA,<br>➢ Comparison of the obtained measuring and calculation information with the maximum permitted values of indicators PI,<br>➢ Coding CD,<br>➢ Data transmission DT, etc.<br>➢ Datatype identification, for example, OD = {PDA, PI, CD, DT,…} |
| Harvest forecasting | ➢ Processing of data arrays PDA,<br>➢ Modeling M,<br>➢ Regression and correlation analysis of RCA,<br>➢ Coding CD,<br>➢ Data transmission DT, etc.<br>➢ Identification of data types D, for example, D = {PDA, M, RCA, CD, DT, ..} |

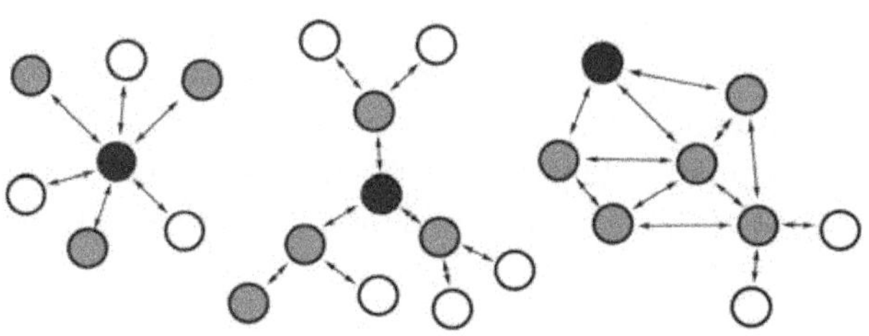

a) b) c)

**Fig. 4.9.** Architecture of wireless sensor network nodes: a) Star type; b) Cluster tree; c) Cell structure.

## 4.2. Cyber-Physical System for Greenhouse Operation

Greenhouses are an important point of agricultural production in all seasons, regardless of the weather. However, for the effective cultivation of each plant variety in the greenhouse requires a specific microclimate. The specificity is that each variety of plants is needed own temperature-humidity-insolation optimum, at which they develop satisfactorily and give a high yield [25-27]. Outside the temperature optimum, plant growth slows down, and with a double deviation from the optimum, it stops altogether. Besides, exceeding the temperature in the greenhouse by 5-10 $^0$C above the required limit leads to extra costs and thus to an increase in the cost of production. One of the ways to solve this problem consists of the development of an automated control system for temperature, humidity, and insolation of the greenhouse using microprocessors to ensure an optimal microclimate with minimal energy.

An important role in plant cultivation is provided by the processes of monitoring the microclimate parameters of the greenhouse. For example, the required accuracy of maintaining the temperature should not exceed ± 1°C. Also, according to the physiological characteristics of plant life, the maintained temperature should be consistent with the level of illumination. The Temperature-humidity regime is maintained by automatic heating, ventilation, irrigation, shading, etc. It is determined by the power of heat sources, as well as the design and technological features of the greenhouse. Therefore, it is important to select correctly the structural elements of the greenhouse. To improve the operation of them, research has to be conducted in the direction of increasing the number of analyzed parameters of greenhouses' microclimatic conditions, the development and improvement of the element base, including sensors, which produce primary information on the particular characteristics in the greenhouse], actuators, microcontrollers for their circuits. Therefore, the urgent task is to analyze the microclimatic conditions in greenhouses, identify the smart sensors for greenhouse design with an optimal statistical error of temperature control $\Delta t_c$ and relative humidity $\Delta \varphi_c$, which would not exceed ± 1°C and ± 3.5% accordingly.

### 4.2.1. Technical Means of the Greenhouse Microclimate Control Subsystem

The principle of operation of the microclimate control system in the greenhouse is predicting the control of the electrical power, f.i. switching in and out, of the equipment based on the processing of a set of different MIs readouts. Namely, the air-heating system has to switch up is started if the air temperature in the greenhouse is lower than set before; the soil-heating system is switching up if the soil temperature in the greenhouse is lower than the set; ventilation is switched up by opening the ventilation transoms if the temperature or humidity in the greenhouse is higher than specified; air exchange with the environment is stopped by closing the ventilation transoms; humidification system is started in case if he humidity in the greenhouse is lower than the set; curtains are opened to ensure the growth of plants, increasing the sun-shining if the illumination of plants in the greenhouse is lower than specified; curtains are closed to reduce the sun irradiating if illumination is higher than specified; irrigation occurs when the soil moisture in the greenhouse is lower than specified; system of providing air with carbon dioxide to accelerate photosynthesis is started if the concentration of carbon dioxide in the air in the greenhouse is lower than specified. Therefore, to ensure the operation of the microclimate control system in the greenhouse, it is necessary to select technical means of the automatic climate control system and optimally locate them in the greenhouse or nearby it. Besides, the characteristics of the external weather conditions and design parameters of the greenhouse have to be measured and analyzed.

To design a CPS for vegetable cultivation, the Arduino Uno board is applied. It can smoothly regulate temperature and humidity while maintaining sufficient accuracy of the supported parameters. The following sensors for CPS's needs were selected: in particular, temperature and humidity sensors (accordingly, DHT21 / AM2301 and DHT22 that are the digital sensors of high accuracy equipped with capacitive humidity sensor and NTC thermistor), illumination sensor BH1750, soil temperature sensor DS18B20 and capacitive soil moisture sensor (not susceptible to corrosion and therefore independent of soil salinity), greenhouse gas contamination sensor MG-811 or MH-Z19B. The system based on the Arduino UNO board allows implementing MIs with microprocessor control, which also yields the electric engines. The functional scheme for microclimate control in greenhouses is demonstrated below (Fig. 4.10).

Temperature-humidity sensors that can be exploited together with Arduino, are of DHT11, DHT21, DHT22, and HTU21 types. Digital temperature-humidity sensor DHT11 yields the capacitive humidity sensor and thermistor. Digital temperature-humidity DHT22 / AM2302 sensor yields the NTC thermistor. The latter differs from the DHT11 sensor by a higher accuracy. The digital temperature-humidity high-accuracy sensor DHT21 / AM2301 is connected to the single-wire interface. The protocol and connection scheme are identical to the DHT22 sensor. The HTU21 temperature-humidity sensor applies an appropriate chip with an I2C interface, which provides the accuracy of temperature measurements ± 0.05°C. The technical characteristics of these temperature and humidity sensors are given in Table 4.10. While using the DHT11 sensor with the Arduino board, we receive less accurate results comparing with other high-mentioned sensors. Therefore, the DHT21 / AM2301 and DHT22 sensors have been chosen as elements of a CPS for vegetable cultivation.

**Table 4.10.** Technical characteristics of temperature and humidity sensors.

| Characteristic | DHT11 | DHT21/AM2301 | DHT22/AM2302 | HTU21 |
|---|---|---|---|---|
| Determination of humidity | 20-90 % ± 5% RH | 0-100 % ±3% RH | 0 – 100 % ±2% RH | 0 – 80 % ± 3% RH |
| Determination of temperature | 0-50 °C ± 2% (max) | -40 ~ +80 °C ±0 .5% | -40 -+80 °C ± 0.5 °C | -10 - +85°C ± 0. 4°C (max) |

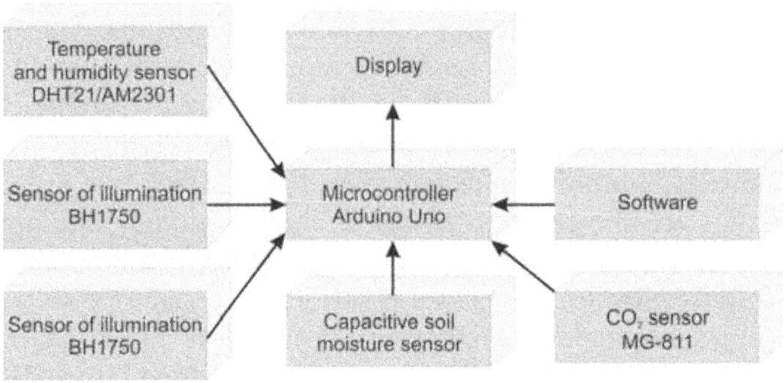


**Figure 4.10.** Functional diagram for climate control in greenhouses.

To measure the illumination has been chosen a digital sensor GY-302 on BH1750 chip, which can work with a microcontroller according to the I2C protocol. A photodiode is used as a light-sensing element. The measuring range is from 0 to 65535 lux (16 bits); the measured wavelength - 560 nm; accuracy in the mode of high resolution is 1 lux; accuracy in the mode of low resolution is 4 lux; measurement period in high-resolution mode is 120 ms.; measurement period in low-resolution mode is 16 ms. Low current consumption and sleep function; filtering of illumination noise 50/60 Hz; ability to select two-chip addresses for the I2C interface (one can connect two sensors to one bus at the same time, which allows distributing sensors in the room for a holistic measurement of light levels). The sensor does not require calibration that is convenient for the project.

To measure soil temperature, it is advisable to use a digital integrated sensor DS18B20 with a unique factory-stitched 64-bit code that can be used by the microcontroller to communicate with a particular sensor on a common bus. Technical characteristics of DS18B20: interface - One-

Wire; temperature range - from -55 °C to +125 °C; operating temperature - from 0 to + 60 °C; measurement accuracy - 0.5 °C; step of indications – 0.0625 °C. In the DS18B20 memory, one can store the boundary temperatures above which the sensor would alter the operation mode into alarm.

Soil moisture is measured by a YL-69 capacitive sensor with a built-in resistor of variable resistance. However, the capacitive soil moisture sensor, unlike resistive moisture sensors, is not prone to corrosion. The output voltage is inversely proportional to soil moisture. The sensor is ideal for monitoring changes in soil moisture, for creating automatic irrigation systems for plants and for monitoring the integrity of the soil pipeline.

To measure the $CO_2$-concentration inside the Greenhouse [25] is applied an eSense, which allows measuring the content in the ambient air in the range up to 10000 ppm and transmit data via analog output. You can also use the sensor of the infrared concentration meter MH-Z19B, designed to quantify the content of carbon dioxide in the air. The sensor operates with two output interfaces, temperature compensation, high linearity, and low power consumption.

- **Study of Temperature Distribution and Relative Humidity in the Greenhouse**

To assess the uniformity of the temperature in the greenhouse and the choice of the location of the device sensors, which were used to study the characteristics of the process, measurements of temperature and humidity were carried out in a stabilized mode; measurements were performed with humidity and temperature sensors DHT11, DHT21 / AM2301, DHT22 / AM2302 in the diagonal direction of the greenhouse every 4 m at a height of 1, 2 and 3 m. The interval between measurements was equal to one minute. The results are shown in Table 4.11.

The mean value is determined by the formula:

$$\bar{x} = \frac{1}{n}\sum_{k=1}^{n} x_k \tag{4.3}$$

Here n is the number of measurements; $x_k$ is the value of the k-th measurement. The standard uncertainty of type A is defined by:

$$u_A(\bar{x}) = \sqrt{\frac{1}{n(n-1)}\sum_{k=1}^{n}(x_k - \bar{x})^2} \quad (4.4)$$

Similar measurements have been fulfilled with the help of the temperature-humidity sensor DHT11. Basing on the received data, it was argued that the DHT11 sensor is worse than the DHT21 / AM2301 sensor. So, we have built the graphs based on measurements made with the help of the latter (Fig. 4.11).

**Table 4.11.** Results of repeated measurements of air temperature at the height and length points of the greenhouse by the temperature-humidity DHT21/AM2301 sensor.

| | **Air temperature in the greenhouse, *t* $^0$C** | | |
|---|---|---|---|
| **Length, m \ Height, m** | 1 | 2 | 3 |
| 0 | 18.56 | 19.62 | 20.66 |
| 0 | 18.55 | 19.61 | 20.68 |
| 0 | 18.52 | 19.61 | 20.69 |
| 0 | 18.49 | 19.57 | 20.69 |
| 0 | 18.55 | 19.56 | 20.61 |
| 0 | 18.51 | 19.61 | 20.66 |
| Mean of multiple measurements, $^0$C | 18.53 | 19.597 | 20.665 |
| Standard uncertainty of type A, $^0$C | 0.01125 | 0.01022 | 0.01232 |
| 4 | 19.08 | 19.83 | 21.73 |
| 4 | 19.07 | 19.89 | 21.71 |
| 4 | 19.05 | 19.86 | 21.73 |
| 4 | 19.03 | 19.88 | 21.74 |
| 4 | 19.1 | 19.86 | 21.7 |
| 4 | 19.07 | 19.85 | 21.77 |
| Mean of multiple measurements, $^0$C | 19.06667 | 19.86167 | 21.73 |
| Standard uncertainty of type A, $^0$C | 0.00999 | 0.00880 | 0.01010 |
| 8 | 19.88 | 20.94 | 23.05 |
| 8 | 19.87 | 20.93 | 23.09 |
| 8 | 19.86 | 20.91 | 23.07 |
| 8 | 19.83 | 20.93 | 23.09 |
| 8 | 19.85 | 20.94 | 23.06 |
| 8 | 19.86 | 20.94 | 23.05 |
| Mean of multiple measurements, $^0$C | 19.85833 | 20.93167 | 23.06833 |
| Standard uncertainty of type A, $^0$C | 0.00703 | 0.00477 | 0.00749 |

**Table 4.11. (Continued)**

| Length, m \ Height, m | Air temperature in the greenhouse, $t$ $^0$C | | |
|---|---|---|---|
| | 1 | 2 | 3 |
| 12 | 18.51 | 20.16 | 21.24 |
| 12 | 18.52 | 20.15 | 21.22 |
| 12 | 18.55 | 20.14 | 21.19 |
| 12 | 18.57 | 20.13 | 21.18 |
| 12 | 18.5 | 20.1 | 21.2 |
| 12 | 18.52 | 20.11 | 21.2 |
| Mean of multiple measurements, $^0$C | 18.52833 | 20.13167 | 21.205 |
| Standard uncertainty of type A, $^0$C | 0.01078 | 0.00946 | 0.00885 |
| 16 | 18.3 | 19.04 | 20.17 |
| 16 | 18.7 | 19.04 | 20.12 |
| 16 | 17.2 | 19.07 | 20.1 |
| 16 | 17.9 | 19.09 | 20.19 |
| 16 | 18.4 | 19.09 | 20.1 |
| 16 | 17.7 | 19.09 | 20.11 |
| Mean of multiple measurements, $^0$C | 18.03333 | 19.07 | 20.13167 |
| Standard uncertainty of type A, $^0$C | 0.22161 | 0.01 | 0.01579 |

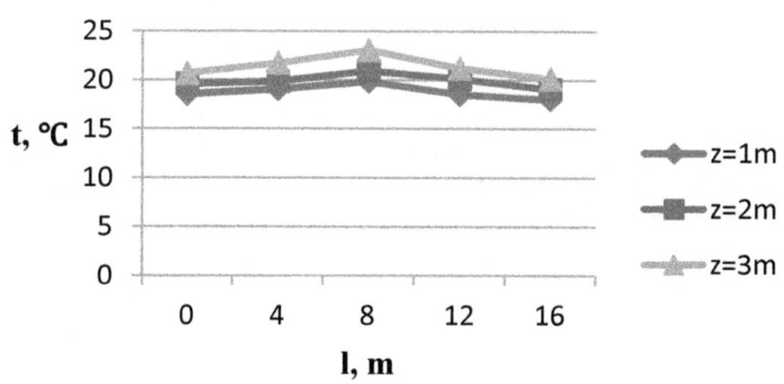


**Figure 4.11.** Change of temperature t of air on height (z) and length (l) of the Greenhouse (measurements are carried out by the temperature-humidity sensor DHT21/AM2301).

Similar measurements have been made for the relative humidity with the help of the same sensors (Table 4.12).

Basing on the measurement results, the graphs of changes in temperature and relative humidity were constructed (Fig. 4.11-4.12). They demonstrate that the air parameters change within the greenhouse space.

Towards the center of the greenhouse, the air temperature rises slightly and the humidity decreases. The temperature falling is accompanied by the increase in the relative humidity of the air near the walls due to the heat exchange with the environment. The significant temperature falling caused by heat leakage was fixed nearby the doors. The maximum temperature and the minimum humidity were recorded in the center of the Greenhouse at a 3 m height. Therefore, this point A has been chosen for the installation of the measuring sensors at the height Z = 2 m. The obtained data give reasons to consider Greenhouse mathematically as an object with concentrated parameters.

**Table 4.12.** The results of repeated measurements of relative humidity on the height and length of the greenhouse by the temperature-humidity DHT21/AM2301 sensor.

| **Length, m \ Height, m** | **Relative humidity in the greenhouse, %** | | |
|---|---|---|---|
| | 1 | 2 | 3 |
| 0 | | | |
| Mean of multiple measurements φ, % | 76.65114 | 83.3241 | 88.62311 |
| Standard uncertainty of type A, % | 0.01125 | 0.01022 | 0.01232 |
| 4 | | | |
| Mean of multiple measurements φ, % | 75.32428 | 81.97186 | 85.96393 |
| Standard uncertainty of type A, % | 0.00999 | 0.00880 | 0.01010 |
| 8 | | | |
| Mean of multiple measurements φ, % | 70.11632 | 79.31843 | 85.96289 |
| Standard uncertainty of type A, % | 0.00703 | 0.00477 | 0.00749 |
| 12 | | | |
| Mean of multiple measurements φ, % | 77.98395 | 81.97106 | 87.29184 |
| Standard uncertainty of type A, % | 0.01077 | 0.00946 | 0.00885 |
| 16 | | | |
| Mean of multiple measurements φ, % | 80.64429 | 84.63285 | 88.62147 |
| Standard uncertainty of type A, % | 0.22161 | 0.01 | 0.01579 |

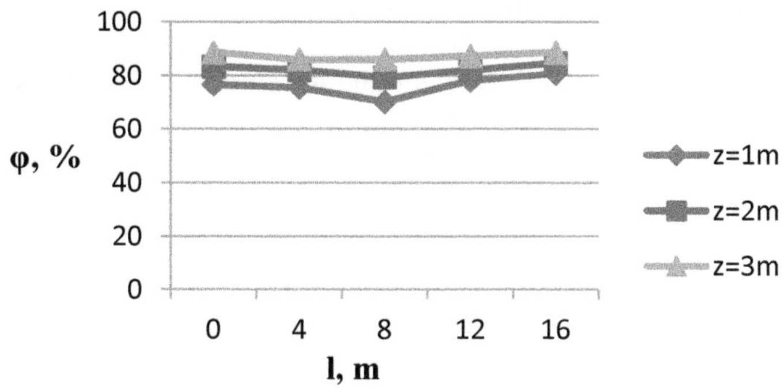


**Figure 4.12.** Change of relative humidity φ of air on height (z) and length (l) of the Greenhouse (measurements are carried out by the humidity and temperature sensor DHT21/AM2301).

The CPS for vegetable cultivation must maintain the specified modes while operating regardless of the impacts on it. The latter can be achieved in control systems resistant to perturbations. It means that small changes in the input signal or any perturbation, initial conditions or parameters would not lead to significant deviations of the output signal. The considered CPS with the temperature-humidity-insolation control yields 3 conjugated subsystems. Each one is equipped with one or more smart sensors and smart actuators relating to a certain physical quantity.

As a result of the study, we justified the choice of smart sensors and their mounting places in the greenhouse. The control function can be provided by the PID controllers, the main task of which are temperature, humidity, and insolation control with minimal error relative to the set mode. The PID controller generates an output control signal, which is fed to the actuators. As a result, the heating element begins to heat or the fan - to cool, the shader changes the insolation mode, and the drip humidification system starts to operate.

To substantiate the parameters of the automatic control system, the modeling of thermal processes dynamics using the equations of heat balance has been considered. The statistical error of air temperature regulation does not exceed ± 1°C at the error of measuring instrument readouts, not higher ± 0.5 °C, the relative humidity of the air – 3.5%, that is that which lies within a zone of the sensitivity of this mean.

## 4.3. Electrical Impedance Spectroscopy

EIS, as an effective analytical technique for electrochemical system, has demonstrated a significant application for food quality and safety assessment recently. Impedance (complex resistance) $Z_X$, as well as the inverse admittance value (complex conductivity) $Y_X = 1/ Z_X$, is a part of the immittance term which is more general. From the formal point of view, impedance and admittance reflect the same physical quantity. Their choice is mainly determined by the convenience of interpreting the primary parameters of the substitution scheme of the object under study: the impedance is used for serial two-element substitution schemes and admittance is used for parallel ones. More complex substitution schemes of the objects under study make equivalent measurements of impedance and admittance, and if necessary, the transition from $Z_X$ to $Y_X$ or vice versa is carried out by recalculation at the results elaboration stage.

The vector (two-dimensional) nature of the impedance is more informative than investigations conducted by using direct current. Since real objects and processes are characterized by certain active and reactive components' dependencies on the frequency, the information about these objects' properties can be obtained by analyzing their frequency response to alternating current. Knowing the dependence on the impedance, a variety of physical quantities, such as humidity or corrosion, can be controlled indirectly by EIS [28-30].

EIS is currently widely used to study objects of non-electric nature, including biological ones. The biomedical measurements [31], the study of materials properties, in particular on micro and nano-level, corrosion monitoring and cancer diagnostics [32], control of the parameters of batteries and elements of electrochemical power sources [30] can be used as the examples.

Nowadays, impedance spectroscopy is the key methodology for solving a lot of problems that currently arise in the field of chemistry, medicine, ecology, semiconductor physics, material science. Impedance spectroscopy (impedance change of frequency) is one of the original methods of studying the researched materials, products, etc., that need quick and reliable results without damaging and destroying the studied samples.

A considered method can be recommended for a lot of areas of science and technology. Relying on considerable theoretical and applied experience in this field, which is reflected in particular by the publication of [30] we analyze below its accomplishment regarding new approaches in particular concerning daily consumed products. This method has proven itself well, in particular in the electrochemical analysis of substances, biological research (impedance microbiology), to control the quality of water, petroleum products, soils, etc.

Simultaneously we develop EIS involving new approaches aimed at further improving the method. Promising in this regard is the application of artificial intelligence to identify patterns' changes of physical properties, including related to the impedance.

Modern impedance analyzers have to yield the analog components, including those that are based on operational amplifiers. For example, the use of an operational amplifier in a measurement impedance-voltage converter provides high sensitivity and performance together with the stability of transfer function and the ability to create specified energy conditions on the studied object. However, the main problem that should be solved is the reduction of so-called frequency errors. This task should be performed with the use of algorithmic correction taking into account the availability of computing facilities.

### 4.3.1. Identification of Fresh and Thawed Meats

Individual differences in livestock cause high variation in the quality of raw meat and fish and their commercialized products. Therefore, to obtain the definite quality information and ensure the quality of each product, a fast and on-line detection technology is demanded to be developed to monitor product processing [33]. The method of EIS for assessing the quality of raw meat is promising. The freshness of the meat is one of the indexes of its quality. However, it is often frozen to be transported and stored. After unfreezing, it is offered to a consumer as a fresh one. Quality of that meat is worse, but the price is, mainly, identical to the cost of meat that wasn't frozen. Therefore, consideration of the method of authentication of both fresh and unfrozen meat seems to be quite important and socially demanded.

If the level of freshness of meat during storage by cooling is determined by at least long-term and time-consuming methods, the identification of meat freezing, except for the organoleptic method (however, it becomes imperfect against the modern methods of adulteration), is not performed yet. Therefore, the proposal to create an express method to detect this counterfeiting of products is relevant.

Analysis of traditional methods of meat quality control has shown that to conclude about the good quality of meat (fresh, questionable fresh, stale) it is studied by several methods that are, subjected to organoleptic, chemical, microscopic, histological, electrical examination.

Regulated methods of evaluation of raw meat are long-lasting, inaccessible to the mass consumer, and do not ensure the effectiveness of measurements. Some studies are quite expensive, focusing only on qualified personnel and special laboratories for product quality control. Some studies are quite expensive, focusing only on qualified personnel and special laboratories for product quality control. For the implementation and wider implementation of this method arises a condition to design the necessary hardware that would meet such criteria as reliability and efficiency (express method), the ability to send information for further processing and storage. However, the implementation of the impedance spectroscopy to monitor the quality of meat products requires an appropriate base sample since the electrical characteristics of it have to be compared with the control sample. It is impossible to ensure invariance to the row of impact factors (temperature, feed, age of the animal, duration of storage in the refrigerated state, etc.).

- **EIS for Quality Control**

The considered method examines the change in the parameters of the impedance, which is fed to the object of control, at fixed frequencies of a wide frequency band. Preferably such parameters are the modulus and phase angle, which characterize the impedance. In this case, compare the measurement results of these parameters of the monitored and base samples.

Let's try to identify fresh and thawed meat by estimating the change in active and reactive components of impedance or admittance. In this case, the components of the immittance (impedance or admittance) of the object are compared with the corresponding components of a similar object of the known quality level at several fixed frequencies of a given frequency band of a sine test signal of a fixed level. Such a comparison can be made by an absolute or relative method, which is simply realized by electrical means. According to the absolute method, we obtain the difference between the values of the impedance and admixture components at fixed frequencies:

$$(X_1 - X_{01})_{f1}, (X_2 - X_{02})_{f2}, (X_3 - X_{03})_{f3}, \ldots, (X_n - X_{0n})_{fn}, \quad (4.5)$$

$$(R_1 - R_{01})_{f1}, (R_2 - R_{02})_{f2}, (R_3 - R_{03})_{f3}, \ldots, (R_n - R_{0n})_{fn}, \quad (4.6)$$

$$(B_1 - B_{01})_{f1}, (B_2 - B_{02})_{f2}, (B_3 - B_{03})_{f3}, \ldots, (B_n - B_{0n})_{fn}, \quad (4.7)$$

$$(G_1 - G_{01})_{f1}, (G_2 - G_{02})_{f2}, (G_3 - G_{03})_{f3}, \ldots, (G_n - G_{0n})_{fn}, \quad (4.8)$$

By the relative method we receive the ratio of the same components:

$$\left(\frac{X_1}{X_{01}}\right)_{f1}, \left(\frac{X_2}{X_{02}}\right)_{f2}, \left(\frac{X_3}{X_{03}}\right)_{f3}, \ldots, \left(\frac{X_n}{X_{0n}}\right)_{fn}, \quad (4.9)$$

$$\left(\frac{R_1}{R_{01}}\right)_{f1}, \left(\frac{R_2}{R_{02}}\right)_{f2}, \left(\frac{R_3}{R_{03}}\right)_{f3}, \ldots, \left(\frac{R_n}{R_{0n}}\right)_{fn}, \quad (4.10)$$

$$\left(\frac{B_1}{B_{01}}\right)_{f1}, \left(\frac{B_2}{B_{02}}\right)_{f2}, \left(\frac{B_3}{B_{03}}\right)_{f3}, \ldots, \left(\frac{B_n}{B_{0n}}\right)_{fn}, \quad (4.11)$$

$$\left(\frac{G_1}{G_{01}}\right)_{f1}, \left(\frac{G_2}{G_{02}}\right)_{f2}, \left(\frac{G_3}{G_{03}}\right)_{f3}, \ldots, \left(\frac{G_n}{G_{0n}}\right)_{fn}, \quad (4.12)$$

where $X_1$, $X_2$, $X_3$, $X_n$ and $B_1$, $B_2$, $B_3$, $B_n$ are the reactive components of thawed meat impedance and admixture, respectively, at frequencies $f_1$, $f_2$, $f_3$, $f_n$; $X_{01}$, $X_{02}$, $X_{03}$, $X_{0n}$ and $B_{01}$, $B_{02}$, $B_{03}$, $B_{0n}$ are the reactive components of impedance and admittance of fresh meat, respectively, at frequencies $f_1$, $f_2$, $f_3$, $f_n$.

The results of comparing the numerical values of the impedance components obtained by (4.5), (4.6), (4.9), (4.10) or the admittance by (4.7), (4.8), (4.11), (4.12) are analyzed. If we construct a graphical dependence of these components on the frequency of the measurements, then analyze the corresponding components of the controlled and base samples. Basing on the analysis results, a decision is made on product identification.

- **Examination of Meat by EIS**

To analyze the ratios of active and reactive components of both impedance and admittance, let's present the studies of chicken meat by EIS. Note that samples of fresh and thawed chicken meat have been studied using a capacitive sensor and a meter of immittance parameters at a given voltage level of the test signal in the frequencies 100 Hz - 100 kHz. Graphical dependencies of active and reactive components on frequency are shown in Fig. 4.13-4.14.

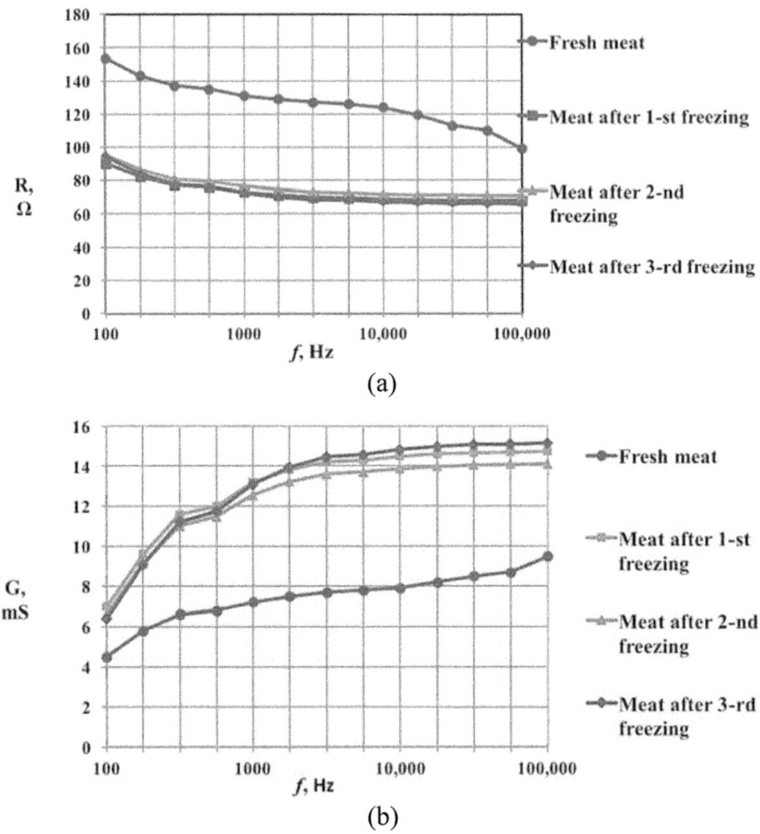


**Figure 4.13.** Dependence of the active components of the impedance and admittance of fresh and defrosted chicken meat on the frequency change of the test signal.

- **Analysis of the Results of the Study**

As the analysis of the obtained dependences of the active component on the frequency demonstrates (Fig. 4.13), after the 1st freezing of meat, the value of the active component of its impedance (Fig. 4.13a) decreases, and the admittance (Fig. 4.13b) increases.

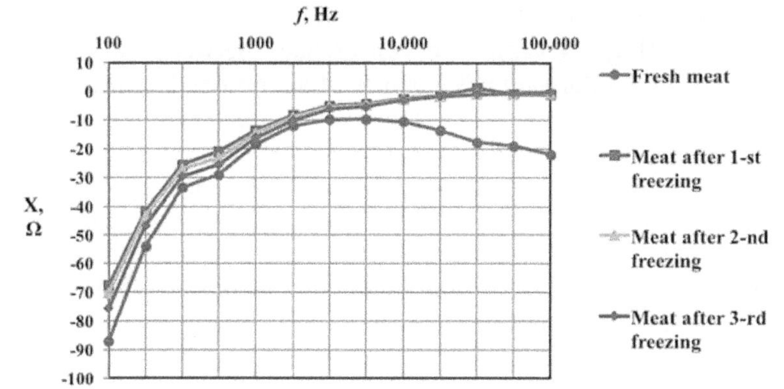


(a)

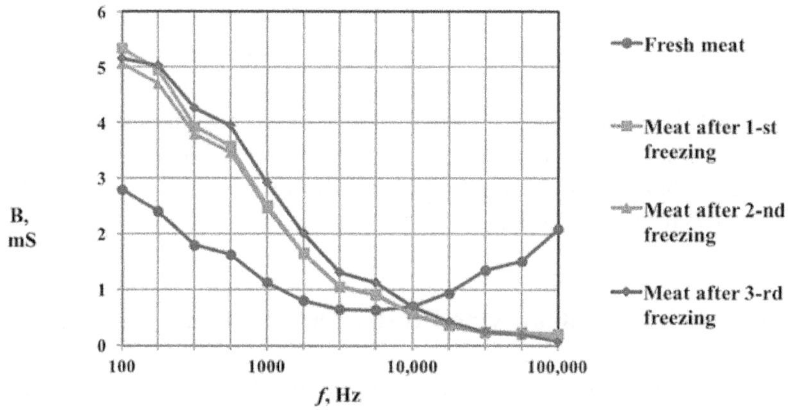


(b)

**Figure 4.14.** Dependence of the reactive components of the impedance and admittance of fresh and defrosted chicken meat on the frequency change of the test signal.

The differences between the results after each of 3 freezes are insignificant, both in the amplitudes of the active ingredients and in the shape of the curves. For the thawed meat, the values of the active component are much higher at each frequency compared with the fresh meat. Such a difference can be an identifying feature for the detection of thawed or fresh meat but under certain measurement conditions and certain requirements for the object.

The reactive components of both impedance and admittance on frequency (Fig. 4.14) for fresh meat differ completely comparing to the thawed meat. If the differences in the active components of fresh meat samples are determined by the amplitude values of the active components (Fig. 4.14a), the difference between the reactive components of similar samples consists of alterations like the curves of the dependences of the corresponding components on frequency (Fig. 4.14b). Such a curve demonstrates a clear extreme value for fresh meat, namely at a certain frequency; here the reactive component of the impedance is maximal and the reactive component of the admittance - minimal. That is, for fresh meat, the reactive component of the impedance increases to a certain value with increasing frequency and then decreases. For thawed meat, it is only increasing in the same frequency range.

We have noticed the similarities in the behavior of the reactive components for fresh and thawed meats, but the direction of their changes is the opposite. Even if the impact factors cause the change in the amplitude of the reactive component, the shape of the curve is permanent. The frequency at which the reactive component reaches an extremum may alter slightly. Since invariance of the results to the amplitude changes is ensured, by such the differences is advisable to identify fresh / thawed meat.

- **Method of Identifying Fresh / Thawed Meat**

Analysis of the curves reflecting the dependencies of the reactive components of impedance and admittance envisages the have a different nature of frequency changes outside the frequency of extremum. For fresh meat, the reactive component of impedance decreases, and for thawed meat, it increases (Fig. 4.14a). The similar applies to the reactive component of admittance (Fig. 4.14b), but for fresh meat, we note increasing and for thawed meat – falling. Therefore, under this condition, you can choose two frequencies of measurement of these components of the impedance or admittance of fresh and thawed meats and compare their values. Frequencies must be selected outside the frequency of the extreme value of the reactive component.

This is the basis of the proposed method of identification of fresh / thawed meat. The illustration of the proposed method based on the analysis of the obtained results (Fig. 4.14) of measuring the impedance (Fig. 4.15a) and the admixture (Fig. 4.15b) components. Here, the curves 1 concern to the thawed meat and curves 2 - the fresh meat.

The proposed method is implemented as follows. The basic then will be the dependence curve of the reactive component of fresh meat (curve 2). At first, measure the reactive component X of the impedance or admittance B at 2 pre-selected frequencies $f_1$ and $f_2$. Obtain 2 values of impedance $X_1^{'}$ and $X_1^{''}$ for frequency $f_1$, and respectively 2 values of admittance $B_1^{'}$ and $B_1^{''}$. Similarly, the results are obtained at the frequency $f_2$ (Fig. 4.15a) that is the impedance values $X_2^{'}$ and $X_2^{''}$ (Fig. 4.15b) as well as the admittance values $B_2^{'}$and $B_2^{''}$. The analysis of dependencies is as follows.

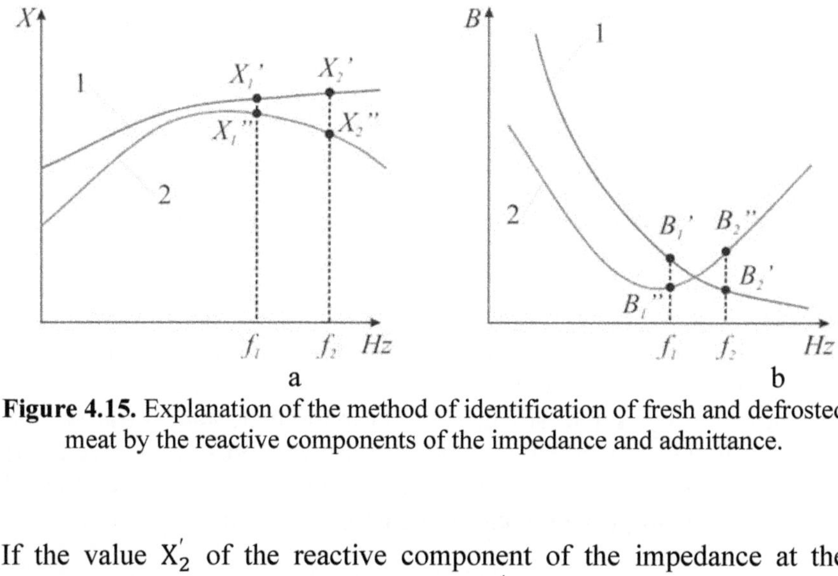


a b

**Figure 4.15.** Explanation of the method of identification of fresh and defrosted meat by the reactive components of the impedance and admittance.

If the value $X_2^{'}$ of the reactive component of the impedance at the frequency $f_2$ (Fig. 4.15a) is bigger than $X_1^{'}$ at the frequency $f_1$, i.e. the condition$(X_2^{'})_{f_2} > (X_1^{'})_{f_1}$ is met, such a slice of meat is identified as thawed. If the condition $(X_2^{''})_{f_2} < (X_1^{''})_{f_1}$ is met, the meat is identified as fresh.

The analysis of the reactive component of admittance is similar. Provided at the condition that $(B_2^{'})_{f_2} < (B_1^{'})_{f_1}$ the meat is identified as thawed and provided that $(B_2^{''})_{f_2} > (B_1^{''})_{f_1}$ it is identified as fresh. Such dependencies can be used as a basis for building a measuring instrument for the identification of fresh / thawed meat.

- **Unconventional Basic Sample for Quality Control**

The basic sample is characterized by a set of non-electrical parameters in the regulatory documentation for products of non-electrical nature. However, analyzing the trend that has developed regarding the development of impedance spectroscopy, we can predict the following. In the future, the technical requirements and regulatory documentation must be supplemented taking into account product quality norms based on impedance spectroscopy. The latter for the identification of fresh / thawed meat have demonstrated the application of an unconventional approach to assessing product quality by the dependencies of their reactive components in the frequency band. Instead of the established values of quality indicators, it is proposed to use a change in the shape of the curves of the reactive component of the impedance or admixture of the monitored object relative to a similar curve of the standard sample. This ensures the invariance of the obtained result to impact factors.

### 4.3.2. Impedance Analyzer Error Adjustment Using Artificial Neural Network

Measurement converters, built based on the OAs based on the auto-balancing circuit, have become widely used in modern impedance measurement instrumentation. Often called as AMCs, they are characterized by some advantages, in particular, the linearity and stability of the conversion function, high sensitivity, and speed, the possibility of creating the specified energy conditions on the investigated object. They also provide measurements both in impedance and in the admittance modes.

The main disadvantage of AMC is the occurrence of frequency errors due to the decrease in the op-amp gain with the increase in the frequency of the probe signal and the increase of impact parasitic shunt capacity at op-amp input. There are several approaches to solve this problem. For example, Agilent, the manufacturer of many types of impedance meters, uses a structural method to reduce frequency errors at frequencies above 100 kHz. An auto-balancing circuit that consists of a zero-detector, in-phase, and quadrature phase-sensitive detectors and a vector modulator is constructed to stabilize the op-amp gain. By nature, the frequency errors of such a measurement converter are determined by the lock-in loop [34]. We have studied another structural method for reducing frequency errors, this time in a static structure of the measurement converter. The composition of the measuring channel, in addition to the specified zero-detectors, phase-sensitive detectors, and a vector modulator, consists of numerical accumulators, which perform the function of integrators.

The above described structural methods for reducing the frequency error rely on the introduction of the hardware redundancy (additional elements) in the measurement channel of impedance analyzers. An alternative solution for expanding the AMC frequency band is the use of algorithmic correction of measurement results, which is based on simple computing operations. Such a "software" approach to improving the metrological characteristics of measurement instrumentations is promising and expedient, considering the availability of computing tools and existing trends for their further improvement.

Special correction algorithms have been developed to reduce the frequency errors of AMC [35]. The algorithmic correction implies:

- Synthesis of correction algorithms based on the mathematical model of AMC, which means, analytical expressions, taking into account the influence of parasitic parameters and destabilizing factors;
- Calculation correction terms using developed algorithms and adding them to the measurement results.

However, this approach is characterized by some difficulties, in particular, it is not always possible to obtain analytical expressions that will ensure that the measurement results are corrected with the required precision, therefore it is necessary to solve a system of complex nonlinear equations for this. In this case, it is expedient to use other variants of algorithmic correction implementation, for example, with the help of machine learning techniques [36].

The purpose of this point is to study the possibility of using ANNs to correct the obtained impedance measurement results, as well as a comparative assessment of the efficiency of this approach to the classical one, which is based on the application of deterministic correction algorithms, obtained analytically.

Regardless of the type of research, we deal with separate or built-in measuring instruments, which must evaluate metrologically autonomous objects with their specifics [37]. Therefore, the approaches here are slightly different from those adopted and are partially described below.

- **Active Measurement Converter Based on Autobalancing Circuit**

AMC is the main element of the impedance analyzers' measurement channel. This is the unit where an impedance (passive value) is converted into a proportional complex voltage under the action of a harmonic probe signal. The general view of the impedance measurement converter based on the auto-balancing circuit is shown in Fig. 4.16.

The essence of the method is to stabilize the current through the investigated object using the reference resistor $R_0$ and to form the complex voltage proportional to the impedance:

$$\dot{U}_Z = -\dot{I}_R \dot{Z}_X = -\frac{\dot{U}_P}{R_O}\dot{Z}_X \tag{4.13}$$

However, given expression does not consider the influence of the real op-amp parameters and the feedback loop. The second one is especially noticed when the probe signal frequency increases and causes errors that can reach tens of percent. These impacts are taken into account by the so-called low-signal model of active measurement converter, which is described by the following expression [38-39]:

$$\dot{H} = -\frac{\dot{U}_Z}{\dot{U}_P} = P + jQ = \dots$$

$$= \frac{-\dfrac{\dot{Z}_X}{R_O} + \left(\dfrac{1}{A_0} + j\dfrac{f}{f_T}\right)\dfrac{R_{OUT}}{R_O}}{1 + \left(\dfrac{1}{A_0} + j\dfrac{f}{f_T}\right)\left[\left(1 + \dfrac{\dot{Z}_X}{R_O} + \dfrac{\dot{Z}_X}{\dot{Z}_P}\right)\left(1 + \dfrac{R_{OUT}}{\dot{Z}_L}\right) + \dfrac{R_{OUT}}{R_O} + \dfrac{R_{OUT}}{\dot{Z}_P}\right]} \tag{4.14}$$

$U_P$ and $U_Z$ are the probe and output (measuring) voltage; P and Q are the in-phase and quadrature AMC output signal components (preliminary results of impedance measurement); $A_0$ is an OA open-loop DC gain; $f_T$ and f are an operational amplifier unity-gain bandwidth and probe signal frequency respectively; $R_{OUT}$ is the OA output resistance; $Z_L$ is the load resistance; $Z_P=Z_D||Z_S$ is the shunt impedance created by differential input impedance $Z_D$ and common-mode impedance $Z_S$ of OA; $R_0$ is the resistance of the converter's reference resistor; $Z_X$ is an impedance of studied object.

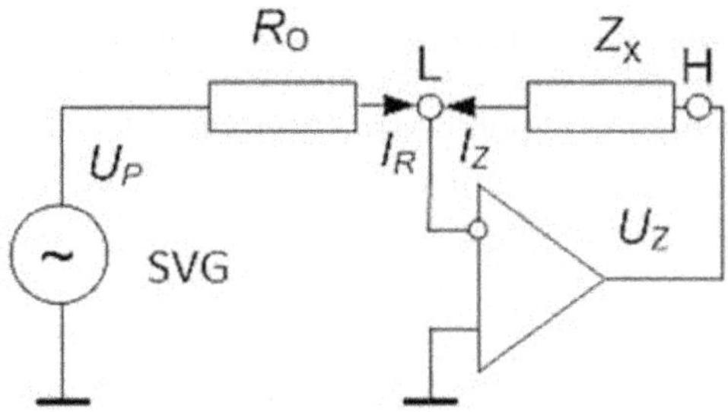


**Figure 4.16.** Active measurement converter based on the auto-balancing circuit.

This equation is bulky, therefore, during the derivation of expressions for algorithmic correction, many simplifications were made, which, as shown in [35], had no significant effect on the adequacy of the model. Since the impedance is a complex value, a separate expression was obtained for each component:

for the active component

$$R = \frac{P + C\frac{P^2+Q^2}{K} - Q\frac{1+D}{K} - \frac{P^2+Q^2}{K^2}}{1 + 2\frac{PC+Q}{K} + \frac{P^2+Q^2}{K^2}\left(1+C^2\right)}, \qquad (4.15a)$$

for the reactive component

$$X = \frac{Q + \frac{P^2+Q^2}{K} + P\frac{1+D}{K} + \frac{D}{K} + C\frac{P^2+Q^2}{K^2}}{1 + 2\frac{PC+Q}{K} + \frac{P^2+Q^2}{K^2}\left(1+C^2\right)} \qquad (4.15b)$$

Here $K = f_T/f$ is the ratio of unity-gain bandwidth and probe signal frequency; $D = R_{OUT}/R_0$ is the ratio of the OA output resistance to reference resistor value $R_0$; $C = 2\pi f C_{IN} R_0$ is an OA time constant by the input capacitance $C_{IN}$ for the reference resistor $R_0$. Each parameter displays one of the three main sources of AMC frequency error: K is the decrease in OA gain; D is the direct passage of the signal due to a nonzero output impedance of the OA; C is the bypassing the op-amp differential and common-mode resistances with the input capacity.

- **Development of Algorithmic Adjustment Based on ANN**

ANN can approximate elementary mathematical functions of almost any level of complexity. Thus, instead of polynomial expression (4.14), ANN-based error correction means matrix multiplication of input data and network's weights. To model more complex dependencies dot products can be passed through non-linear activation function and after that multiplied with the weights of the next layer. The last layer is usually named as output and the others are called hidden.

Experiments have been made in Matlab using Neural Network Toolbox. Design flow of ANN-based algorithmic correction can be split into the following stages:

- Data preparation;
- Initialization;
- Training (calibration);
- Testing (accuracy evaluation).

At the first stage train and test data sets should be prepared. Both sets have identical structures and include pairs of input and target (output) vectors, also called samples. Each input vector consists of six quantities f, P, Q, $f_t$, $C_{in}$, $R_0$, and additional constant that equals one (typical machine learning trick for better performance). Each target vector consists of two quantities R, X. Train set includes 10 batches with 10 thousand samples in each. The test set includes one batch with 1 million samples. Quantities f, $f_t$, $C_{in}$, $R_0$, R, X are randomly generated using a uniform distribution. Quantities P, Q have been calculated according to (4.14).

At the initialization stage, ANN of the given type and architecture is being created. The initial values of weights are random. All necessary parameters for training and testing should also be set.

Experiments have been performed using feedforward type ANN (multi-layer perceptron). This type has a linear activation function in the output layer and non-linear activation function (sigmoid) in hidden layers. Two ANNs have been trained separately for resistance and reactance correction. Networks have the same structure as two hidden layers with 50 neurons in each one (Fig. 4.17).

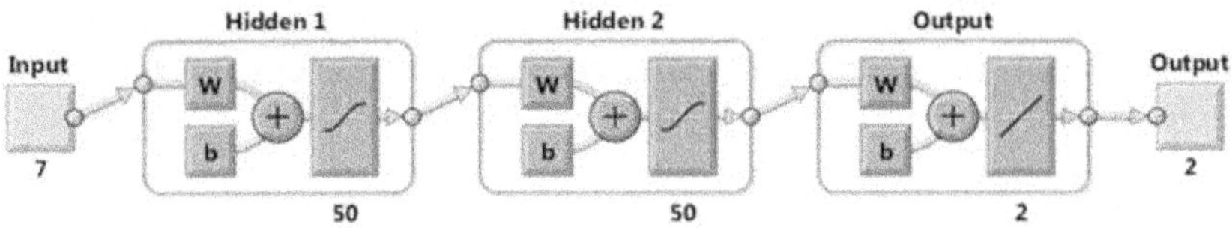


**Figure 4.17.** Feedforward ANN with two hidden layers.

Following training and testing parameters were chosen:

- Training algorithm: Bayesian regularization [40];
- Number of iterations: 500 for each batch;
- Maximal cross-validation error: 10;
- Minimal gradient: $10^{-8}$;
- Training accuracy: MSE.

Other parameters were set as default.

Training stage long for ~18 hours on CPU Intel Core i7-5500 (Windows operating system, 8 GB RAM).

- **Results Analysis and Comparison**

The next step after ANN models training is performance estimation on the test set. Also, correction accuracy should be compared with analytical models. Two types of errors have been chosen to report the results: maximum absolute error and MSE. As correction error is normally distributed, the MSE is estimated as standard deviation multiplied by factor 3. This allows ensuring the probability of 0.997. The results of both types of algorithmic correction are presented in Table 4.13.

**Table 4.13.** Comparison of algorithmic correction results.

| | **MSE ($P$ = 0.997), %** | | **Maximum absolute error, %** | |
|---|---|---|---|---|
| | ANN model | Analytical model | ANN model | Analytical model |
| $R$ | 0.0022 | 0.0037 | 0.0274 | 0.0244 |
| $X$ | 0.0016 | 0.0032 | 0.0128 | 0.0155 |

As follows from Table 4.13, ANN tends to show slightly better results. Application of a more complex ANN model (more layers and neurons) as well as input feature preprocessing (e.g. Z-scoring or logarithmic scale for frequency) theoretically allows achieving even more significant improvement in correction efficiency. Nevertheless, proportional improvement of overall measurement accuracy on the real-life data is not expected. The reason for that is that impact of instrumentation errors ($\Delta_P$ and $\Delta_Q$) will remain uncompensated.

For both cases, correction takes the few dozens of microseconds. Both models are based on simple mathematical operations – multiplication and addition. Thus, they can be easily implemented on DSP or FPGA.

- Artificial neural networks can be applied for algorithmic correction of errors while implementing the impedance spectroscopy for product quality studies. For the algorithmic correction development, feedforward type ANN with two hidden layers of 50 neurons in each one was used. The training set contained 10 batches with 10 thousand samples in each, and test set - 1 batch with a million samples. The training algorithm was Bayesian regularization, the number of iterations for each package was 500. During the research, the following results were obtained:
- The maximum absolute error does not exceed 0.027 % for the active component and 0.013 % for the reactive;
- MSE for 0.997 confidence intervals are 0.0022 % and 0.0016 % for the active and reactive components correspondingly.

For comparison, using the analytical algorithmic correction approach based on a measuring transducer mathematical model, the error values are as follows:

- Maximum absolute error for the active component – 0.024 % and for the reactive – 0.015 %;
- MSE – 0.0037 % and 0.0032 % for the specified components.

### 4.3.3. Analysis of Product Quality by Relative Electric Parameters

The quality of products is determined by the compliance of its indicators with the requirements established in the regulatory and technical documentation. According to the range of changes in the values of indicators (their number) or the presence of additional indicators, products are classified into different levels of quality. To establish such levels, a differential method of quality assessment is used under standard [41]. It provides for any qualimetric object a relative indicator by which the level of its quality is assessed. Obtain a relative indicator based on the results of comparing single or complex indicators of the quality of the object, the quality level of which is set, and the sample, taken as a baseline, with a known level of quality. The comparison of the indicators of these objects can be carried out directly through the measurement of their relations or by comparing the obtained results of measuring the indicators of the controlled object with the indicators of the basic sample specified in the regulatory documentation. The availability of MIs for measuring the relative quality indicators would directly simplify the procedure for assessing the level of product quality.

- **Differential Method of Quality Monitoring by Electrical Parameters and Methods of Comparing Imitation Parameters as Quality Indicators**

The application of a measuring method to compare objects of a non-electrical nature is based on the reflection of quality indicators of such objects with adequate electrical parameters. So, one can compare two electrical quantities, which give us the possibility to ensure the efficiency and objectivity of measurement, storage, and transfer of information about the level of quality, as well as ease of implementation of the differential method by electrical parameters. There are two ways to realize the mentioned method.

The first one is the electrical measurements of individual product quality indicators of known and unknown quality levels and comparing them with each other. To do this, it is necessary to have standard samples of a known level of quality of characterized products and through measurements to obtain their electrical parameters that correspond to each of the unit indicators. Quality assessment is performed by comparing the measured respective electrical parameters of the examined and base samples.

Carrying out such measurement requires several MIs for characterizing individual quality indicators, as the indicators are mostly inhomogeneous quantities, and the methods of measurement are different mostly. This complicates the quality assessments. Therefore, one or two indicators are chosen for comparison (octane number of gasoline, milk fat content, water hardness, or salt content in water, electrolyte concentration, etc.). This simplifies the quality assessment procedure but does not take into account other individual quality indicators.

The second method involves the implementation of a differential method based on comparisons of generalized indicators that can represent the objects of qualimetry. Then the object must be represented by an electrical quantity whose parameters would reflect any change in the unit values. This quantity may be the complex resistance (impedance) or complex conductivity (admittance), i.e. the immittance of the objects of comparison. Generally, the immittance parameters depend on the internal structure of the object, the applied signal level, the signal frequency, types of polarization and conductivity, and much more. In this case, the parameters of comparison of one electrical physical quantity can be a lot, which allows detecting the changes in many single indicators simultaneously. Here the difference from the previous method consists of the application of one measuring instrument only, which is the immittance analyzer. The relations of the parameters of multi-element bipolar schemes, which represent the comparison objects [41-42], are analyzed. If the structure of such two bipolar schemes is the same as the same are the quantitative characteristics of the unit quality indicators, then the parameters of their immittances, measured at the row of frequencies must be the same. Any change in the parameters of the monitored object indicates an alteration in its quality indicators. That is, the electrical parameters can be used to identify products as corresponding to a given level of quality.

Qualimetry objects with unknown bipolar substitution schemes which represent the monitored and basic objects can be identified by comparing their immittances at frequencies of a given band, namely by ratios of impedance $N_Z = \left(\frac{Z_X}{Z_0}\right)_{\omega_i}$ or admittance $N_Y = \left(\frac{Y_X}{Y_0}\right)_{\omega_i}$. Immittances, as vector quantities, are given through active $\mathrm{Re}(Z_x) = R_x$, $\mathrm{Re}(Y_x) = G_x$ and reactive $\mathrm{Im}(Z_x) = X$, $\mathrm{Re}(Y_x) = G_x$ components as:

$$Z_x = R_x - jX_x,\ Z_0 = R_0 - jX_0 \tag{4.16}$$

$$Y_x = G_x + jB_x,\ Y_0 = G_0 + jB_0 \tag{4.17}$$

or through the modules and phases:

$$Z_x = |Z_x| \cdot e^{j\phi_0},\ Z_0 = |Z_0| \cdot e^{j\phi_0} \tag{4.18}$$

$$Y_x = |Y_x| \cdot e^{j\phi_x},\ Y_0 = |Y_0| \cdot e^{j\phi_0} \tag{4.19}$$

Here, the two bipolar schemes that characterize the monitored and base objects are compared while measuring the active and reactive components of (4.16)-(4.17) or modules and phase differences of (4.18), (4.19) of the immittances of the comparison objects, as well as defining their appropriate ratios at fixed frequencies. Therefore, the electrical relative values at any frequency $\omega_i$ of the test signal (where $i = 1,2,3,\ldots,n$ are the selected frequency of a given frequency band) for the impedance and admixture are recorded as:

$$\left(\frac{\mathrm{Re}(Z_x)}{\mathrm{Re}(Z_0)}\right)_{\omega_i}, \left(\frac{\mathrm{Im}(Z_x)}{\mathrm{Im}(Z_0)}\right)_{\omega_i}, \left(\frac{\mathrm{Re}(Y_x)}{\mathrm{Re}(Y_0)}\right)_{\omega_i}, \left(\frac{\mathrm{Im}(Z_x)}{\mathrm{Im}(Z_0)}\right)_{\omega_i} \qquad (4.20)$$

$$\left(\frac{|Z_x|}{|Z_0|}\right)_{\omega_i}, \left(\varphi_{Z_x} - \varphi_{Z0}\right)_{\omega_i}; \left(\frac{|Y_x|}{|Y_0|}\right)_{\omega_i}, \left(\varphi_{Y_x} - \varphi_{Y_0}\right)_{\omega_i} \qquad (4.21)$$

According to the obtained relative electrical indicators, you can set the appropriate quality levels. The direct measurements of immittance parameters and definitions of the relative indicators are carried out by dividing the measurement results of each of the parameters. Also, it is possible to record changes in product performance by the parameters of the immittance ratio, in particular by measuring the active and reactive components of the immittance ratio for instance by the modulus of their ratio or phase shift. That is, the active and reactive components of the relationship (the modulus of the relationship between the phases of the relationship vector) are measured directly, namely:

$$\mathrm{Re}\left(\frac{Z_x}{Z_0}\right)_{\omega_i}, \mathrm{Im}\left(\frac{Z_x}{Z_0}\right)_{\omega_i}, \mathrm{Re}\left(\frac{Y_x}{Y_0}\right)_{\omega_i}, \mathrm{Im}\left(\frac{Y_x}{Y_0}\right)_{\omega_i} \qquad (4.22)$$

$$\left|\frac{Z_x}{Z_0}\right|_{\omega_i}, \left(\varphi_Z\right)_{\omega_i}; \left|\frac{Y_x}{Y_0}\right|_{\omega_i}, \left(\varphi_Y\right)_{\omega_i} \qquad (4.23)$$

Let's consider in detail each of these methods of identification. To analyze the electrical parameters of the bipolar schemes in the frequency range, which can be used to assess the level of quality, it is advisable to study the vectors of immittances of these schemes, which represent the monitored and basic objects.

Fig. 4.18,a demonstrates the vectors of impedances $Z_x$, $Z_0$ and Fig. 4.18, b - admittances of the mentioned samples at the same frequency.

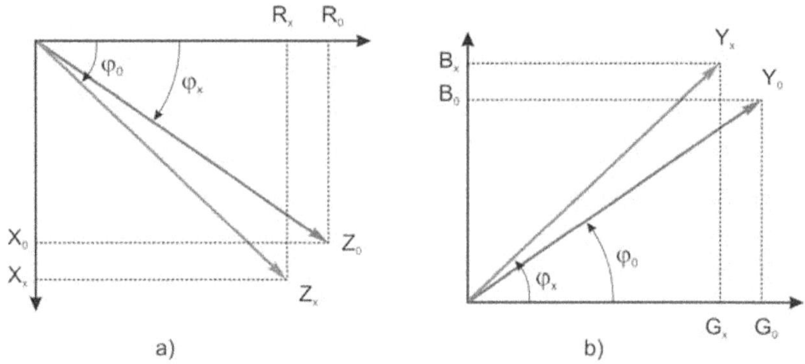


**Figure 4.18.** Impedance and admixture vectors of comparison objects at the same frequency.

The mismatch of immittance vectors, as shown in Fig. 4.18, for the monitored and basic products, indicates a different level of their internal structure, and accordingly of quality indicators. Quantitative evaluation is carried out by measuring and comparing the parameters that characterize these vectors, namely: module and phase, active and reactive components, or other combinations of these parameters. The above methods are used to assess the quality of qualimetric objects, the substitution schemes of which are unknown. The generalized quality indicator, in this case, can be given by one of the sets of parameters, namely on relations of modules of impedances $|Z_x|$, $|Z_0|$, or admixtures $|Y_x|$, $|Y_0|$ and differences of phases $\varphi_x$, $\varphi_0$, of the electric signals at fixed frequencies $\omega_1, \omega_2, \ldots, \omega_n$:

$$P_1' \supset \left\{ \begin{matrix} \left(\frac{|Z_x|}{|Z_0|}\right)_{\omega 1}, \left(\frac{|Z_x|}{|Z_0|}\right)_{\omega 2}, \ldots, \left(\frac{|Z_x|}{|Z_0|}\right)_{\omega n} \\ (\varphi_x - \varphi_0)_{\omega 1}, (\varphi_x - \varphi_0)_{\omega 2}, \ldots, (\varphi_x - \varphi_0)_{\omega n} \end{matrix} \right\}, \tag{4.24}$$

$$P_1'' \supset \left\{ \begin{matrix} \left(\frac{|Y_x|}{|Y_0|}\right)_{\omega 1}, \left(\frac{|Y_x|}{|Y_0|}\right)_{\omega 2}, \ldots, \left(\frac{|Y_x|}{|Y_0|}\right)_{\omega n} \\ (\varphi_x - \varphi_0)_{\omega 1}, (\varphi_x - \varphi_0)_{\omega 2}, \ldots, (\varphi_x - \varphi_0)_{\omega n} \end{matrix} \right\}, \tag{4.25}$$

To implement such measurements, it is necessary to apply the meter of amplitude values of informative parameters and the meter of phase shifts. While assessing quality by the ratio of a component of impedances or admittances at selected frequencies, a similar quality indicator is defined as:

$$P_2' \supset \left\{ \begin{matrix} \left[\frac{\mathrm{Im}(Z_x)}{\mathrm{Im}(Z_0)}\right]_{\omega 1}, \left[\frac{\mathrm{Im}(Z_x)}{\mathrm{Im}(Z_0)}\right]_{\omega 2}, \ldots, \left[\frac{\mathrm{Im}(Z_x)}{\mathrm{Im}(Z_0)}\right]_{\omega n} \\ \left[\frac{\mathrm{Re}(Z_x)}{\mathrm{Re}(Z_0)}\right]_{\omega 1}, \left[\frac{\mathrm{Re}(Z_x)}{\mathrm{Re}(Z_0)}\right]_{\omega 2}, \ldots, \left[\frac{\mathrm{Re}(Z_x)}{\mathrm{Re}(Z_0)}\right]_{\omega n} \end{matrix} \right\}, \quad (4.26)$$

$$P_2'' \supset \left\{ \begin{matrix} \left[\frac{\mathrm{Im}(Y_x)}{\mathrm{Im}(Y_0)}\right]_{\omega 1}, \left[\frac{\mathrm{Im}(Y_x)}{\mathrm{Im}(Y_0)}\right]_{\omega 2}, \ldots, \left[\frac{\mathrm{Im}(Y_x)}{\mathrm{Im}(Y_0)}\right]_{\omega n} \\ \left[\frac{\mathrm{Re}(Y_x)}{\mathrm{Re}(Y_0)}\right]_{\omega 1}, \left[\frac{\mathrm{Re}(Y_x)}{\mathrm{Re}(Y_0)}\right]_{\omega 2}, \ldots, \left[\frac{\mathrm{Re}(Y_x)}{\mathrm{Re}(Y_0)}\right]_{\omega n} \end{matrix} \right\} \quad (4.27)$$

The identical is considered the objects for which the ratios of modules or components at every frequency of the band are equal to each other, and the phase shifts also are equal to zero. Let's evaluate the results of direct measurements of the components of the immittance ratio. To fulfill this, write the ratio of impedances, taking into account the components of the monitored and base objects:

$$\frac{Z_x}{Z_0} = \frac{R_x - jX_x}{R_0 - jX_0}, \quad (4.28)$$

From here the active and the reactive components are described by the equations:

$$\mathrm{Re}\left(\frac{Z_x}{Z_0}\right) = \frac{R_x R_0 + X_x X_0}{R_0^2 + X_0^2}, \quad (4.29)$$

$$\mathrm{Im}\left(\frac{Z_x}{Z_0}\right) = \frac{R_x X_0 - X_x R_0}{R_0^2 + X_0^2} \quad (4.30)$$

Similarly, the ratio of admittances, taking into account the components of the monitored and base objects, is:

$$\frac{Y_x}{Y_0} = \frac{G_x + jB_x}{G_0 + jB_0}, \quad (4.31)$$

therefrom the similar components are derived:

$$\mathrm{Re}\left(\frac{Y_x}{Y_0}\right) = \frac{G_x G_0 + B_x B_0}{G_0^2 + G_0^2}, \tag{4.32}$$

$$\mathrm{Im}\left(\frac{Y_x}{Y_0}\right) = \frac{B_x G_0 - G_x B_0}{G_0^2 + B_0^2} \tag{4.33}$$

Analysis of expressions (4.29, (4.32) envisages that the active component is always of the same sign and any change in the components of the immittance of the monitored object leads to an adequate change in its value. However, from expressions (4.30), (4.33) it is obvious that the reactive components of the ratio of impedances and admittances can change their sign to the opposite under certain conditions and can be equal to zero. Graphic interpretation of the relationship specified in the form of vectors is shown in Fig. 4.19.

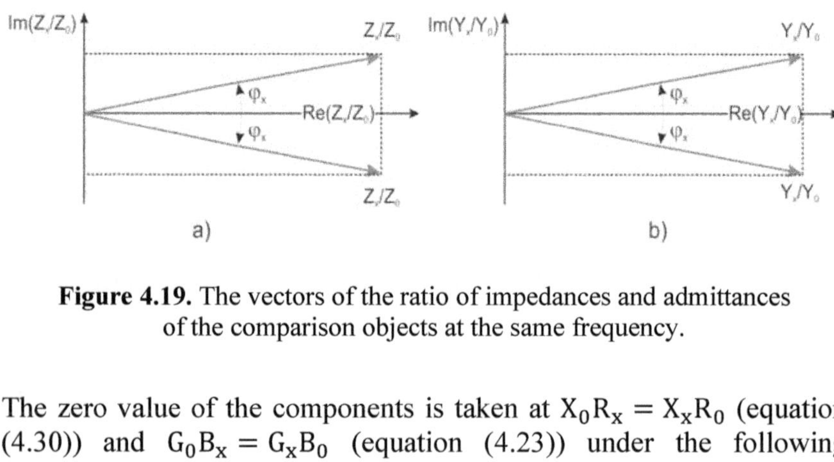


**Figure 4.19.** The vectors of the ratio of impedances and admittances of the comparison objects at the same frequency.

The zero value of the components is taken at $X_0 R_x = X_x R_0$ (equation (4.30)) and $G_0 B_x = G_x B_0$ (equation (4.23)) under the following conditions:

- For the reactive component of the impedance ratio:
If $R_x = R_0$ at $X_x = X_0$;
If $R_x > R_0$ and if $X_x > X_0$, besides it is necessary that $R_x - R_0 = X_x - X_0$;
If $R_x < R_0$ and if $X_x < X_0$, besides it is necessary that $R_0 - R_x = X_0 - X_x$;

- For the reactive component of the ratio of admittances:
If $G_x = G_0$ at $B_x = B_0$;
If $G_x > G_0$ and $B_x > B_0$, besides it is necessary that $G_x - G_0 = B_x - B_0$;
If $G_x < G_0$ and $B_x < B_0$, besides it is necessary that $G_0 - G_x = B_0 - B_x$.

Thus, in one of the three cases, the equality of the reactive component of the ratio can be established that the quality levels of the compared objects are the same. However, the reactive components of the ratio of impedances and admittances change the sign to the opposite at $X_0R_x < X_xR_0$ and at $G_0B_x < G_xB_0$. This is achieved under the following conditions:
If $X_0 < X_x$ at $R_x = R_0$;
If $R_x < R_0$ at $X_x = X_0$;
If $R_x < R_0$ and $X_x > X_0$, besides it is necessary that $R_0 - R_x = X_x - X_0$.

In other cases, the reactive components are of the same sign as the active components. That is, the level of quality of monitored products can be judged by the characteristic features of the change of the reactive component of the admittance, as well as the dependencies of the components on the quality, but only under the above conditions. Relative values (4.26)-(4.27) can be calculated by solving (4.29)-(4.30) or (4.32)-(4.33), provided that the values of the components for different frequencies of the base sample are known.

The generalized quality index $P_3$ in the case of quality assessment by the modulus of ratio and phase is determined for impedances and admittances as:

$$P_3' \supset \left\{ \begin{matrix} \left|\dfrac{Z_x}{Z_0}\right|_{\omega 1}, \left|\dfrac{Z_x}{Z_0}\right|_{\omega 2}, \left|\dfrac{Z_x}{Z_0}\right|_{\omega 3}, \ldots, \left|\dfrac{Z_x}{Z_0}\right|_{\omega n} \\ (\phi_x)_{\omega 1}, (\phi_x)_{\omega 2}, (\phi_x)_{\omega 3}, \ldots, (\phi_x)_{\omega n} \end{matrix} \right\}, \tag{4.34}$$

$$P_3' \supset \left\{ \begin{matrix} \left|\dfrac{Y_x}{Y_0}\right|_{\omega 1}, \left|\dfrac{Y_x}{Y_0}\right|_{\omega 2}, \left|\dfrac{Y_x}{Y_0}\right|_{\omega 3}, \ldots, \left|\dfrac{Y_x}{Y_0}\right|_{\omega n} \\ (\phi_x)_{\omega 1}, (\phi_x)_{\omega 2}, (\phi_x)_{\omega 3}, \ldots, (\phi_x)_{\omega n} \end{matrix} \right\} \tag{4.35}$$

and for the active and reactive components of the ratio of the same parameters, a similar indicator is described as:

$$P_4' \supset \left\{ \begin{array}{l} \left[ \mathrm{Im}\left( \dfrac{Z_x}{Z_0} \right) \right]_{\omega 1}, \left[ \mathrm{Im}\left( \dfrac{Z_x}{Z_0} \right) \right]_{\omega 2}, \ldots, \left[ \mathrm{Im}\left( \dfrac{Z_x}{Z_0} \right) \right]_{\omega n} \\ \left[ \mathrm{Re}\left( \dfrac{Z_x}{Z_0} \right) \right]_{\omega 1}, \left[ \mathrm{Re}\left( \dfrac{Z_x}{Z_0} \right) \right]_{\omega 2}, \ldots, \left[ \mathrm{Re}\left( \dfrac{Z_x}{Z_0} \right) \right]_{\omega n} \end{array} \right\}, \quad (4.36)$$

$$P_4'' \supset \left\{ \begin{array}{l} \left[ \mathrm{Im}\left( \dfrac{Y_x}{Y_0} \right) \right]_{\omega 1}, \left[ \mathrm{Im}\left( \dfrac{Y_x}{Z_0} \right) \right]_{\omega 2}, \ldots, \left[ \mathrm{Im}\left( \dfrac{Y_x}{Z_0} \right) \right]_{\omega n} \\ \left[ \mathrm{Re}\left( \dfrac{Y_x}{Y_0} \right) \right]_{\omega 1}, \left[ \mathrm{Re}\left( \dfrac{Y_x}{Y_0} \right) \right]_{\omega 2}, \ldots, \left[ \mathrm{Re}\left( \dfrac{Y_x}{Y_0} \right) \right]_{\omega n} \end{array} \right\} \quad (4.37)$$

- **Ways to Ensure the Effectiveness of Determining the Parameters of Non-Electric Objects**

The creation of CPSs for monitoring parameters and quality management of non-electrical objects, such as the soils, significantly improve the environmental situation, the balanced use of natural resources, including water. The CPS could streamline the process of collecting and forecasting real-time land condition parameters and linking them to spatial coordinates, as well as improve the procedure for their certification and technical supervision of the use of certified soils. The formation of a developed information base for monitoring the state of soil pollution based on the CPS ensures an appropriate level of management.

To optimize the process of collecting information on the quality indicators of soils, it is desirable to use the admittance method, according to which the object of control of non-electrical nature, f. e. the soil, placed in an AC circuit is considered as a complex bipolar object, parameters of which contain information about the relevant physical and chemical information on it. The results of measuring the RLC parameters of such a scheme determine the relationships between the electrical and the corresponding physical & chemical characteristics of the monitored object. With this method it is most expedient to monitor important integral quality indicators of soils, that is the levels of acidity, salinity, humidity, etc. This encourages the creating the admittance maps of soils and their real-time usage, which provide representative information on changes in soil properties and soil heterogeneity parameters comparing to surveys performed by classical methods. Prompt response to heterogeneous conditions of plant growth and development allow the agronomist to increase yields at fixed production costs and / or reduce these costs without reducing yields. The operation of a mobile device for admittance mapping can be based on the capacitive method. For example, when using the contact method, the electrodes made in the form of disks can be attached to a vehicle moving at a constant speed in the field, contact with the ground and determine the admittance in several layers of soil at the different depths. In this way, the amount of data multiplies. GPS can help in information provision. The obtained soil admittance map highlights the contrasting areas of the soil structure. Although the absolute values of the admittance increase with the soil moisture, the relative values remain consistently adequate within time. To ensure the reliability of the measurement results of RLC-meters for CPS's purposes it is necessary to verify the metrological characteristics of MIs directly while operation.

- **Concept of Express Control of Admittance MIs Performance for Analytical Purposes**

Based on the analysis, it is proposed to calibrate the measuring channel of RLC-meters by intelligent code-operated measures. First of all, it needs to prepare a row of the standard samples that serve as a measure of the studied property of the defining object in the entire range of its alteration. Next, for a given frequency band of the impedance (admittance) parameters are processing the received results of the mentioned samples parameters with the help of RLC-meter. The latter is carried out by forming a matrix $N_i$ of digital codes for the descriptive measure $Z_M$, corresponding to the certain parameter of the object (Fig. 4.20).

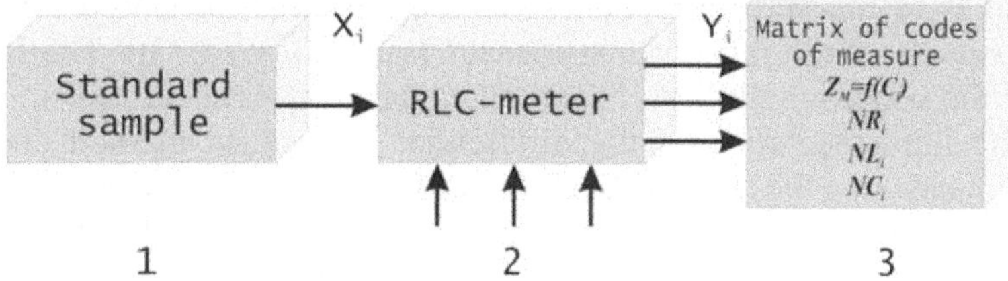


**Figure 4.20.** The general structure of determining the parameters of the electrical measure for calibration of FTA analytical purposes.

Henceforward, it is advisable to measure the characteristics of the admittance MIs by substitution method, or by differential or compensatory methods of measuring resistance. For example, when measuring the impedance $Z_X$ by the substitution method of bridge impedance meters, the measurement result can be given as:

$$Z_X = Z_M \, Z_{thresX} / Z_{thresM} \tag{4.38}$$

Here $Z_{thresX}$ is the value of the resistance of the comparison arm $Z_X$ when included in the resistance circuit; $Z_{thresM}$ is the value of the resistance of the comparison arm when included in the resistance circuit $Z_M$. Assuming that the errors of the resistances of the comparison arms $Z_X$; $Z_M$ are rather the same: $\delta_{ZthresX} \approx \delta_{ZthresM}$, the error of measuring the impedance $Z_X$ is equal to the sum of the error of the measure and the double discreteness error $\delta_{ZX} = \delta_{ZM} + 2\delta_{discr}$.

- **Device for Analyzing the Product Quality by Relative Electric Parameters**

Differential method of quality assessment consists in comparison with the base sample through direct measurement of the same parameters of the standard sample and monitored sample. Herewith, the comparison of the measured parameters is carried out for the specified parameters of samples regarding the standard sample parameters responsible for a certain level of quality. In this case, two options are possible. The first is to measure the modulus and phase differences or their components for each of the comparison objects. The relative indicators (4.24)-(4.27) are defined from the obtained results. This requires 2 measuring channels containing primary transducers, immittance-voltage converters, meters of the selected parameters equipped with the microcontroller. The 2nd option is to measure the modulus of the ratio and the phase shift or components of the ratio of immittances (see (4.34); (4.37)). The latter gives the wider possibilities for product quality assessment and simpler for the implementation of the differential assessment method due to direct measuring the relationship between the relevant components of the compared objects at different frequencies (expressions (4.18)-(4.19)). Such a measurement can be implemented with the help of MI, the functional scheme of which is in Fig. 4.21.

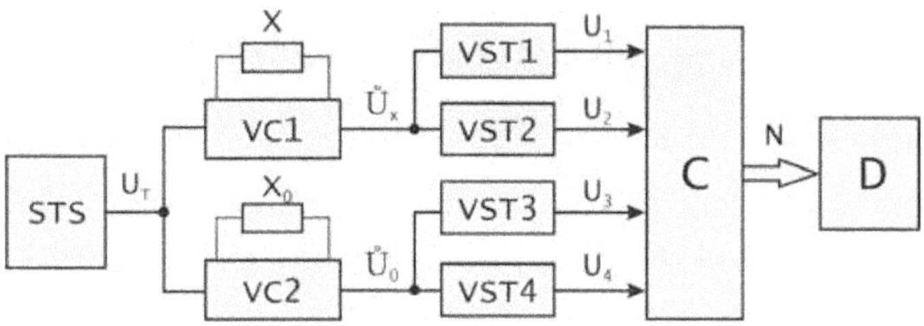


**Figure 4.21.** Block diagram of measuring the ratio of the components of the immittances of the comparison objects.

The circuit contains an STS of fixed voltage level with the frequencies $\omega_1, \omega_2, \dots, \omega_n$, vector converters "immittance-voltage" VC1 and VC2 to convert the immittance of the monitored and base objects to the appropriate voltages $U_x$ and $U_o$ [43]. The VST1, VST3, and VST2, VST4 distinguish the active and reactive components of complex voltages, respectively. The obtained DC voltages U1, U2, U3, U4, are fed to the controller C, which according to a given program produces the codes that correspond to the ratio of active and reactive components of the immittances of the comparison objects. These codes are visualized on the display D. So, the output voltages of the VC1 and VC2 at the particular frequencies of the test signal can be written as:

$$\dot{U}_{x1} = U_T k_1 X\,,\ \dot{U}_{x2} = U_T k_2 X\,,\ \dot{U}_{xc} = U_T k_c X \tag{4.39}$$

$$\dot{U}_{01} = U_T k_1' X_0\,,\ \dot{U}_{02} = U_T k_2' X_0\,,\ \dot{U}_{0c} = U_T k_c' X_0\,, \tag{4.40}$$

where $k_1, k_2, \dots, k_n$ та $k_1', k_2', \dots, k_n'$ are the factors of immittances converting in the voltages $\dot{U}_{x1}, \dot{U}_{x2}, \dots, \dot{U}_{xn}$ та $\dot{U}_{01}, \dot{U}_{02}, \dots, \dot{U}_{0n}$ , respectively, performed by convertors VC1 and VC2 on the frequencies $\omega_1, \omega_2, \dots, \omega_n$. The voltages at the outputs of the vector-scalar converters are described by expressions:

$$U_{11} = U_T k_1 b_1 \operatorname{Re}(U_{x1}),\ U_{12} = U_T k_2 b_2 \operatorname{Re}(U_{x2}),\ U_{1c} = U_T k_c b_c \operatorname{Re}(U_{xc}) \tag{4.41}$$

$$U_{21} = U_T k_1 c_1 \operatorname{Im}(U_{x1}),\ U_{22} = U_T k_2 c_2 \operatorname{Im}(U_{x2}),\ U_{2c} = U_T k_c c_c \operatorname{Im}(U_{xc}) \tag{4.42}$$

$$U_{31}=U_T k_1' b_1' \operatorname{Re}(U_{01}),\; U_{32}=U_T k_2' b_2' \operatorname{Re}(U_{02}),$$
$$U_{3c}=U_T k_c' b_c' \operatorname{Re}(U_{0c}) \tag{4.43}$$

$$U_{41}=U_T k_1' c_1' \operatorname{Im}(U_{01}),\; U_{42}=U_T k_2' c_2' \operatorname{Im}(U_{02}),$$
$$U_{4c}=U_T k_c' c_c' \operatorname{Im}(U_{0c}) \tag{4.44}$$

where $b_1, b_2, \ldots, b_n$, $b_1', b_2', \ldots, b_n'$ and $c_1, c_2, \ldots, c_n$, $c_1', c_2', \ldots, c_n'$ are the converting factors for VST1, VST3, and VST2, VST4 which transform the voltages of vector converters in the appropriate voltages proportional to the components on the separate frequencies. At the controller output we get:

$$N_1=\frac{U_{11}}{U_{31}}=\frac{k_1 b_1 \operatorname{Re}(U_{x1})}{k_1' b_1' \operatorname{Re}(U_{01})},\; N_2=\frac{U_{12}}{U_{32}}=\frac{k_2 b_2 \operatorname{Re}(U_{x2})}{k_2' b_2' \operatorname{Re}(U_{02})},$$
$$N_n=\frac{U_{1c}}{U_{3c}}=\frac{k_c b_c \operatorname{Re}(U_{xc})}{k_c' b_c' \operatorname{Re}(U_{0c})} \tag{4.45}$$

$$N_1'=\frac{U_{21}}{U_{41}}=\frac{k_1 c_1 \operatorname{Im}(U_{x1})}{k_1' c_1' \operatorname{Im}(U_{01})},\; N_2'=\frac{U_{22}}{U_{42}}=\frac{k_2 c_2 \operatorname{Im}(U_{x2})}{k_2' c_2' \operatorname{Im}(U_{02})},$$
$$N_1'=\frac{U_{2c}}{U_{4c}}=\frac{k_c c_c \operatorname{Im}(U_{xc})}{k_c' c_c' \operatorname{Im}(U_{0c})}, \tag{4.46}$$

which is adequate to the relative indicators characterized by the ratios of active and reactive components of the comparison objects at fixed frequencies. Assessment of the level of product quality by the components of the relationship can be implemented by MI shown in Fig. 4.22.

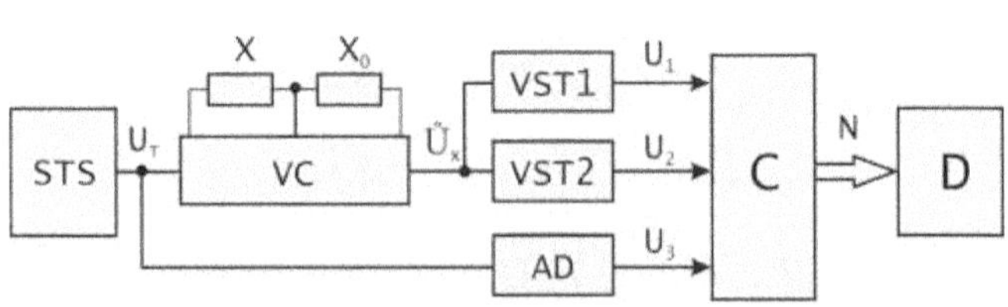


**Figure 4.22.** Block scheme of measuring the components of the ratio of immittances of the comparison objects.

The circuit contains similar to the previous scheme nodes, as well as an additional amplitude detector BP to ensure the invariance of the measurement result to the voltage of the STS. So, at certain frequencies of the test signal at the output of the VC we obtain:

$$\dot{U}_{x1} = U_T k_1 \frac{X}{X_0},\ \dot{U}_{x2} = U_T k_2 \frac{X}{X_0},\ \dot{U}_{x3} = U_T k_3 \frac{X}{X_0},$$
$$\dot{U}_{xc} = U_T k_c \frac{X}{X_0} \tag{4.47}$$

Accordingly, at the outputs of VST1 and VST2 we receive:

$$U_{11} = U_T k_1 b_1 \,\mathrm{Re}\left(\frac{X}{X_0}\right), U_{12} = U_T k_2 b_2 \,\mathrm{Re}\left(\frac{X}{X_0}\right),$$
$$U_{1c} = U_T k_c b_c \,\mathrm{Re}\left(\frac{X}{X_0}\right) \tag{4.48}$$

$$U_{21} = U_T k_1 c_1 \,\mathrm{Im}\left(\frac{X}{X_0}\right),\ U_{22} = U_T k_2 c_2 \,\mathrm{Im}\left(\frac{X}{X_0}\right),$$
$$U_{2c} = U_T k_c c_c \,\mathrm{Im}\left(\frac{X}{X_0}\right) \tag{4.49}$$

and at the AD output we get:

$$U_{31} = a_1 U_T,\ U_{32} = a_2 U_T,\ U_{3n} = a_n U_T \tag{4.50}$$

That is, the results of measuring the reactive and active components of the ratio of immittances of the comparison objects are obtained. For known values of the components of the basic sample, it can be computed the relative values using the dependences (4.29)-(4.30) or (4.32)-(4.33). Ratios of (4.47) and (4.48) to (4.49) ensure the invariance of the measurement result to the parameters of the test signal.

- **Development of the Code-Operated Measure**

The analysis envisages that the main requirements while designing the code-operated impedance measure are lowering the impact of input and output impedances of DACs, of the number of unambiguous impedance measures and mutual independence of conduction reproduction processes of active, capacitive and inductive nature. Taking them into account, there was developed the design of a code-operated measure of complex conductivities (Fig. 4.23). It contains 4 OAs, 3 unambiguous measures of impedance, DAC1-DAC3, 4 VR1 - VR4. This impedance circuitry makes it possible to generate currents flowing through every unambiguous impedance measure. The values of these currents are set by external codes.

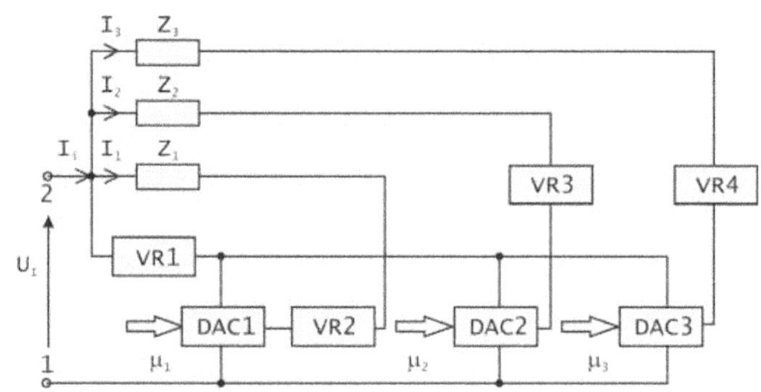


**Figure 4.23.** The structure of the code-operated measure of complex conductivity.

Due to the use of buffer elements, simultaneously is reduced the impact of input and output resistances, is expanded the frequency band of the measure, is eliminated the possibility of self-excitation of the code-operated measure in the whole range of reproducible values. A voltage $U_i$ is applied to inputs 1, 2 of external sources; it causes currents $I_1$, $I_2$, $I_3$ to pass through the measures $Z_1$, $Z_2$, and $Z_3$, respectively, and the outputs of the VR2 - VR4, respectively. Due to the high input resistance of the VR1, its input current can be neglected compared to the currents $I_1$, $I_2$, $I_3$. The input voltage Ui of the simulator is repeated at the output of the VR1 and is applied simultaneously to the information inputs of DAC1 - DAC3. Codes $\mu_1$ - $\mu_3$ from an external source are fed to their control inputs. The output voltages DAC1 - DAC3 are repeated at the outputs of the VR2 - VR4, through which pass the appropriate currents $I_1$, $I_2$, $I_3$. Their values are determined from the relations: $I_1=U_i(1-\mu_1)G_1$, $I_2=U_i(1-\mu_2)G_2$, $I_3=U_i(1-\mu_3)G_3$, where $G_1=1/Z_1$, $G_2=1/Z_2$, $G_3=1/Z_3$ are the conductivities of unambiguous impedance measures $Z_1$, $Z_2$, $Z_3$; $U_{VR2} = \mu_1 U_i$, $U_{VR3} = \mu_2 U_i$, $U_{VR4} = \mu_3 U_i$ are the output voltages of

VR2 - VR4. The input conductivity $G_i$ of the code-operated conductivity simulator is determined from:

$$G_i = I_i / U_i = (1-\mu_1)G_1 + (1-\mu_2)G_2 + (1-\mu_3)G_3 \qquad (4.51)$$

Here $I_i = I_3 + I_2 + I_1$ is an input current of the code-operated conductivity simulator. If, for example, $G_3 = \frac{1}{R} = G$, $G_2 = \frac{1}{j\omega L}$, and $G_1 = j\omega C$, then the code-operated simulator reproduces the resistive-capacitive-inductive conductivity $G_{iRLC} = (1-\mu_3)G + \frac{(1-\mu_2)}{j\omega L} + (1-\mu_1)j\omega C$. F.i., if $G_2 = \frac{1}{R} = G$, $G_3 = \frac{1}{j\omega L}$ and $\mu_3 = 1$, then the code-operated simulator reproduces the resistive-inductive conductivity $G_{iRL} = (1-\mu_2)\cdot G + \frac{(1-\mu_1)}{j\omega L}$.

The code-operated measure of complex conductivity makes it possible to separately and mutually independently control the reproduction of each of the components of complex conductivity - active, capacitive, and inductive. The reproduction errors of the conductivity components are determined almost exclusively by the errors of unambiguous measures of resistance, capacitance, and inductance and code-operated voltage dividers. Due to the large values of the transmission factors of OAs, the static error can be reduced to the thousands of percent, which is significantly smaller comparing to the errors of unambiguous impedance measures (typical values of one hundredth to tenths of a percent) in a wide frequency band (up to tens of MHz). The structure of the proposed code-operated measure of complex conductivity is potentially stable in the frequency band for all reproducible conductivity values.

## 4.4. Dielectric Moisture-Metry of Bulk Materials

Measuring the moisture content of bulk materials and designing express moisture meters is an urgent task, as the major direction of the economy is the production of cereals and products of its processing. Precise control of the moisture content in the grain helps to reduce the costs associated with drying and storage. According to the Food and Agricultural Organization at the UN, annual grain losses account for more than 10% of total production, and in some less developed countries 30-50%. To the above, it should be added that most often these losses occur at high humidity and high grain temperature.

The ability to accurately determine the moisture content of grain at the right time, and especially during harvest - is the factor that separates profits from losses. Express moisture meters are used when receiving grains at elevators, to control the preservation during grain processing. The ability to determine the humidity not only of cereals and legumes but also the products of their processing (flour, etc.), allows significantly expand the range of moisture meters.

In contrast to some known methods for measuring the moisture content of bulk products, we focused on the dielectric method for measuring humidity. The latter is best suited to determine the moisture content of biomaterials and products [44].

Dielectric method of moisture determination involves the assessment of moisture content by measuring the dielectric constant and the tangent of the angle of dielectric loss of wet materials and substances in a wide range of frequencies - from sound to microwave. Measurement of humidity by the dielectric method is based on the difference between the values of the dielectric constant of the solid base, air, and water. The main characteristics of the test substance are the dependence of dielectric constant and dielectric loss on moisture content, frequency of the alternating field, temperature, density, etc.

Compared with the conductometric method, dielectric one is inherent in many advantages: lower temperature sensitivity, the ability to compensate for the impact of significant losses (up to $tg\delta = 100$) with the appropriate choice of measurement scheme, reduced moisture distribution by volume of the test object, and high accuracy measurements. Taken together, it leads to the fact that, despite the variety of existing methods for measuring the moisture content of bulk materials, today preference is given to the dielectric method.

The main characteristics used in the considered method are the dependence of the specific electrical parameters of the material on its humidity, properties, and condition, as well as on the measurement conditions:

a) Humidity characteristics of the electrical parameters, i.e. the dependence of $\varepsilon$ and $tg\delta$, or other, including complex parameters on humidity W. Humidity characteristics must be measured at the permanent values of electric field frequency f, material temperature $\theta$, degree of compaction (for dispersed materials) and other quantities that affect the measurement results;

b) Frequency characteristics - functions $\varepsilon(f)$ and $tg\delta(f)$ at constant humidity W, temperature $\theta$ and other values that characterize the state of the material;
c) Temperature characteristics - functions $\varepsilon(\theta)$ and $tg\delta(\theta)$ at constant humidity W, frequency f and other factors;
d) Characteristics that describe the dependence of the electrical parameters of the object on its composition, density, moisture distribution, etc.

Humidity characteristics are the basis for calibration of moisture meters and largely determine their metrological properties. If the electrical parameters change according to the proportional-additive law, the considered characteristics are described by the equations:

$$\varepsilon = \varepsilon_c + AW, \quad (4.52)$$

$$tg\delta = tg\delta_c + BW, \quad (4.53)$$

where $\varepsilon_c$ and $tg\delta_c$ are the parameters of absolutely dry material, A and B are factors that depend on the properties of the material and the measurement conditions. Fig. 4.24 and 4.25 envisage the dependencies of $\varepsilon$ and $tg\delta$ on the humidity for different materials.

Only some materials have linear moisture characteristics. These include some liquid dielectrics - oil, fuel oil, and other petroleum products, in which a linear dependence of $\varepsilon$ on the humidity in the range up to 10–12% has been established. In the vast majority of materials $\varepsilon$ and $tg\delta$ increase with humidity faster than linearly; the slope of the characteristic $tg\delta$ (W) is usually greater than the characteristic $\varepsilon$ (W). The shape of the moisture characteristics strongly depends on the frequency; therefore, the combined frequency-humidity characteristics are considered more frequently for certain materials.

The shape of the frequency characteristics of wet materials is associated with the inhomogeneity of the inner field of these materials due to the difference in the electrical properties of their components. The influence of frequency on the electrical properties of an inhomogeneous dielectric is determined by its unequal impact on different types of polarization. In the low-frequency region, most materials demonstrate a significant increase in $tg\delta$ due to raising the relative value of the through conductivity. For materials in which the losses are mainly due to through conductivity, the dependence of tgδ (f) is close to hyperbolic, that is, with increasing frequency tgδ fall, moving to a certain limit.

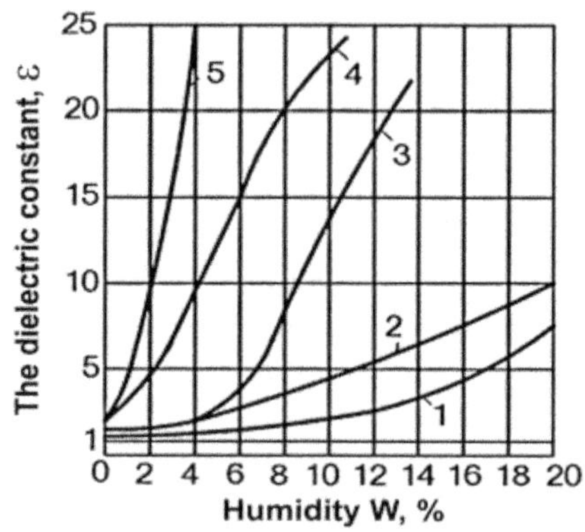


**Figure 4.24.** Dependence of dielectric constant on humidity:
1 – grain; 2 - activated alumina; 3 - coal; 4 - sand; 5 – gravel.

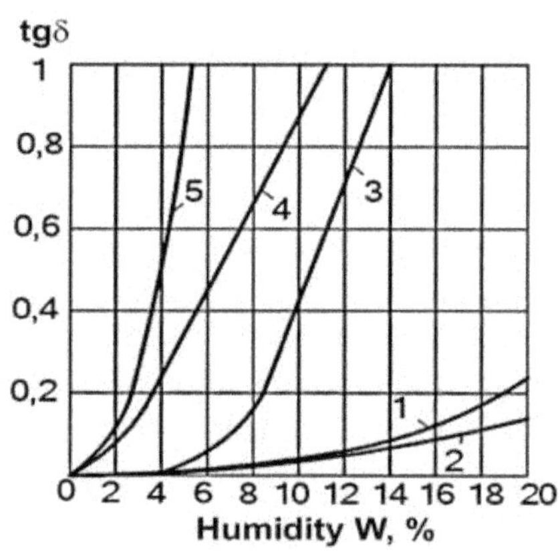


**Figure 4.25.** Dependence of tgδ on humidity:
1 – grain; 2 - activated alumina; 3 - coal; 4 - sand; 5 – gravel.

As for the dielectric constant, in the wet materials with increasing frequency, there is noted a gradual decrease in ε. These general properties of the frequency characteristics may vary depending on the degree of inhomogeneity of the dielectric, in particular, the presence of conductive inclusions, etc. The most important of the factors influencing the nature of the frequency curves is the humidity of the material. In this regard, it is advisable to consider the frequency-humidity characteristics, i.e. the families of curves ε (f) and tgδ (f) for different values of W = const.

- **Factors Affecting the Humidity of Bulk Materials and Temperature Errors Adjustment while Measuring**

The humidity of bulk materials significantly depends on many factors, such as temperature, particle size distribution (shape, size), grain structure, biochemical composition, etc. In this case, the temperature of the environment has to be controlled during the measurement. To correct the temperature error, it is enough to define the so-called temperature coefficients.

The chemical composition of organic origin materials, such as grain, significantly depends on the region and climatic conditions of its cultivation, on mineral fertilizers (organic and inorganic), which are added to the soil to obtain a better harvest. The dielectric constant of bulk materials of inorganic origin also depends on similar factors, in particular humidity and temperature. In general, it is difficult to take into account all these factors when developing secondary MIs, but during the measurement influence of their impacts should be corrected.

Among such non-informative parameters, the most significant are the temperature, as well as the biochemical composition of studied materials. The temperature of the medium can be additionally defined during the measurement and it is enough to know the temperature coefficients for the temperature error adjustment. In contrast to temperature, it is extremely difficult to control the biochemical composition of materials during measurement; therefore, it is almost impossible to correct errors caused by the ambiguity of composition by conventional methods. To perform it, we must use other methods, which are considered just below.

- **Effect of Temperature and Humidity of Bulks on the Result of Measuring the Capacity of the Primary Transducer**

Dielectric humidity meters are capacitive. They carry out the same conversion algorithm, which can be represented as Wd → εd → Cd → Bd, where Wd ; εd ; Cd ; Bd are respectively, the measured humidity, dielectric constant, electrical capacity of the measured material in the primary transducer and the readout of the MI. Thus, the output value is proportional to the measured capacitance of the transducer, filled with the measured material. Then the capacity becomes a simplified form for all structural particles in it: Cd = εd K; here K is the design parameter of the primary transducer, which depends on the shape and size of both the electrodes and the interelectrode space in which the bulk material is located.

The dielectric constant of bulk materials substantially depends on their temperature. In general, the change in temperature introduces an additional error as a result of measuring the capacity of the transducer by the dielectric method. To adjust, it is necessary to know the temperature coefficients of the studied material. The latter may not be given in references. So, to determine, it seems to be necessary to conduct experimental studies.

Previously, we have prepared several samples of wheat grain taken from one field, i.e. material was of the same type. These samples were moistened in different ways. The humidity of the samples was determined by thermogravimetric method (drying and weighing method). As a result, we obtained samples of wheat grain with the following moistures: 10.87%; 14.63%; 19.30%; 23.42%. After that, the samples of different humidity were alternately filled in primary transducer and located in a thermal chamber type TV 1000, where the temperature was set from +10°C to +40°C in steps of 5°C with an error ±0.2°C The output capacity of the primary transducer was measured with pre-calibrated LCR E7-12 meter. At the same time, it was measured the temperature by a non-contact method with the help of an infrared pyrometer, taking into account the time-unstable coefficient of emissivity of the test material [45].

The obtained results are presented in Table 4.14 and Fig. 4.26.

**Table 4.14.** Values of capacitance (pF) at different temperatures and humidity.

| Humidity / Temperature | 10.87 % | 14.63 % | 19.30 % | 23.42 % |
|---|---|---|---|---|
| 10 °C | 17.52 | 42.65 | 73.42 | 100.94 |
| 15 °C | 36.39 | 73.87 | 120.08 | 161.01 |
| 20 °C | 55.38 | 104.54 | 166.29 | 219.56 |
| 25 °C | 74.19 | 136.21 | 211.67 | 278.83 |
| 30 °C | 93.50 | 166.53 | 259.17 | 339.01 |
| 35 °C | 112.03 | 197.83 | 305.48 | 400.02 |
| 40 °C | 131.28 | 228.79 | 350.25 | 458.36 |

These dependencies are linear of different inclination and displacement. To establish the parameters of the dependence C = f (W, θ) and further analysis, we apply the method of least squares, which allows defining a bilinear mathematical model for two factors: humidity and temperature in the form:

$$C_{ij} = b_0 + b_1 W_i + b_2 \theta_j + b_3 W_i \theta_{jj}, \quad (4.54)$$

where $b_0$, $b_1$, $b_2$, $b_3$ are the model factors; W is the humidity of material; θ is the temperature of it.

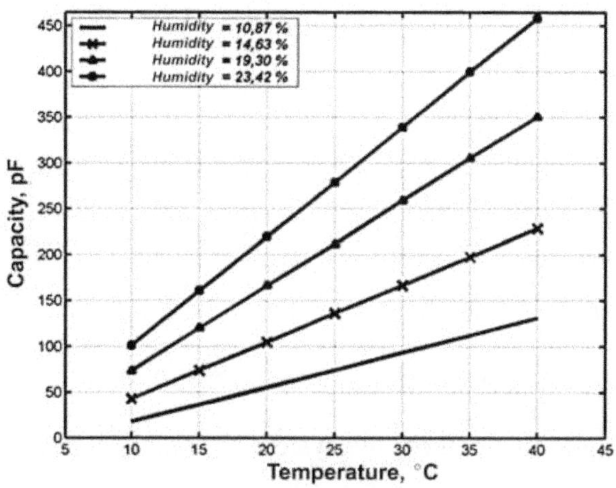


**Figure 4.26.** Influence of material's temperature on the capacity of the primary transducer at different humidity of studied material.

- **Determination of the Parameters of the Mathematical Model of Primary Transducer by the Method of Least Squares**

The basis for determining the coefficients of dependence (4.54) is the results of the experiment (see Table 4.14). We have obtained 28 different values of capacity for a combination of 4 values of humidity (i = 4) and 7 values of temperature (j = 7), i.e.

$$b_0 + b_1 W_i + b_2 \theta_j + b_3 W_i \theta_j = y_{ij} + \Delta_{ij}, \quad (4.55)$$

where i = 1...4, j = 1...7, Δij error of modelling. The considered method's criterion consists in minimization of the sum of error squares:

$$\sum_{i=1}^{4}\sum_{j=1}^{7}\Delta_{ij}^{2}=\min, or \sum_{i=1}^{4}\sum_{j=1}^{7}\left(b_0+b_1W_i+b_2\theta_j+b_3W_i\theta_j-y_{ij}\right)^2=\min \quad (4.56)$$

If denote the matrix of the primary system of equations (4.55) as:

$$\mathbf{A}=\begin{vmatrix} 1 & W_1 & \theta_1 & W_1\theta_1 \\ 1 & W_1 & \theta_2 & W_1\theta_2 \\ & & \vdots & \\ \cdots & \cdots & \cdots & \cdots \\ 1 & W_2 & \theta_1 & W_2\theta_1 \\ 1 & W_2 & \theta_2 & W_2\theta_2 \\ & & \vdots & \\ \cdots & \cdots & \cdots & \cdots \\ 1 & W_3 & \theta_1 & W_3\theta_1 \\ 1 & W_3 & \theta_2 & W_3\theta_2 \\ & & \vdots & \\ \cdots & \cdots & \cdots & \cdots \\ 1 & W_4 & \theta_1 & W_4\theta_1 \\ & & \vdots & \\ 1 & W_4 & \theta_7 & W_4\theta_7 \end{vmatrix}, \quad (4.57)$$

and the vector of measurement results as:

$$Y^T = (y_1; y_2; y_3; y_4; \ldots; y_{28}), \quad (4.58)$$

then according to the method, the system of equations is as following:

$$(A^TA)\cdot B = A^TY, \quad (4.59)$$

here B is the vector of required coefficients:

$$B^T = (b_0; b_1; b_2; b_3) \quad (4.60)$$

The solution of (4.59) is:

$$B = (A^TA)^{-1}\cdot A^TY \quad (4.61)$$

From the point of the errors following analysis, simpler results are obtained when both input factors (W, θ) are shifted to their average value, i.e.

$$\begin{cases} W_i' = W_i - \overline{W} \\ \theta_j' = \theta_j - \overline{\theta} \end{cases} \quad (4.62)$$

$$\overline{W} = \frac{1}{4}\sum_{i=1}^{4} W_i; \qquad \overline{\theta} = \frac{1}{7}\sum_{j=1}^{7} \theta_j. \tag{4.63}$$

In this case, the matrix A of the system (4.55) is modified to A′, resulting in a matrix

$$M' = (A'^{T} \cdot A') \tag{4.64}$$

and for a system of linear equations it becomes diagonal:

$$\mathbf{M}' = 28 \cdot \begin{vmatrix} 1 & 0 & 0 & 0 \\ 0 & \overline{W'}^2 & 0 & 0 \\ 0 & 0 & \overline{\theta'}^2 & 0 \\ 0 & 0 & 0 & \overline{W'}^2\,\overline{\theta'}^2 \end{vmatrix}, \tag{4.65}$$

where $\overline{W'}^{2} = \frac{1}{4}\sum_{i=1}^{4}(W_i - \overline{W})^2$; $\overline{\theta'}^{2} = \frac{1}{7}\sum_{j=1}^{7}(\theta_j - \overline{\theta})^2$. Let's define the values of the elements of the vector

$$C = A'^{T} \cdot Y, \;\; C_0 = 28 \cdot \overline{y}; \tag{4.66}$$

$$C_1 = 7\left[\begin{array}{l} W_1 \dfrac{y_1 + y_2 + \ldots + y_6 + y_7}{7} + W_2 \dfrac{y_8 + y_9 + \ldots + y_{13} + y_{14}}{7} + \\ + W_3 \dfrac{y_{15} + y_{16} + \ldots + y_{20} + y_{21}}{7} + W_4 \dfrac{y_{22} + y_{23} + \ldots + y_{26} + y_{28}}{7} \end{array}\right];$$

$$C_2 = 4\left[\begin{array}{l} \theta_1 \dfrac{y_1 + y_8 + y_{15} + y_{22}}{4} + \theta_2 \dfrac{y_2 + y_9 + y_{16} + y_{23}}{4} + \\ + \theta_3 \dfrac{y_3 + y_{10} + y_{17} + y_{24}}{4} + \ldots + \theta_7 \dfrac{y_7 + y_{14} + \ldots + y_{21} + y_{28}}{4} \end{array}\right];$$

$$C_3 = 7\left[\begin{array}{l} W_1 \dfrac{(\theta_1 y_1 + \theta_2 y_2 + \ldots + \theta_6 y_6 + \theta_7 y_7)}{7} + W_2 \dfrac{(\theta_1 y_8 + \theta_2 y_9 + \ldots + \theta_6 y_{13} + \theta_7 y_{14})}{7} + \\ + W_3 \dfrac{(\theta_1 y_{15} + \theta_2 y_{16} + \ldots + \theta_6 y_{20} + \theta_7 y_{21})}{7} + W_4 \dfrac{(\theta_1 y_{22} + \theta_2 y_{23} + \ldots + \theta_6 y_{27} + \theta_7 y_{28})}{7} \end{array}\right].$$

Then the values of the required coefficients of the mathematical model (4.54) with an offset by the average value are:

$$b_0' = C_0; b_1' = C_1 / \overline{W}'^2;$$
$$b_2' = C_2 / \overline{\theta}'^2; b_3' = C_3 / \left(\overline{W}'^2 \cdot \overline{\theta}'^2\right).$$

Using the software package Mathcad we obtain the following values of the coefficients:

$b'_0 = 175.529$; $b'_1 = 16.366$; $b'_2 = 7.791$; $b'_3 = 0.649$.

- **Analysis of the Characteristics of the Errors of the Defined Factors**

The variance of the dependence coefficients (4.55) can be found based on the residual squares of the deviations of the defined linear dependence on the experimental points:

$$S_y^2 = \frac{\sum_{i=1}^{28} (y_i - y_{\text{mod}})^2}{28 - 4} = 0.227 \tag{4.67}$$

and diagonal coefficients of the so-called dispersion matrix of the equations system

$$\text{MD} = (\text{A}'^{\text{T}}\,\text{A}')^{-1}, \tag{4.68}$$

which due to the diagonality is the following:

$$\mathbf{MD} = \frac{1}{28} \cdot \begin{vmatrix} 1 & 0 & 0 & 0 \\ 0 & \frac{1}{\overline{W}'^2} & 0 & 0 \\ 0 & 0 & \frac{1}{\overline{\theta}'^2} & 0 \\ 0 & 0 & 0 & \frac{1}{\overline{W}'^2 \cdot \overline{\theta}'^2} \end{vmatrix} = \tag{4.69}$$
$$= \begin{vmatrix} 0.03571 & 0 & 0 & 0 \\ 0 & 1.593 \cdot 10^{-3} & 0 & 0 \\ 0 & 0 & 3.571 \cdot 10^{-4} & 0 \\ 0 & 0 & 0 & 1.593 \cdot 10^{-5} \end{vmatrix}$$

That is, estimates of the variances of the model coefficients are calculated as follows:

$$S_{b_0}^2 = S_y^2 \mathrm{MD}_{0,0} = 8.124 \cdot 10^{-3}; S_{b_1}^2 = S_y^2 \mathrm{MD}_{1,1} = \frac{S_y^2}{\overline{W}'^2} = 3.623 \cdot 10^{-4};$$

$$S_{b_2}^2 = S_y^2 \mathrm{MD}_{2,2} = \frac{S_y^2}{\overline{\theta}'^2} = 8.24 \cdot 10^{-5}; S_{b_3}^2 = S_y^2 \mathrm{MD}_{3,3} = \frac{S_y^2}{\overline{W}'^2 \overline{\theta}'^2} = 3.623 \cdot 10^{-6}. \tag{4.70}$$

Due to the diagonality, the dispersion of the matrix, the values of these coefficients are not mutually correlated, so the estimate of the variance of the model (4.55) for any value of humidity ($W_i$) and temperature ($\theta_j$) (in the studied range of temperature and humidity) is described by:

$$S_{y\,\mathrm{mod}}^2(W_i', \theta_j') = S_{b_0}^2 + S_{b_1}^2 W_i'^2 + S_{b_2}^2 \theta_j'^2 + S_{b_3}^2 W_i'^2 \theta_j'^2 = \\ = S_y^2 \left[ 1 + \frac{W_i'^2}{\overline{W}'^2} + \frac{\theta_j'^2}{\overline{\theta}'^2} + \frac{W_i'^2 \cdot \theta_j'^2}{\overline{W}'^2 \cdot \overline{\theta}'^2} \right], \tag{4.71}$$

$$S_{y\,\mathrm{mod}}^2(W_i, \theta_j) = S_y^2 \left[ 1 + \frac{(W_i - \overline{W})^2}{\overline{W}'^2} + \frac{(\theta_j - \overline{\theta})^2}{\overline{\theta}'^2} + \frac{(W_i - \overline{W})^2 \cdot (\theta_j - \overline{\theta})^2}{\overline{W}'^2 \cdot \overline{\theta}'^2} \right]., \tag{4.72}$$

Assessments of the standard deviation of the defined coefficients and the approximating model are the square roots of the variance estimates:

$$\sigma_{b_0} = \sqrt{S_{b_0}^2} = 0.09013; \quad \sigma_{b_1} = \sqrt{S_{b_1}^2} = 0.01903; \\ \sigma_{b_2} = \sqrt{S_{b_2}^2} = 9.013 \cdot 10^{-3}; \quad \sigma_{b_3} = \sqrt{S_{b_3}^2} = 1.903 \cdot 10^{-3}, \tag{4.73}$$

$$\sigma_{y\,\mathrm{mod}}(W_i, \theta_j) = \sqrt{S_{y\,\mathrm{mod}}^2(W_i, \theta_j)} = \\ = S_y \sqrt{\left[ 1 + \frac{(W_i - \overline{W})^2}{\overline{W}'^2} + \frac{(\theta_j - \overline{\theta})^2}{\overline{\theta}'^2} + \frac{(W_i - \overline{W})^2 \cdot (\theta_j - \overline{\theta})^2}{\overline{W}'^2 \cdot \overline{\theta}'^2} \right]}. \tag{4.74}$$

- **Estimation of Essentiality of Calculated Coefficients of the Approximating Model**

To make sure that the found coefficients adequately describe the mathematical model, let's check their significance. The estimation will be performed using the t-Student's test, which is based on the fact that the obtained coefficients $b_i$ are normally distributed, and their mathematical expectation coincides with the values of the coefficients M ($b_i$) = $b_i$, and assessments of their standard uncertainty are obtained from the covariance matrix $\sigma_{b_i} = S_y \cdot \sqrt{M_{i,i}}$. Therefore, to check the significance of the values of the coefficients $b_i$ by the t-Student's test, define the parameter of the ratio:

$$g_t = \frac{|b_i|}{\sigma_{b_i}} = \frac{|b_i|}{\sqrt{S^2_{b_i}}}, \tag{4.75}$$

which has a Student's distribution with v = n - m degrees of freedom. From the tables of Student's quantiles for a given level of essentiality $\alpha$ find the critical value $t_{1-\alpha/2}(n-m)$. If the condition is met:

$$g_t \geq t_{1-\alpha/2}\,(n-m), \tag{4.76}$$

then the coefficient is considered significant. So, it is given a zero value and the calculation is repeated. So, in our case n = 28, m = 4, from Student's tables for $P_{trust}$ = 0.95 and v = n – m = 24, critical value $t_{0.95}$ (24) = 2.064.

Respectively,
$g_{b_0} = 367.985$; $g_{b_1} = 162.0396$; $g_{b_2} = 162.3125$; $g_{b_3} = 64.9$.
As can be seen, condition (4.76) is satisfied, i.e. the values of the coefficients of the approximating model are significant (Fig. 4.27).

- **Evaluation of the Effectiveness of Temperature Error Adjustment**

First, we determine the boundary value of the humidity measurement error under the impact of temperature. From the function

$$C_x = b_0 + b_1(W_x - \overline{W}) + b_2(\theta - \overline{\theta}) + b_3(W_x - \overline{W})\cdot(\theta - \overline{\theta}) \tag{4.77}$$

the required humidity is determined as:

$$W_x = \overline{W} + \frac{C_x - b_0 - b_2(\theta - \overline{\theta})}{b_1 + b_3 \cdot (\theta - \overline{\theta})} \tag{4.78}$$

Hence the error from the influence of temperature (θ) is:

$$\Delta W(\theta) = W_x(\theta) - W_x(\overline{\theta}) = \frac{-(\theta - \overline{\theta})\left[b_1 b_2 + b_3(C_x - b_0)\right]}{b_1\left[b_1 + b_3 \cdot (\theta - \overline{\theta})\right]} \tag{4.79}$$

Dependences (4.79) for different values of humidity are given in Fig. 4.28.

As you can see, when the temperature deviates from the average value of 25 °C, the error changes linearly and can reach from ±4% for humidity 10.87% to ±11% for humidity 19.30%. Due to temperature control and known coefficients $b_i$, the humidity value is adjusted according to (4.77). However, the coefficients of approximation and temperature are known with certain errors, which are calculated above. Let's estimate the unadjusted residual error from the temperature impact:

$$\Delta W_x = \frac{\partial W_x}{\partial b_0}\Delta b_0 + \frac{\partial W_x}{\partial b_2}\Delta b_2 + \frac{\partial W_x}{\partial b_1}\Delta b_1 + \\ + \frac{\partial W_x}{\partial b_3}\Delta b_3 + \frac{\partial W_x}{\partial \theta}\Delta \theta. \tag{4.80}$$

Accordingly,

$$\begin{cases} \dfrac{\partial W_x}{\partial C_x} = \dfrac{1}{b_1 + b_3 \cdot (\theta - \overline{\theta})}; \quad \dfrac{\partial W_x}{\partial b_0} = -\dfrac{1}{b_1 + b_3 \cdot (\theta - \overline{\theta})}; \\ \dfrac{\partial W_x}{\partial b_1} = -\dfrac{C_x - b_0 - b_2(\theta - \overline{\theta})}{(b_1 + b_3 \cdot (\theta - \overline{\theta}))^2}; \\ \dfrac{\partial W_x}{\partial b_3} = -\dfrac{(C_x - b_0 - b_2(\theta - \overline{\theta})) \cdot (\theta - \overline{\theta})}{(b_1 + b_3 \cdot (\theta - \overline{\theta}))^2} \\ \dfrac{\partial W_x}{\partial \theta} = \dfrac{-b_1 b_2 - (C_x - b_0) b_3}{(b_1 + b_3 \cdot (\theta - \overline{\theta}))^2}. \end{cases} \tag{4.81}$$

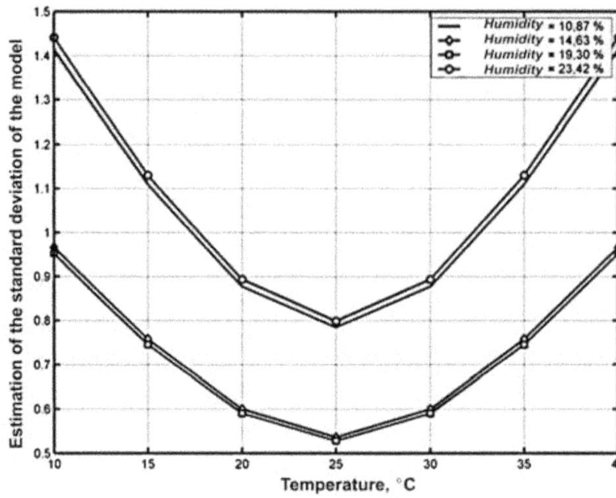


**Figure 4.27.** Estimation of the standard deviation of the approximating model.

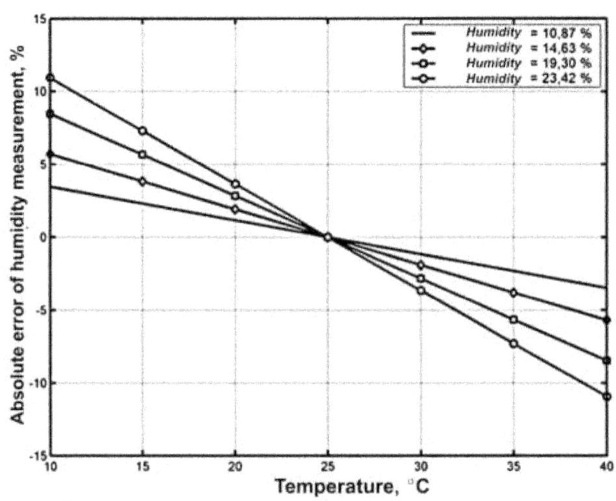


**Figure 4.28.** The absolute error of measurement of the humidity of the material caused by the temperature impact (without adjustment).

Assuming that the components of equation (4.79) are not mutually correlated, the estimate of the variance of the unadjusted residual temperature error is described by:

$$S_{W_x}^2(\theta)=\frac{S_{b_0}^2+(\theta-\bar{\theta})^2\cdot S_{b_2}^2+C_x^2\cdot S_{b_1}^2+C_x^2\cdot(\theta-\bar{\theta})^2\cdot S_{b_3}^2}{(b_1+b_3\cdot(\theta-\bar{\theta}))^2}+ \\ +\frac{\dfrac{(b_1b_2+(C_x-b_0)b_3)^2}{(b_1+b_3\cdot(\theta-\bar{\theta}))^2}S_\theta^2}{(b_1+b_3\cdot(\theta-\bar{\theta}))^2} \tag{4.82}$$

Estimation of the standard deviation of unadjusted residual temperature error:

$$S_{W_x}(\theta)=\sqrt{S_{W_x}^2(\theta)}=\frac{1}{\left|b_1+b_3\cdot(\theta-\bar{\theta})\right|}\cdot \\ \cdot\left[\sqrt{S_y^2\begin{bmatrix}\mathrm{M}_{0,0}+(\theta-\bar{\theta})^2\cdot\mathrm{M}_{2,2}+\\+C_x^2\cdot\mathrm{M}_{1,1}+C_x^2\cdot(\theta-\bar{\theta})^2\cdot\mathrm{M}_{3,3}\end{bmatrix}+ \\ +\frac{(b_1b_2+(C_x-b_0)b_3)^2}{(b_1+b_3\cdot(\theta-\bar{\theta}))^2}\cdot\frac{\Delta\theta_{\text{гр}}^2}{3}}\right] \tag{4.83}$$

The confidence limits of the unadjusted residual temperature error at $R_{conf} = 0.95$ are determined from the expression:

$$\Delta_{\text{conf}}(W(\theta))=1.96\cdot S_{W_x}(\theta) \tag{4.84}$$

The efficiency of the correction of the error temperature component is determined (Fig. 4.29) as the ratio of the maximum value of the temperature error without correction $|\Delta W(\theta)|_{max}$ to the confidence value of the unadjusted residual temperature error $|\Delta_{conf}(W(\theta))|$, that is

$$E_{W_i,\theta_j}=\frac{\left|\Delta W(\theta)\right|_{\max}}{\left|\Delta_{\text{conf}}(W(\theta))\right|} \tag{4.85}$$

The efficiency of adjusting the temperature component of the error by (4.85) is envisaged in Fig. 4.30.

So, as can be seen, the significant temperature impact entails the need to measure it simultaneously with measuring the capacitance by the primary transducer. To fulfill this, it can be used some methods similar to well-known electrical measurements of other physical quantities, for instance, the complicated method of studying an energy spectrum of stochastic signals caused by the measured variations of electrical resistance [46].

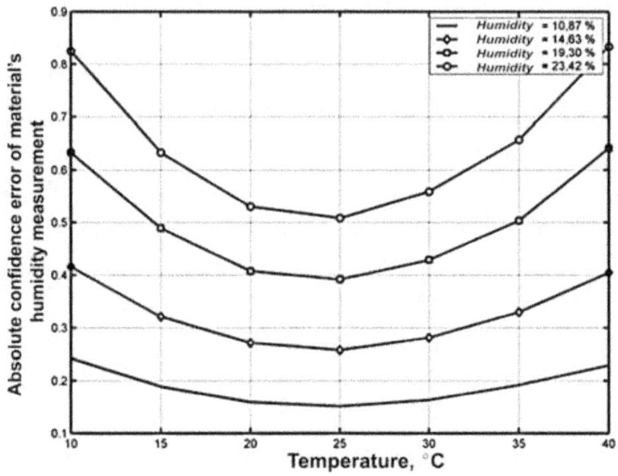


**Figure 4.29.** Absolute confidence error of material's humidity measurement due to temperature impact (adjusted).

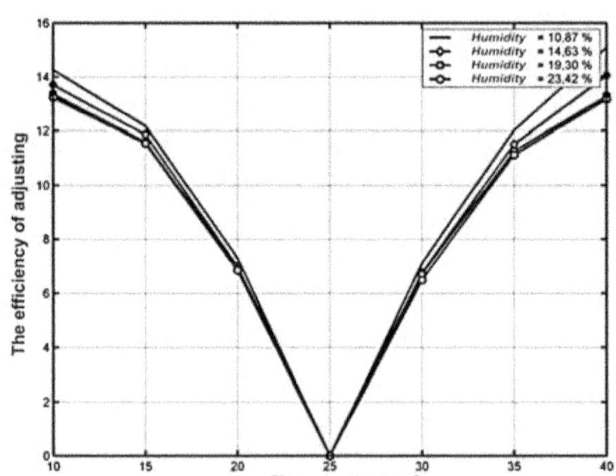


**Figure 4.30.** The efficiency of adjusting the error temperature component while measuring the bulk's humidity.

## 4.5. Conclusions

1. To ensure unity and compatibility in the operation of agricultural CPSs, it is necessary to apply the standard requirements for their construction, maintenance, and support.

2. Based on the conducted research, the urgency of the development of CPS vegetable cultivation with the regulation of a temperature-moisture-insolation mode is substantiated. Usually, the CPS's vegetable cultivation necessitates 2 control subsystems that are the systems of the air temperature-control and the relative humidity-control in a confined space of Greenhouse. These subsystems determine moisture in the soil. The most effectively seems to implement additional the 3$^{rd}$ soil moisture-control subsystem as a corrective one.

3. Studies of Impedance Spectroscopy for economy purposes have demonstrated the following results:

- Investigation of active and reactive constituents of impedance and admittance of fresh and unfrozen chicken meat in the frequency range of 100 Hz – 100 kHz have revealed that the difference between fresh and thawed meats consists of reactive constituents of impedance or admittance on the fixed frequencies of the chosen frequency band. Concerning performed studies, it was offered a method of authentication of the multiple freezing the meat. Note the first freezing did not change the characteristics of the meat. The obtained results can be used to develop the means of express meat quality measurement.
- Artificial neural networks can be applied for algorithmic correction of errors while implementing the impedance spectroscopy for product quality studies. For the algorithmic correction development, feedforward type ANN with two hidden layers of 50 neurons in each one was used. The obtained results were the next: maximum absolute error does not exceed 0.027% for the active component and 0.013% for the reactive; MSE for 0.997 confidence intervals are 0.0022% and 0.0016% for the active and reactive components correspondingly.
- For comparison, using the analytical algorithmic correction approach based on a measuring transducer mathematical model, the error values were as following: maximum absolute error for the active component – 0.024% and for the reactive – 0.015%; MSE – 0.0037% and 0.0032% for the specified components.

4. It is proved to construct the means of realization of a differential method of estimation of the quality of production by the measurements of relative electric parameters of the monitored and base samples. To fulfill this, it is advisable to use, as informative parameters of the comparison objects, the module, and phase or the active and reactive components of the immittances, measured at several frequencies of the test. The identification of the qualimetric objects is carried out by the ratio of the measured parameters of the monitored and base samples or by analyzing the same parameters for the measured immittances' ratio. MI with parallel "immittance-voltage" conversion makes it possible to obtain directly the ratio of active and reactive components of the immittances of the comparison objects. There is no need to define previously the relevant parameters of the basic sample. MI with simultaneous transformation provides direct measurement of the components of the ratio of immittances of the mentioned samples by analyzing the characteristic changes in which one can assess the quality level.

5. The humidity measuring of the bulk materials involving the method of impedance spectroscopy gives possibilities to enhance significantly, at 7 – 14 times, the research accuracy, depending on the temperature at which measurements are fulfilled. This means that it may be recommended to implement the conjugated monitoring of a few electric parameters of bulk materials (temperature, humidity, etc.) with in-situ processing the obtained results and self-adjusting the processed results on the error value.

## References

[1]. Living in a networked world. Integrated research agenda Cyber-Physical Systems (agendaCPS), E. Geisberger, M. Broy (Eds.), *Acatech STUDY*, March 2015.

[2]. Cyber-Physical Systems. Metrological Issues, Ed. by S. Yatsyshyn, B. Stadnyk, Barcelona, *IFSA Publishing*, 2016.

[3]. D. Brunner, A. Winton, The key principles of vehicle cyber security for connected and automated vehicles, UK, 2017, https://www.gov.uk/government/publications/principles-of-cyber-security-for-connected-and-automated-vehicles/the-key-principles-of-vehicle-cyber-security-for-connected-and-automated-vehicles

[4]. National Institute of Standards and Technology, Cyber Physical Systems Public Working Group 'Framework for Cyber-Physical Systems Release 1.0', May 2016, pp. 236–238.

[5]. A. J. C. Trappey, C. V. Trappey, U. H. Govindarajan, J. J. Sun and A. C. Chuang, A Review of Technology Standards and Patent Portfolios for Enabling Cyber-Physical Systems in Advanced Manufacturing, *IEEE Access*, Vol. 4, 2016, pp. 7356-7382.
[6]. Law of Ukraine: About the basic principles of ensuring cyber security of Ukraine, *Law of Ukraine*, 2017, pp.1–2.
[7]. ISO/IEC 7498–1:2004 Information Technology. The interconnection of open systems. Basic reference model. Part 1. The reference model, 2004.
[8]. ISO 7498–2:2004 Information processing systems. The interconnection of open systems. Basic reference model. Part 2. Information security architecture, 2004.
[9]. ISO/IEC 27033–1:2009 Information technology. Security techniques. Network security. Part 1: Overview and concepts: ISO/IEC JTC 1/SC 27 Information security, cybersecurity and privacy protection, 2009.
[10]. ISO/IEC 27033–2: 2012 Information technology. Security techniques. Guidelines for the design and implementation of network security (Information Technology. Security Methods. Guidelines for Developing and Implementing Network Security). Information security, cybersecurity and privacy protection, 2012.
[11]. J. Lee, B. Bagheri, H. Kao, A Cyber-physical Systems architecture for Industry 4.0-based manufacturing systems, *Manufacturing Letters,* Vol. 3, 2015, pp. 18–23.
[12]. H. Dat Dac, H.-Y. Paik, C.-K. Kim, Service-oriented middleware architectures for cyber-physical systems, *Int. Journ. Comp. Sc. and Network Sec.*, Vol. 12.1, 2012, pp. 79–87.
[13]. Wu Fang-Jing, Yu-Fen Kao, Yu-Chee Tseng, From wireless sensor networks towards cyber physical systems, *Pervasive and Mob. Comp.,* Vol. 7.4, 2011, pp. 397–413.
[14]. S. Teodora, L. Miclea, Cyber-Physical Systems–Concept, Challenges and Research Areas, *Journal of Control Eng. and Appl. Informatics*, Vol. 14.2, 2012, pp. 28–33.
[15]. A. Melnyk, Cyber-physical Systems: Problems of creation and direction of development, *Bull. Lviv Polytech. Nat. Un.: Comp. Systems and Networks,* Vol. 806, 2014, pp. 154–161.
[16]. P. Kumar, M. Rosegrant, P. Hazen, Cereals prospects in India to 2020, ResearchGate, Jan. 1995, (https://www.researchgate.net/publication/ 5055789_Cereals_prospects_in_India_to_2020).
[17]. C. Lin, S. Zeadally, T. Chen, Enabling Cyber Physical Systems with Wireless Sensor Networking Technologies, *Int. Journ. Distributed Sensor Networks,* May 2012, 21 p.
[18]. I. Akyildiz, Y. Sankarasubramaniam, Wireless sensor networks: a survey, *Computer Networks*, Vol. 38, 2002, pp. 393–422.
[19]. O. Palagin, V. Romanov, I. Galelyuka, Data acquisition systems for precision farming, *Information Technologies & Knowledge*, Vol. 5, Issue 2, 2011, pp. 103–109.

[20]. V. Romanov, O. Palagin, I. Haleluka, O. Voronenko, Wireless sensor network for precision farming and environmental, *Comp. Tools, Networks and Systems*, Vol. 13, 2014, pp. 53–62.
[21]. S. Lindsey, C. Raghavendra, Power-efficient gathering in sensor information systems, in *Proceeding of the IEEE Aerospace Conference (AERO '2002)*, Big. Sky, USA, 9–16 March 2002, pp. 924–935.
[22]. C. Perkin, E. Belding-Royer, S. Das, Ad hoc On-Demand Distance Vector Routing, in *Proceeding of the Workshop on Mobile Computing Systems and Applications (WMCSA '99)*, New Orleans, USA, 25–26 February 1999, pp. 90–100.
[23]. Y. Xu, J. Heidemann, D. Estrin, Geography-informed energy conservation for ad hoc routing, in *Proceeding of the ACM/IEEE International Conference on Mobile Computing and Networking (MobiCom '01)*, New York, USA, 16–21 July, 2001, pp. 70–84.
[24]. L. Subramanian, R. Katz, An architecture for building self-configurable systems, in *Proceeding of Mobile Ad Hoc Network Comput. Workshop,* Boston, USA, 11 August 2000, pp. 63–73.
[25]. T. Stuhlfauth, H. Fock, Effect of whole season $CO_2$ enrichment on the cultivation of a medicinal plant, Digitalis lanata, *J. Agronomy & Crop Sc.*, Vol.164, Issue 3, 1990, pp. 168–173.
[26]. J. Famiglietti, M. Rodell, Water in the Balance, *Env. Sc.,* Vol. 340, 6138, 14 June 2013, pp. 1300–1301.
[27]. J. McNutty, Solar greenhouses generate electricity and grow crops at the same time, UC Santa Cruz study reveals, *USC News center*, Santa Cruz: University of California. Retr. 6 Nov. 2017.
[28]. Ye. Barsukov, R. MacDonald, Electrochemical Impedance Spectroscopy, in Characterization of Materials, ed. E. Kaufmann, *John Wiley & Sons* 2012, (https://jrossmacdonald.com/jrm/wp-content/uploads/258ElectrochemicalIS.pdf)
[29]. M. Orazem, B. Tribollet, Electrochemical Impedance Spectroscopy, *Wiley*, 2017 (https://www.wiley.com/en-us/Electrochemical+Impedance+Spectroscopy)
[30]. V.V. Khoma, Y. Pokhodylo, Problem-oriented tools of immittance measurement. Theory and practice, *Publ. House of Lviv Polytechnic Nat. Un.*, Ukraine, 2013 (in Ukrainian).
[31]. N. Jaffrezic-Renault, Label-Free Affinity Biosensors Based on Electrochemical Impedance Spectroscopy, *Springer Protocols*, 2013 (https://experiments.springernature.com/articles/10.1007/978-1-62703-370-1)
[32]. Z. Haeri, M. Shokoufi, M. Jenab, R. Janzen, F. Golnaraghi, Electrical impedance spectroscopy for breast cancer diagnosis: Clinical study, *Integr Cancer Sci Therap.*, 3, 2016.
[33]. X. Zhao, H. Zhuang, S.-C. Yoon, Y. Dong, W. Wang, W. Zhao, Electrical Impedance Spectroscopy for Quality Assessment of Meat and Fish: A Review on Basic Principles, Measurement Methods, and Recent

Advances, *New Food Processing Technologies and Food Safety,* Special Issue, Vol. 2017, Art. ID 6370739, 16 p.
[34]. V. Khoma, M. Wrzusczak, Compensation impedance meter, *Electrotechnical Review*, Vol. 84, Issue 5, 2008, pp. 33-36 (in Polish).
[35]. K. Santhosh, B. Roy, An Improved Intelligent Temperature Measurement by RTD using Optimal ANN, in *Proceedings of the Int. Conf. on Advances in Comp., Electronics and El. Eng.*, 2012, pp. 82-86.
[36]. J. Roj, Neural Network Based Real-time Correction of Transducer Dynamic Errors, *Meas. Sc. Rev.*, Vol. 13, Issue 6, 2013, pp. 286 – 291.
[37]. M. Mykyychuk, R. Kochan, O. Kochan, S. Jun, B. Stadnyk, P. Stolyarchuk, Increasing metrological autonomy of in-plant measuring systems, *Adv. in Sc. and Techn. Res. Journ.*, Vol. 10, Issue 322016, pp.193 - 197.
[38]. W. Jung, Op Amp Applications Handbook, *Elsevier*, 2005.
[39]. S. Smith, Digital Signal Processing: A Practical Guide for Engineers and Scientists, *Elsevier Science*, Burlington, MA, USA, 2003.
[40]. F. Burden, D. Winkler, Bayesian regularization of neural networks, *Methods in Molecular Biology (Clifton, N.J.),* 01 January 2008, 458, 25-442008.
[41]. Y. Pohodylo, P. Stolyarchuk, M. Chyrka, PC-Based Devices for Immittance Control of Multidimensional Objects, *IEEE Trans. on Instr. & Meas.*, Vol. 51, Oct. 2002, pp.1133-1136.
[42]. Y. Pohodylo, P. Stolyarchuk, Immitance control of quality, Lviv, Ukraine, *Publ. House of Lviv Polytechnic Nat. Un.*, 2012 (in Ukrainian).
[43]. Y. Pohodylo, V. Khoma, CLR meters with imitation voltage conversion, Lviv, Ukraine, *Publ. House of Lviv Polytechnic Nat. Un.*, 2012 (in Ukrainian).
[44]. L. Nollet, F. Toldra, Handbook of Seafood and Seafood Products Analysis, *CRC Press*, 2009 (https://www.routledge.com/Handbook-of-Seafood-and-Seafood-Products-Analysis /Nollet-Toldra/p/book/9781420046335)
[45]. S. Yatsyshyn, B. Stadnyk, P. Skoropad, M. Mykyychuk, Chapter 8. Energy Audit and Contactless Temperature Measurement, in *Advances in Sensors: Reviews, Vol 7: Physical and Chemical Sensors: Design, Applications & Networks*. Ed. by S. Yurish, 2020, pp. 179-196.
[46]. Z. Kolodiy, B. Stadnyk, A. Kolodiy, Energy Spectrum of Stochastic Signals Caused by Variations of Electrical Resistance, *Autom. Control and Comp. Sc.*, Vol. 52, No. 4, 2018, pp. 311–316.

# Chapter 5

# Metrology 4.0 for Health-Care and Human Being Identification

**Bohdan Stadnyk, Svyatoslav Yatsyshyn, Yuriy Khoma and Orest Kochan**

## Definition

- **Biometrics** is the science that uses statistical methods to recognize the human identity based on the physiological and behavioral attributes of the individual such as fingerprints, face, and voice. It includes the measurement and statistical analysis of people's unique physical and behavioral characteristics and refers to metrics related to human characteristics.

This chapter aims to consider some estimates and insights on how robust and reliable biometric information and especially markers are and how to develop smart devices to meet the contemporary challenges for nowadays. To accomplish this task that is simultaneously the metrological issue and to figure out some common trends and limitations, studies have been performed for a few directions, a common feature of which was for ECG-based technology. The latter includes human identification as well as human rehabilitation. It can be emphasized that common is not only the heart-produced information including the ECG-orientation but also the use of smart devices, which is the basis of CPSs.

## 5.1. Biometrics

The main idea of Biometrics is to operate matching a given biomarker value with the reference(s) value that represents an individual [1-2]. As biomarkers, some physiological and behavioral attributes that are unique for each human being are considered. Modern biometric systems rely on various biomarkers, but most common are fingerprints, face, voice, iris, hand geometry, finger vein.

Main requirements that biomarker should fulfill are the following [1]:
- Universal (present for all individuals);
- Stability over time;
- Easy to measure/acquire;
- Low sensitivity to other physiological factors (stress, fatigue);
- Unique for each person.

There are a few additional requirements that are considered as a big plus:
- Fraud resistance (difficult to fake);
- Continuous nature (always available to measure);
- Liveness indication (present only from live humans).

### 5.1.1. ECG Signal as Robust and Reliable Biometric Marker: Datasets and Algorithms Comparison

As it was shown in [3-5] electrocardiogram signal is a promising biometric marker. Historically, ECG was mostly used for medical purposes, but recent progress in the fields of consumer electronics and information technologies already enabled its applications in biometric systems [6-7]. On the other hand, ECG-based identification systems are still not widely used in commercial and government services, as most of them are provided as research prototypes or very new commercial products that have just appeared on the market [8-9]. In this situation, some questions remain open, for example, ECG signal reliability and reproducibility over time, its behavior on a scale in real-world applications, the impact of the measurement process, hardware configuration, and matching algorithms on identification performance, etc.

- **Biometric System Architecture**

ECG-based identification process typically consists of three main stages: data acquisition, data processing (filtering, normalization, feature extraction), and template matching (classification) [4, 6-7, 10]. Data acquisition requires an analog front-end (two- or three-electrode measuring circuit, typically based on instrumental amplifiers) followed by ADC. Digitized data being streamed to MCU/PC. Data processing includes filtering (low-pass to remove offsets and respiration, high pass to remove noise, 50 Hz coupling, movement artifacts), heartbeats segmentation, and normalization. To split ECG waveform into separate heartbeats, R peaks need to be detected. There are multiple algorithms developed for this purpose, but in our case, a third-party implementation of the Hamilton algorithm was chosen from the library. After segmentation, each heartbeat was normalized to a range of [-1; 1]. Afterward data points on the ends of each heartbeat were dropped, so only the point of the central part (around 60% of entire heartbeat) was involved in the further identification process.

Another important transformation of the data processing stage is outlier correction. It is expected that the ECG signal has regular nature and beats tend to be a similar one to another. However, for some of the beats, strong deviations were observed. There might be different kinds of their origin, e.g. muscle noise, respiration, non-stable contact impedance, movements of electrodes, etc. To detect and correct these corrupted segments (outliers) special algorithm proposed in [11] was applied.

Data processing is followed by the classification stage. The classification model should recognize some user-specific patterns in the processed ECG signal and perform matching with one of the corresponding classes (users). This is the last step of the identification flow. The entire ECG-based identification flow is presented in Fig. 5.1.

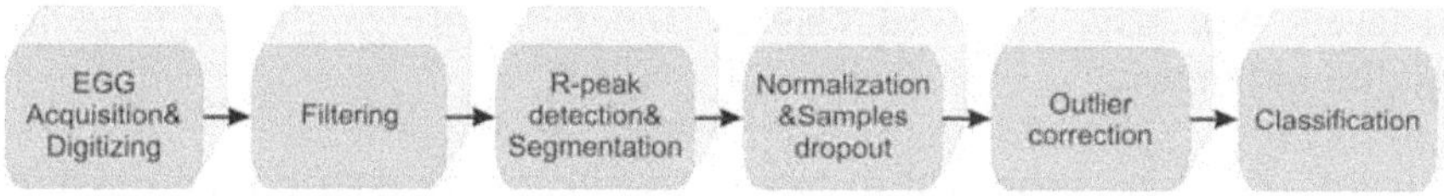


**Figure 5.1.** ECG-based identification flow.

- **The Methodology of the Experiment**

The whole idea of the study is to check whether the ECG signal is sustainable as a biometrics marker. To perform this task the following methodology is proposed:

First, different datasets from different sources (both self-collected and publicly available on the Internet) should be included in the study. It's expected that these datasets should have different origins and internal structure. The main parameters that should be taken into consideration are the number of users, the total number of records, mean, the minimal and maximal number of records per user, records length, etc. There are two constraints related to dataset selection. The first one is that the dataset should contain only records of healthy people with normal rhythm [12]. The second constraint assumes that data should be recorded from the same scheme of electrodes placed on the patient's body, called lead in cardiology. The reason is that the ECG waveform highly varies when measured from different parts of the body.

Second, different algorithms should be used for records classification. In our experiment, we expect to use just one heartbeat for human identification. Thus, no sequential algorithm will be analyzed here. Only simple supervised machine learning techniques that map the multidimensional input vector (samples of ECG heartbeat) to a dedicated output vector (number of users). Algorithm comparison is required for two reasons. Firstly, to get some basic intuition on how complex non-linear and simpler linear algorithms will behave on such kind of data. And secondly, to ensure that there is no bias in different datasets. It means that algorithms should demonstrate similar behavior on different datasets.

Third, one of the most important stages while designing machine learning experiments is to select appropriate performance metrics. In our case, we expect to do multiclass classification. In case the classes are balanced it might be enough to use simple accuracy. In case some datasets will have screwed classes, the F1 score might be used as additional metrics.

Finally, it is also proposed to have two-alternative data pre-processing algorithms in place. The first one is described in the section above. That means filtering, normalization, and outlier correction. Another one proposes the use of dimensionality reduction on the top. This trick is commonly used in machine learning and might help to improve classification performance. It was decided to use PCA as one of the simple and commonly used, efficient compression algorithms.

- **Datasets Description**

Taking into consideration the requirements formulated above (see section 3), the following four datasets were chosen for the current study, namely Lviv Biometric Dataset (self-collected), Physionet ECG-ID dataset, Physionet QT dataset (only some records with normal sinus rhythm) and Physionet MIT-BIH Normal Sinus Rhythm. A short description of each dataset as well as a comparison of the main parameter are presented below (see Table 5.1) [13].

- Lviv Biometric Dataset

This is a self-collected dataset available at [14]. All the records were acquired using the eHealth Arduino extension board [15]. More details on the measurement procedure can be found at [7].

- Physionet ECG-ID

This dataset was created for human identification purposes, as a part of the master thesis. The records were acquired from 44 men and 46 women aged from 13 to 75 years old. For some users, there are only a few records available, which means they were recorded during one day. For other users, there are over 20 records collected periodically over 6 months period. The Physionet ECG-ID database is available at [16].

- Physionet QT-Database

Dataset was designed for the evaluation of ECG heartbeat segmentation algorithms. It has an annotation to each record with the boundaries of each heartbeat. This dataset includes not only records of healthy people, but also patients with cardiological disorders. Because of this, all annotations were manually reviewed to pick up records with normal ECG rhythm. The following records were selected: sel103, sel117, sel123, sel16265, sel16272, sel16273, sel16420, sel16483, sel16539, sel16773, sel16786, sel17152, sel17453, sel301, sel302, sel306, sel307, sel310, sele0111, sele0124, sele0133, sele0210. The Physionet QT database is available at [17].

- Physionet MIT-BIH Normal Sinus Rhythm

The database includes ECG records obtained by the Arrhythmia Laboratory at Boston's Beth Israel Hospital. Records originate from healthy people with no significant arrhythmias in the age between 20 to 50 years old. Physionet MIT-BIH Normal Sinus Rhythm database is available at [18].

**Table 5.1.** Basic parameters of ECG datasets.

| Parameter | LBDS | ECG-ID | QT | Normal Sinus Rhythm |
|---|---|---|---|---|
| Lead | modified I-lead (from fingers of the right and left hand) | I-lead | I-lead | I-lead |
| Number of users | 53 | 90 | 22 | 18 |
| Total number of records | 545 | 310 | 22 | 18 |
| Records per user | from 3 to 15 | from 1 to 22 | 1 | 1 |
| Sampling rate | 277 Hz | 500 Hz | 250 Hz | 125 Hz |
| Average record time | ~ 10 seconds | 20 seconds | 15 minutes | ~ 10:20 hours (from 8:00 to 13:50 hours) |

- **Results and Discussions**

Datasets selection and ECG signal pre-processing stages have already been described in previous sections. The final stage is down to user identification. It can be considered as a classification task because the identification algorithm must match each ECG record to one of the existing users (classes). In general, the classification is done using machine learning techniques. The machine learning approach requires selecting an appropriate algorithm that is powerful enough to model complex internal data relations and dataset splitting for correct estimation of classifier performance in real-world applications [19].

Machine learning algorithms have different natures, are based on different ideas and mathematical frameworks, and typically are used in different applications. These factors should be taken into consideration while selecting an algorithm suitable for ECG identification. The following seven algorithms have been chosen as the most promising: Logistic Regression, SVM, LDA, Naive Bayes, KNN, Neural Networks or MLP, Extreme Gradient Boosting (xGboost), Random Forest [19-20].

Dataset splitting requires their division into two subsets: training and test set. Samples from the training set are used to fit a classification model, while samples from the test set are used to provide an unbiased evaluation of the model performance. The test set must be carefully prepared as it should realistically represent the real-world data that the classification model would operate on.

As ECG-ID and LBDS have multiple records per user, we will split the test and training set based on the records level. Some records will be randomly selected as the training set, remaining will go for the test. Experiments will be conducted for training and test set ratio of 0.7 and 0.3, correspondingly. To achieve more realistic identification performance, dataset split was done 5 times, and then mean values for each subset were calculated. For MIT-BIH Normal Sinus Rhythm and Qt database, just one record per user is available. However, these records are of quite a substantial length. The idea is to use the time split for training and test set. In this case, training and test ratio were also conducted as 0.7 and 0.3.

Classification models have been trained for two different scenarios: with and without PCA compression. The only exception here is neural networks because they are complex non-linear models, which can learn efficient data compression in the first hidden layer. Thus, it makes no sense to use PCA for them. Experiment results are presented below (see Table 5.2).

As one can see in Table 5.2, algorithms seem to behave similarly across all datasets. Simple algorithms, like KNN and linear models (logistic regression, LDA, SVM) work surprisingly well. Other simple algorithms like Naive Bayes, gradient boosting, and random forest shows relatively poor performance. Neural networks show very high accuracy also, which was expected to take into consideration their complex non-linear nature and modeling capacity. PCA compression might slightly improve accuracy on some datasets while decreasing it on the other. So, it seems there is no need to include PCA in the data preprocessing pipeline. The best accuracy was achieved by LDA and MLP over all four datasets. KNN shows high results on all datasets, except MIT-BIH Normal Sinus Rhythm. Taking into consideration that this database is much larger compared to others it is not clear if KNN will scale well for a larger number of users and records. MLP and xGboost were the most time-consuming algorithms to train. Logistic regression and LDA were among the fastest algorithm.

**Table 5.2.** ECG identification results.

| | Physionet ECG-ID | LBDS | Physionet QT | MIT-BIH Normal Sinus Rhythm |
|---|---|---|---|---|
| Logistic Regression | 0.8286 | 0.9417 | 0.8809 | 0.7492 |
| SVM classifier | 0.8817 | 0.9599 | 0.9174 | 0.7707 |
| LDA classifier | 0.9328 | 0.9831 | 0.9659 | 0.9017 |
| KNN classifier | 0.8903 | 0.9746 | 0.9686 | 0.7967 |
| Naive Bayes | 0.7003 | 0.9587 | 0.9034 | 0.6607 |
| Random Forest | 0.8362 | 0.9546 | 0.9278 | 0.8192 |
| xgboost classifier | 0.7352 | 0.9126 | 0.9191 | 0.8591 |
| MLP (1 hidden layer) | 0.8933 | 0.9711 | 0.9162 | 0.8925 |
| MLP (2 hidden layer) | 0.8976 | 0.9464 | 0.9478 | 0.8744 |
| MLP (3 hidden layer) | 0.8406 | 0.92373 | 0.9294 | 0.8808 |
| PCA+Logistic Regression | 0.8286 | 0.9383 | 0.8465 | 0.7335 |
| PCA+SVM classifier | 0.8865 | 0.9593 | 0.8832 | 0.7472 |
| PCA+LDA classifier | 0.9536 | 0.9833 | 0.9481 | 0.8798 |
| PCA+KNN classifier | 0.8913 | 0.9758 | 0.9675 | 0.7957 |
| PCA+Naive Bayes | 0.6211 | 0.9511 | 0.8915 | 0.6681 |
| PCA+Random Forest | 0.7782 | 0.9199 | 0.8947 | 0.7418 |
| PCA+xgboost classifier | 0.6723 | 0.8911 | 0.9460 | 0.7305 |

Another important observation based on the results from Table 5.2 is that hardware parameters (e.g. measurement instrumentation, lead type, sampling rate) do not impact on the identification results significantly. The lowest accuracy was achieved for the ECG-ID database (potentially because of highly skewed classes and the larger number of users) and MIT-BIH Normal Sinus Rhythm (potentially difficult to scale on a much bigger number of samples).

### 5.1.2. Biometric Identification from Raw ECG Signal Using Deep Learning Techniques

In modern society, there is a pronounced need for reliable and robust human recognition techniques in critical applications such as secure access control, international border crossing, and law enforcement. Biometric systems operate under the premise that many of the physical or behavioral characteristics of humans are distinctive to an individual, and that they can be reliably acquired via appropriately designed sensors and represented in a numerical format that enables automatic decision-making in the context of identity management.

There are two basic operating modes in biometrics. First, verification (or authentication) mode in which the system performs the one-to-one comparison of captured features with a specific template stored in a biometric database to verify if the individual is the person - he/she - claims to be. Second is an identification mode in which the system performs a one-to-many comparison against a biometric database in an attempt to establish the identity of an unknown individual. The system will succeed in identifying the individual if the comparison of the biometric sample to a template in the database falls within a previously set threshold [1]. Human identification based on ECG is a relatively new and fast-developing approach [1, 3]. The ECG signal is created by electrical impulses coming from the brain to our heart. Each of these pulses is stimulating various parts of the heart to make a complete beat. Many factors create ECG variants among normal persons, like anatomy, position, and size of the heart, chest geometry, age, sex, weight.

Comparing to common commercial biometric systems such as fingerprints, hand geometry, face recognition ECG identification has several advantages [1, 10]:

1. More reliable and robust, because it is an internal but not external biometric, and due to this is difficult to mimic and forge;
2. Possibility to provide fresh biometric reading continuously;
3. Good accuracy even in abnormal cases, low sensitivity to noise;
4. Relatively easy to acquire: ECG signals acquisition can be done at the fingers and hand palms with one lead sensor or using textile electrodes.

The process of ECG recognition consists of such phases: data acquisition, pre-processing, feature extraction, feature reduction, and classification [4, 6, 10]. Data acquisition is necessary to register the ECG signals from the person. Besides conventional means of signals reading, nowadays due to recent advances in biomedical instrumentation, ECG signals acquisition can be done on the chest using a shirt with textile embedded electronics, at the neck using a necklace with a pendant, at the fingers and hand palms with one lead sensor or using textile electrodes, and in these last one, it is not necessary to be applied on the body of the person as previous approaches [21-22]. Phase of pre-processing is intended to remove various distorting effects and to keep useful information in the input signal. Digital band-pass filters or wavelet transform can be used for noise reduction, power-line suppression, baseline-wandering removal, etc. Also, on this stage segmentation and normalization of ECG signals are performed. Thus, the single heartbeat signal lends itself to further processing [23-24]. The next stage of the recognition process is linked to extracting features. Features are the ECG represents attributes allowing to recognize the specific person using inter-subject variability. The selection of appropriate feature subsets is a critical step in pattern recognition problems. Today, there are many approaches to perform feature selection which can be divided into two categories: fiducial or non-fiducial. Both approaches have advantages and disadvantages.

Fiducial techniques can be further subdivided into temporal, amplitude, and morphological. These methods are based on locating specific anchor points on the ECG recordings, namely fiducials, such as wave's peaks, boundaries, slopes, and others. Detecting fiducial points is a challenging process due to the high variability of the signal. Further, the fiducials are used for generating the feature set. These features can be extracted with adaptive thresholds, wavelet transform, and other means [10-12]. Non-fiducial methods don't use the characteristic points, instead, they are extracting features directly from the ECG signal or its fragments using this trait of the signal as a quasi-periodicity. Non-fiducial approaches deal with a large amount of redundant feature set that needs to be reduced. In this case, the challenge is to remove this information in a way that the intra-subject variability is minimized and the inter-subject is maximized. A literature overview has shown that non-fiducial methods may be subdivided into 3 main categories: autocorrelation based, phase-space based, and frequency-based analyses [3, 25].

The last stage of the ECG recognition process is classification. On this stage selected feature subsets of signals are applied as classifier inputs. Depending on how accurately and appropriately have been chosen this features the decision classifier will make the correct or wrong decision.

Classification methods that have been proposed during the last years include Bayesian Networks, Kalman Filtering, Hidden Markov Models, Linear Discriminant Analysis, Genetic Algorithm, Decision Trees, k-Nearest-Neighbour, Self-Organizing Map, Fuzzy Logic Algorithm, Support Vector Machine, Artificial Neural Network [3, 14-15]. Each approach exhibiting its advantages and disadvantages.

- **Data Acquisition**

Arduino Uno and e-Health Sensor Platform V2.0 were chosen for ECG measurements. Arduino Uno is a microcontroller board based on the ATmega328P, with 16 MHz quartz crystal and USB port for programming, debugging and data transfer [30]. The e-Health Sensor Platform V2.0 extends Arduino Uno and allows to implement biometric and medical applications. The body monitoring can be done by 10 different sensors: pulse, oxygen in the blood, airflow (breathing), body temperature, electrocardiogram, glucometer, galvanic skin response (sweating), blood pressure, patient position (accelerometer) and muscle/electromyography sensor [28].

Data acquisition was made using differential OpAmp schema followed by an 8-bit ADC operating at a 277 Hz sampling rate. ADC data have been transferred to PC via COM-port using PySerial library. Each measurement took approximately 10 seconds, which means that user records typically contain about 10 or more heartbeats.

It was decided to use a modified Lead I scheme for electrodes placed on the body surface to record the ECG tracing. Modified schema requires, that the user should touch the electrodes with two fingers on the left hand and one finger on the right hand as shown in the Figure below. It is a very convenient method that can be applied in everyday life for user authorization. In our experiments to minimize preprocessing efforts users were in sitting position and in the neutral emotional state during the measurement, and minimal or without body and hands movements (Fig. 5.2). Each record additionally has passed manual visual quality analysis.

- **Signal Preprocessing**

The resulted signal should be subjected to preprocessing which includes filtering, segmentation, and normalization. The preprocessing stage prepares the input signal for classification.

ECG signal is influenced by multiple factors. The recording is made through electrodes, which capture more than just the electrical activity of the heart.

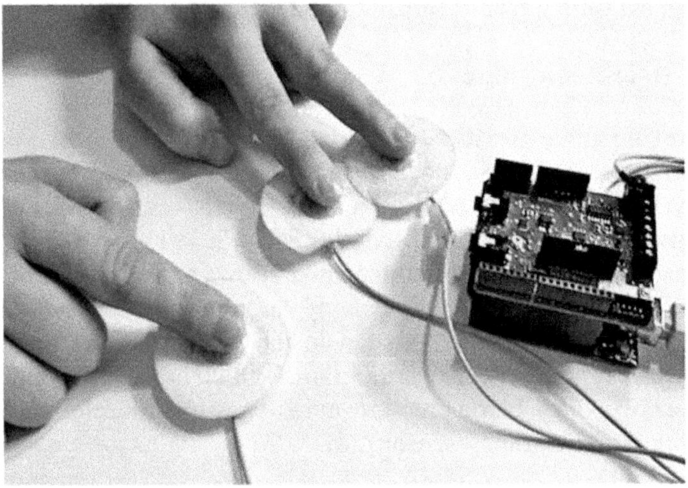

**Figure 5.2.** ECG recording hardware.

Common noise/distortion sources for ECG signal measurements are:

1. Baseline wander (low-frequency noise caused by perspiration that affects electrode impedance, respiration, body movements, for example, finger movements on the electrode);
2. Electromagnetic fields from power lines can cause 50 Hz sinusoidal interference and its harmonics;
3. Muscle artifacts including respirations movements;
4. Noise generated by electronic devices used in signal acquisition circuits.

The informative part of the ECG signal locates between 5 and 35 Hz. In to remove all of the distortions mentioned above, a bandpass IIR-filter was applied. Filter parameters were taken as following: polynomial type - Butterworth, filter order - 7, sampling rate - 277 Hz, stopband - below 1 Hz and above 50 Hz, passband - between 4 and 35 Hz, stopband gain - 20 dB, passband gain - 1 dB. The filter was designed using the SciPy library. The filtered signal was normalized from the raw ADC digits to the range from -1 to +1. The next step is to split the signal into separate heat beats. To detect R-peaks, Hamilton algorithm from the BioSPPy library was used [15]. Afterward, the signal is split up into individual heartbeats, using an assumption that an informative part of the common heartbeat doesn't exceed 270 samples (80 samples to the left of R-peak and 170 samples to the right). For shorter beats signal was accomplished with zeros at the end.

Also, ECG preprocessing included the selection of appropriate beats and remove various artifacts. The sliding window approach has been used for uniformity and outliers' detection. This works as following: ECG beats are split into fixed-length windows (windows may overlap). And, if the standard deviation of at least one sample exceeds some selected threshold, then all samples within the window is recognized as outliers. Samples outliers are replaced by the mean values obtained by averaging corresponding sampled from other segments. This procedure is done iteratively until no outliers are detected. Afterward, samples have been averaged across all heartbeats for each record. Details of ECG signal preprocessing are shown in Fig. 5.3-5.4.

- **Classification**

Supervised machine learning uses the concept of learning by example. A system infers a set of rules from a set of input instances (training set). Once the model is generated, the built model can be used to classify previously unseen instances (testing set). There is a wide range of classification algorithms and the choice of one or another is determined by the nature of the problem, the dataset characteristics, and the application where it will be used.

A deep feedforward neural network, which was chosen as the basic architecture for this study. The details are the following: 270 neurons in the input layer (corresponding to 270 preprocessed features), three hidden layers with 70, 50, 30 neurons, and 19 neurons in the output layer (corresponding to several classes). The Rectified Linear Unit was selected as an activation function for hidden layers and softmax as the activation function for the output layer. Training algorithm - Adagrad, number of training epochs - 3000, learning rate - 0.05, L1 regularization rate - 0.001. After the training stage is finished, the system performance should be evaluated. Accuracy was used to a classification performance metric. The data set was randomly split into train sets (70%) and test set (30%). To prevent skewed class problems, this split was done for each class separately, thus each class is proportionally represented in both train and test data.

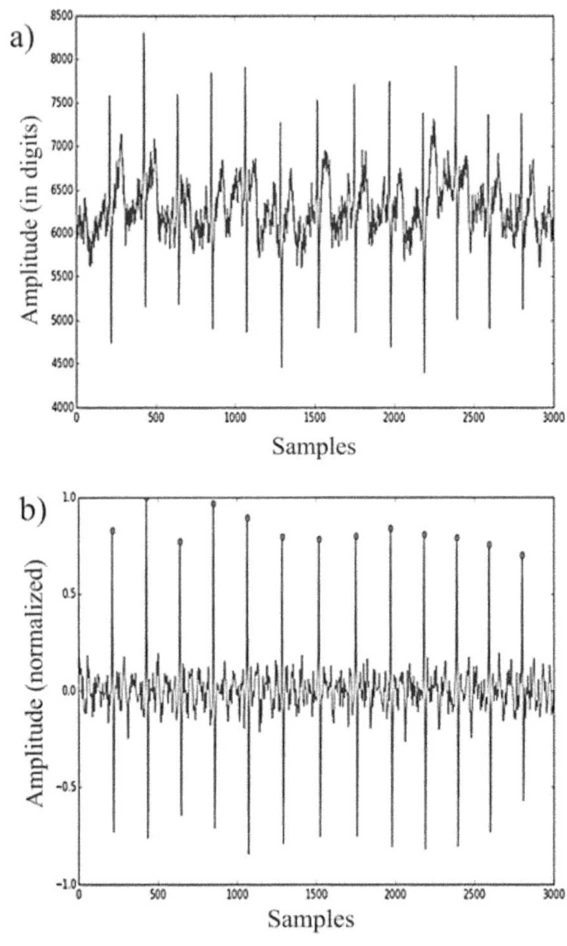


**Figure 5.3.** Raw ECG signal (a) and ECG signal after filtering and normalization with detected R peaks (b).

- **Experiments and Results**

For current research 147 ECG records of 18 unique persons have been selected from Lviv Biometric Data Set. The minimal number of records per person is 3. The train set consists of 88 records, test set 49 [13]. Experiments have been made in Python 2.7. The following frameworks and libraries have been used skflow, scipy, numpy, matplotlib, sci-kit learn. All deep learning algorithms were implemented on the top of the Tensorflow framework. The source code of the project can be found here [31-32]. Experiments have been performed using the next parameters and configurations. Training stage long for around 10 minutes on the machine with the following parameters: CPU - Intel Core i7-5500, operating system - Ubuntu 14.04, 8 GB RAM.

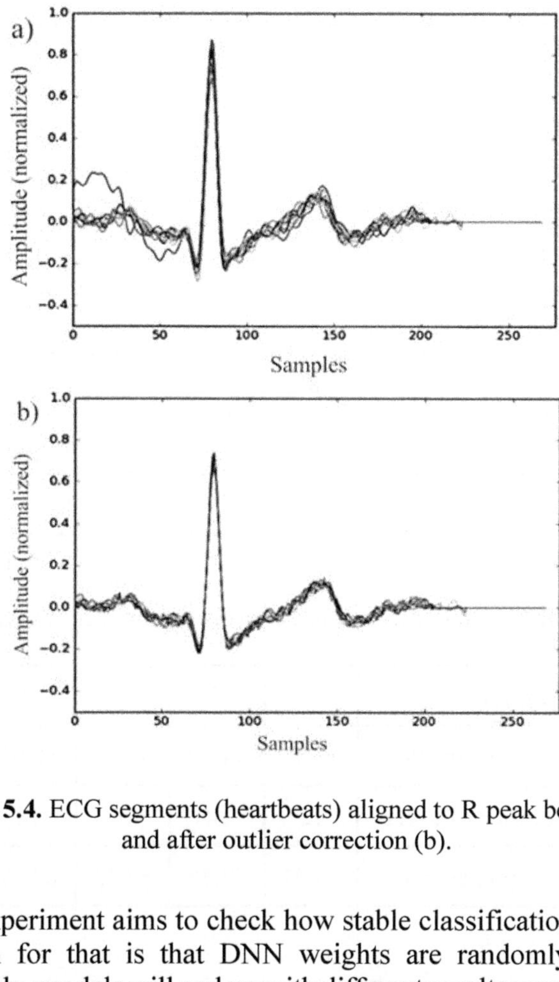


**Figure 5.4.** ECG segments (heartbeats) aligned to R peak before (a), and after outlier correction (b).

The first experiment aims to check how stable classification results are. The reason for that is that DNN weights are randomly initialized. Consequently, models will end up with different results even if they were trained under the same conditions (data and hyperparameters). For this purpose, different (selected) DNN architectures were trained iteratively for 100 times (Table 5.3). The results are shown using the table below. As follows from the results, the deviations are quite big for all architectures. So, the most reasonable approach is to run the same model multiple times and pick-up the best one. Classification accuracy does not vary strongly from architecture to architecture. For further experiments in this research three hidden layers with 70, 50, 30 neurons were chosen.

The second performed experiment has investigated how the number of classes impacts overall classification accuracy. The results are shown in Table 5.4 and Fig. 5.5 below. As we can see adding more classes do not impact system accuracy significantly.

**Table 5.3.** Accuracy dispersion for different DNN architectures.

| Neurons in hidden layers | Mean accuracy, % | Accuracy standard deviation, % |
|---|---|---|
| [100 70 50 30] | 88.50 | 3.35 |
| [70 50 30] | 88.97 | 2.81 |
| [50 50 20] | 84.14 | 5.28 |
| [70 20] | 88.84 | 4.23 |

**Table 5.4.** Classification accuracy vs the number of users.

| **Number of users** | 3 | 5 | 7 | 10 |
|---|---|---|---|---|
| **Accuracy, %** | 97.03 | 94.69 | 93.26 | 88.97 |
| **Number of users** | 12 | 14 | 16 | 18 |
| **Accuracy, %** | 88.97 | 92.78 | 88.43 | 88.97 |

By default, the DNN classifier assigns records of the unknown user to one of the existing classes, which is completely wrong for identification. However, the confidence of such a decision is quite low. To handle this issue rejection option is required, which allows skipping identification if DNN predicted probability is lower than a certain threshold. On the other hand, this can lead to a problem that some records of the known user will be rejected also (e.g. because of poor signal quality). The third experiment aimed to select the best rejection threshold to get an optimal relation between classification accuracy and identification rate. The results are shown in the Figure below. As follows from the presented plot the optimal threshold is around 70%. For this threshold classification accuracy achieves ~96%, while the acceptance rate remains ~90%, which relatively high.

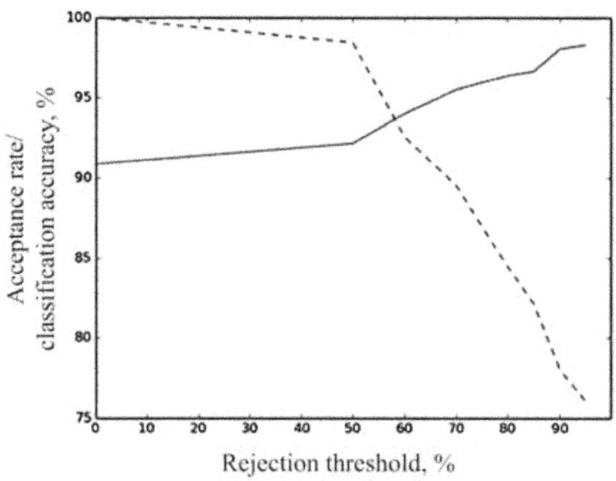


**Figure 5.5.** Classification accuracy (solid line) and identification rate (dotted line).

- **Considerations**

Biometric identification is key to solving many problems related to such areas as information security and access control, authorization, digital and online transactions, etc. Using biosignals in addition to common biometric techniques such as fingerprints, iris, face recognition for multi-factor identification is a quite promising approach. Currently, we focus on human identification using ECG data.

In recent years, deep learning has demonstrated tremendous results in solving commercial and scientific problems. We combine deep learning techniques with an ECG signal for human identification. This approach was promising because of two different reasons. Firstly, deeply learning typically overperforms most of the other classification techniques in terms of accuracy, which can lead to more reliable and robust identification results. Secondly, deep learning allows performing identification using raw ECG data, without feature preparation step, which is required for most of the other techniques and is quite challenging to implement both from algorithmic and computational points of view.

Below, an embedded MCU device with a differential amplifier circuit was used for ECG recording. Signal preprocessing and classification was done on the PC side. To make data recording procedure more user-friendly Lead I was modified and data was collected not from the human chest, but fingers of both hands (two fingers on the left and one finger on the right hand). This approach shows that ECG biometric systems can be miniaturized and integrated with existing biometric systems or other consumer electronics gadgets.

Recorded data have been used to create Lviv Biometric Data Set which currently contains 137 ECG records of 18 unique persons and is publicly available on the Internet.

Main results of the study are the following:

- Number of users to identify, as well as the number of neurons and hidden layers, do have a significant impact on identification accuracy compared to other factors;
- Identification accuracy can vary strongly for the same model structure and training hyperparameters, probably because of random weights initialization and non-convex cost function. To achieve higher accuracy same experiment should be run multiple times and model with the best results should be selected for further usage;
- To prevent wrong identification in case of unknown user rejection threshold was proposed to ignore identification results that have a low confidence level. The optimal threshold value at 70% was selected to maximize the ratio between classification accuracy and acceptance rate.

Achieved results were not as satisfactory as expected. A possible reason is ECG distortions that remain after filtering and a relatively small number of records per class. Identification accuracy can be potentially improved by using another solution that can be other DNN architectures, more advanced outlier correction algorithms, as well as data augmentation using generative models (generative adversarial networks, variational autoencoders) or other techniques. These hypotheses require further research and investigations.

### 5.1.3. Autoencoder Neural Networks for Outlier Correction in ECG-Based Biometric Identification

Biometrics is the field of study focused on recognition of human identity using some unique biological attributes of a person. These attributes, also known as biometric markers, can be of physiological or behavioral origin, for example, fingerprint, iris, voice, face, hand geometry, signature, keyboard typing manner, etc.

Typically, biometric markers should fulfill the following requirements [1]:

103. Universal (present for all individuals);
104. Stability over time;
105. Easy to measure;
106. Low sensitivity to other physiological factors (stress, fatigue);
107. Unique for each person.

There are also additional biomarker features that are not strictly required, but their availability is considered as a big plus:

108. Fraud resistance (difficult to mimic);
109. Liveness indication (present only from live humans);
110. Continuous nature (always available to measure).

The rapid growth of digital and information technologies has made information security and data protection one of the most vital problems on the market, both from technical and business points of view. Thus, there is a strong demand for robust and reliable biometric techniques for human recognition. Classical approaches operate under the assumption that various physiological or behavioral patterns are individual and unique and with minimal or no variability over time. They are mostly based on direct biomarker matching with the reference from the database (e.g. fingerprint or face) which do not always fulfill existing needs. So, the investigation of new alternative biometric techniques is considered as a promising and emerging field of study [2].

Biometric systems operate in one of two basic modes: authentication or identification. The first one assumes that the system should verify, whether the person is the one he/she claims to be. On the other hand, identification requires from biometric system to recognize who exactly is asking for access (from the predefined list of users). Formally both modes can be treated as a classification task (binary for authentication and multiclass for identification) [1].

Authentication can be implemented in two different ways. The first way is called one-to-one matching. It's based on comparing user record with the reference record(s) made earlier on the system calibration stage. Authentication is accomplished if some similarity measure (e.g. Euclidean distance) does not exceed a certain threshold. The other way called one-vs-many matching. It requires a representative database with a big amount of biomarker patterns with different variations. For this scenario, authentication will be passed only in case the biometric system was able to classify users correctly among the big number of users with similar biometric patterns.

Identification mode requires to assign new user record to most similar entities among those from the database. The list of users is predefined by the organization (typically all employees that have appropriate access levels). This constraint is the main difference with one-vs-many matching authentications, where we can use any personal records and any number of persons. Another specific moment for identification applications is the so-called rejection option. Unlike pure classification where the system always assigns to the new record to one of the available classes, identification mode allows the system to reject the deciding case of low confidence or small difference between biometric patterns.

One of the most novel and interesting concepts in biometric research is related to biosignals. ECG, EEG, PPG, biological impedance analysis and others are well known and widely used in medical diagnostic. However, there are also multiple efforts to use them as an alternative source of biometric markers [1], [33-34].

For now, ECG signals prospective in the biometric application look more attractive than others. EEG is a mixture of oscillation waves (alpha-, beta-, gamma-rhythms, etc.) and non-regular event-related potentials, with a very low signal level. In comparison, ECG has a much better signal-to-noise ratio and pure regular behavior. Also, ECG can be acquired from fingers [3], which is more user-friendly compared to EEG that can be measured only from the scalp. Unlike the other biological information which can change significantly depending on other physiological factors (e.g. bodyweight variations, skin sweating, etc.), ECG waveform is predefined by anatomy, position, and size of the heart, which makes it extremely stable overtimes (excluding cases of some cardiovascular diseases). PPG has just one peak per heartbeat and is less informative than ECG with three peaks per beat [34].

One of the most important parts of the ECG biometric system design is to obtain signal acceptable quality. This might be problematic as the informative waveform is mixed with various distortions (noise, respiration and muscle artifacts, etc.). Despite the regular nature some heartbeats can be significantly corrupt and might be treated as anomalies or outliers [37], [40-41]. In measurement theory, the most common solution to this problem is to detect abnormal samples and drop them out. In [42] it was proposed to use even corrupted heartbeats, after some correction to provide classification models with better statistics.

- **Biometric System Design**

The process of biometric identification typically includes the following stages: data acquisition, preprocessing, feature engineering, and feature selection, and finally pattern recognition (classification) [39-40]. This is just a very basic description and different modifications are possible depending on the application details. In this paper, we have reused the system architecture from our previous study [42]. Some brief description is given below. The block diagram of such a human identification system based on ECG is shown in Fig. 5.6. The system consists of the following components: analog measuring circuit, typically based on instrumental amplifiers circuit with two or three electrodes [35]. The Analog part is followed by analog-to-digital conversion. Then digitized data is streamed to MCU/PC and saved to the file system. In our study, we have been using previously recorded data so this part was missing. More details on this can be also found in [43]. An important moment is that all records have been measured from the I-lead (the difference between bio-potentials on the chest) [36]. Raw ECG signal records were filtered using two Butterworth filters: low-pass to remove offsets and respiration and high pass to remove noise, 50 Hz, movement artifacts. Filter signal has been also normalized to a range of [-1; 1]. ECG waveform segmentation was performed using Hamilton algorithms for R peak detection. Algorithm of implementation was taken from [44].

The feature engineering stage was very straight forward. Only samples from the central part of each heartbeat were involved in the further identification process, while samples on the ends were simply dropped out. The reason for such a decision is that the most informative P-, T-wave, and QRS complex are located in the center of the heartbeat. They are surrounded by isolines, which consist of more noise than information. Afterward outlier detection and correction have been performed followed by the PCA compression algorithm [45-46]. The final stage is pattern recognition or classification. Similar to [42], the classifier model was built using a Linear Discriminant Analysis algorithm [47]. It's computationally simple, easy to train and there are lots of ready to use implementations [48]. Also as shown in [40], it demonstrates satisfactory identification accuracy in ECG biometric applications.

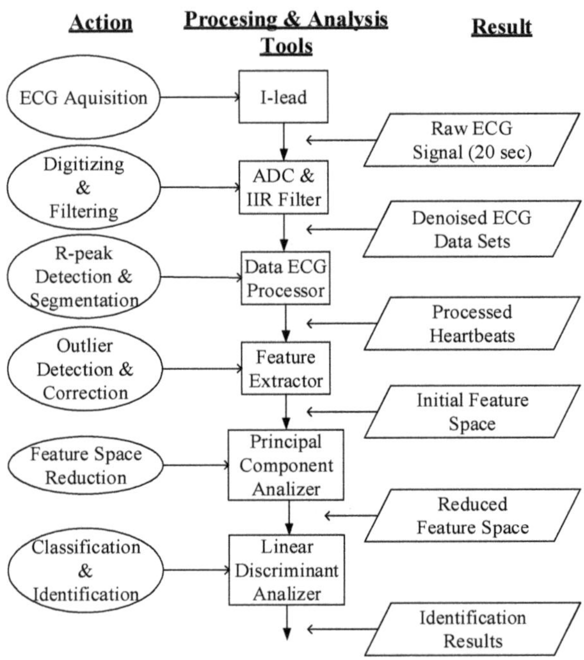


**Figure 5.6.** The block diagram of the human identification system based on ECG.

- **Outlier Correction**

As mentioned higher, the ECG signal might be seriously affected by various reasons, such as muscle noise, respiration movements, electrodes displacement, etc. In many cases, the waveform of some ECG heartbeat will be seriously distorted compared to the other neighboring beats and can be defined as anomalies or outliers. A good way for understanding specific ECG outliers is visual estimation using heartbeat alignment. It assumes that raw signal waveform should be split into separate heartbeats followed by R peaks alignment. ECG outlier example is shown in Fig. 5.7.

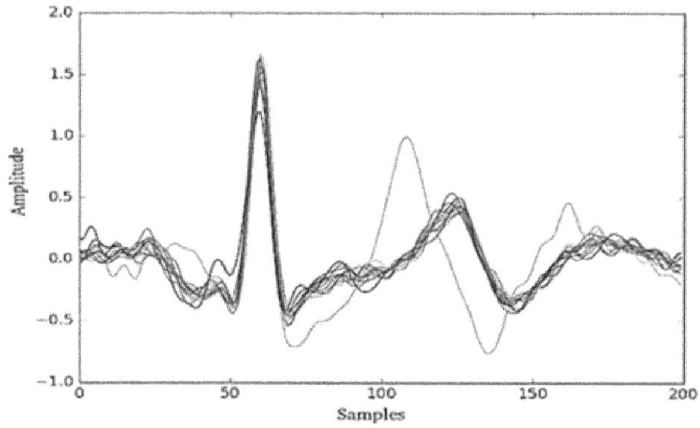


**Figure 5.7.** Abnormal ECG heartbeats (outliers): small (black curve) and sizable (blue curve).

The classical anomaly detection strategy is based on the Euclidean distance estimation between samples of interest. If some sample falls out from confidence interval it is marked as erroneous and is being dropped. The main disadvantage of such an approach is that it relays on an integral estimator, which does not always take into consideration anomalies localization. In many cases, small deviations are grouped in one part of the heartbeat. So, their total impact on Euclid distance is not as big, but they can create a so-called adversarial example, which can easily confuse the classification model and thus lead to wrong identification results.

In work [42], authors have proposed a new method that was capable to detect small and big anomalies simultaneously. Furthermore, it was also proposed to correct corrupted segments of the ECG waveform without the elimination of the entire heartbeat. The idea behind that method was based on the sliding window that moves over the heartbeat. The statistical distribution of samples within each window has been analyzed; standard deviation and confidence interval were calculated. If it turned out that some samples fall out the interval they were replaced with the average values. The procedure was run recursively until no outlier was left for a given window.

The confidence interval is calculated as a product of standard deviation with some gain. Gain and window length are hyperparameters of the algorithm, which values are selected empirically. As shown in [42], outlier correction it's no straightforward task to find optimal values of hyperparameters that will satisfy all the cases. The process is not easy to automate and it requires lots of human intuition. This is one of the major disadvantages of the described method.

An alternative approach is to use machine learning algorithms for outlier detection instead of analytical expressions. There are multiple techniques for this task, but the most promising is to use autoencoders [25]. An autoencoder is a special type of neural networks for which input should coincide with the output. Commonly autoencoders are applied for feature selection and extraction. Formally, they can be described as the following:

$$g[f(x)]=x \tag{5.1}$$

Here f is an encoder function, g is the decoder function, x is an input vector. Despite a complicated name, autoencoder neural networks are a relatively simple machine learning algorithm. Autoencoders are trained using an unsupervised learning algorithm based on backpropagation. Internally it is split into parts encoder and decoder. The encoding algorithm compresses the input values into a latent-space representation ("code"), while the decoder tries to reconstruct the input from the latent-space. Running this procedure iteratively, the model will find some optimal parameters (weights) that will keep output (reconstructed) values as similar to input as possible. This approach works because the latent-space vector is much smaller than the original input one. Thus, just most important, regular information is encoded, while noise and artifact have been automatically removed. After decoding the output vector is quite similar to the input one, but almost without distortions. There are lots of different modifications of autoencoders (e.g. variation, denoising, sparse, etc.), but we use the quite simple architecture so-called vanilla autoencoder shown in Fig. 5.8 [21, 25].

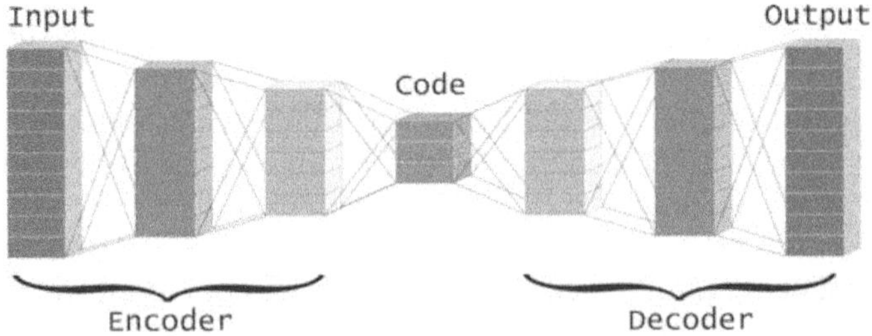


**Figure 5.8.** Basic autoencoder architecture.

The main advantages of autoencoder neural networks are they are adaptive to the data, simple to implement, easy to train, and also computationally efficient. However, the following notes should be considered:

111. Autoencoders are specific to data. The model will work correctly only vs data it was trained on or similar;
112. Autoencoders perform lossy compression. In other words, decompressed data will be degraded compared to the original.

Nowadays, autoencoders are a very popular class of neural networks in many applications. They are widely used for advanced non-linear data filtering and denoising, dimensionality reduction and embedding engineering, etc.

Proposed outlier correction algorithm works as follows:

1. The Autoencoding model is being trained on each heartbeat waveform. While training it learns weights that will be able to minimize total reconstruction on the entire dataset;
2. Passing data through the trained model and comparing the difference between original and reconstructed samples values will indicate potential abnormal samples;
3. If abnormal samples follow each other sequentially the segment is treated as an outlier;
4. All outlier segments are replaced with mean values of the rest segments using the same procedure as in the previous version of the outlier correction algorithm [42].

It is important to notice that the autoencoder based approach requires hyperparameters to perform the correction, but it's expected to be more robust to their variations. Some examples of ECG heartbeats with outlier before and after correction are visualized in Fig. 5.9. According to our research, the most frequent corrections are done for P and T waves. The QRS complex is corrected not so often. The number of corrections varies for different users (from no correction for one user to dozens for another). Also as follows from the plots, correction makes heartbeats more similar, but lots of differences remain. This potentially should be a big plus in real-life applications, since keeping heartbeats similar but still unique will allow us to train more robust classifiers and reduce the error rate.

- **Experiments and Results**

It is reasonable to compare the proposed method with the previously developed method based on standard deviation, which is described in [42]. The latter was conducted on the open Physionet ECG-ID database.

This dataset was created for human identification purposes. The records were acquired from 90 users (44 men and 46 women). Some users have only a few records available recorded during one day. At the same time, other users have over 20 records that were collected periodically for over 6 months.

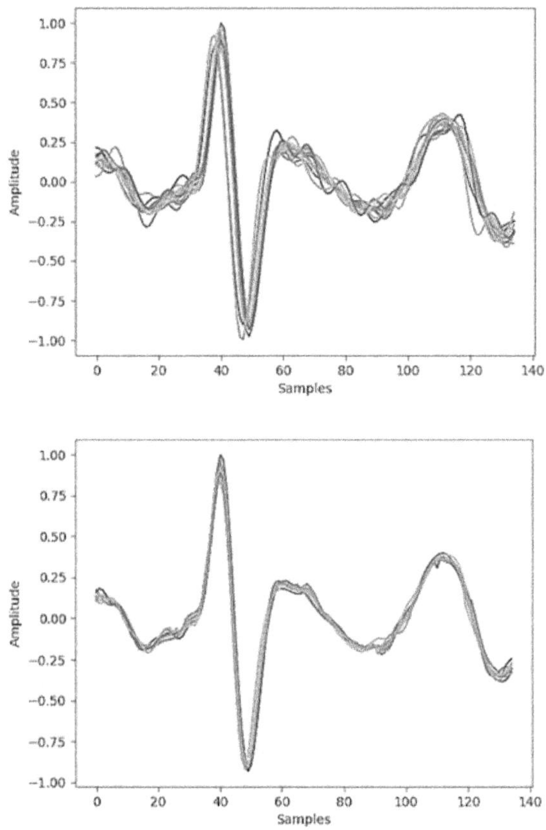


**Figure 5.9.** Results of the operation of the outlier correction based on autoencoder: ECG heartbeats before (higher) and after correction (lower).

Experiment details were completely reproduced from [42], except for the outlier correction part. LDA was used as a classification algorithm. The dimension of the input data was reduced from 270 to 30 features (samples) using PCA transformation (Principal Component Analysis). All experiments have been carried out using Python 2.7. Furthermore, the following frameworks and libraries have been used: SciPy, NumPy, matplotlib, sci-kit learn. The source code of the project can be found here [44]. The autoencoder model was constructed using the $H_2O$AutoEncoderEstimator module from the h2o library. In the experiment, the selection of optimal activation function, the number of layers and neurons for autoencoder based detection stage, as well as hyperparameters for correction stage (windows length and threshold) has been conducted. The choice of hyperparameters was based on the results of classifier training on the ECG-ID dataset. To estimate the classification performance the accuracy score was used [49]. The results are gathered in Table 5.5.

**Table 5.5.** Identification accuracy after outlier correction (in %).

| **Model hyper-parameters** | **Outliers threshold** | **Window length, samples** | | | |
|---|---|---|---|---|---|
| | | **5** | **10** | **20** | **30** |
| Activation = "rectifier with dropout", hidden=[100, 100, 100, 100] | 0.75 | 93.58 | 92.22 | 91.45 | 76.74 |
| | 0.80 | 92.42 | 93.28 | 96.17 | 82.09 |
| | 0.85 | 91.19 | 91.59 | 92.85 | 86.91 |
| | 0.90 | 88.93 | 89.99 | 92.54 | 90.19 |
| | 0.95 | 87.25 | 87.85 | 90.86 | 91.75 |
| Activation = "tanh with dropout", hidden= [100, 100, 100, 100] | 0.75 | 88.90 | 87.76 | 80.49 | 75.91 |
| | 0.80 | 90.07 | 90.22 | 82.63 | 80.00 |
| | 0.85 | 87.43 | 89.78 | 86.82 | 82.19 |
| | 0.90 | 88.58 | 90.27 | 91.73 | 87.33 |
| | 0.95 | 86.09 | 89.23 | 88.07 | 89.25 |

The results of the classification are presented for the test sets. A grid search algorithm has been chosen for hyperparameters selection. Experiments have been conducted for using autoencoders with four hidden layers with 100 neurons in each layer. Rectifier linear unit and hyperbolic tangent were selected as activation functions.

As follows from Table 5.5, the model based on the rectifier linear unit provides higher accuracy of classification. The best outlier correction results were achieved for a window length of 20 samples with an 80% threshold. The Autoencoder method slightly outperforms one described in [42], which was based on standard deviation (maximal achieved accuracy of 95.16 %).

### 5.1.4. ECG-based Artificial Neural Network for Human Being Identification

Biometrics is both, a science domain as well as technology that can be used to determine human identity based on the analysis of physiological and behavioral characteristics such as fingerprints, voice patterns, etc. However human being identification based on his electrocardiograph is a relatively new and intensively developing domain – within a few last years, there have been some publications in this domain [3-4, 35].

To guarantee unerring people identification, the biometric features must satisfy some requirements: universality (each person must have such a feature), uniqueness (each person's feature must be unique), durability (the feature should be time-invariant). Furthermore, the feature must be measurable, and the implementation of the biometrical method – efficient. Equally important there are also requirements regarding acceptability (ergonomic), robustness to forgery (identity fraud), and low cost of the identification system. ECG method satisfies the majority of those requirements and is a good response to the question related to advantages [1].

From the implementation viewpoint, the ECG method is computationally demanding but is also know from providing good results only when it is supported by some artificial intelligence methods, which is a reason why it becomes available nowadays [4, 10]. One can find descriptions for various scenarios when the biometry based on ECG can be used, e.g. for drivers' identification in vehicles, for identity proof in financial institutions, for verification of patients in institutions with remote monitoring, for authentication of passengers at the airports and even for identifying personnel in troops.

So, we present below some interesting research results concerning how the ECG signal parameters and volume of their records in the ECG database impact on the accuracy of human identification for selected machine learning algorithms.

- **Structure and operation of the ECG-based identification system**

In Fig. 5.10 one can see a generic structure of a human being identification system based on an analysis of his ECG features. Yere, we can admit three functional blocks performing appropriate core functions:

ECG signal processing, ECG data conditioning, and smart analysis based on multi-parameter classification.

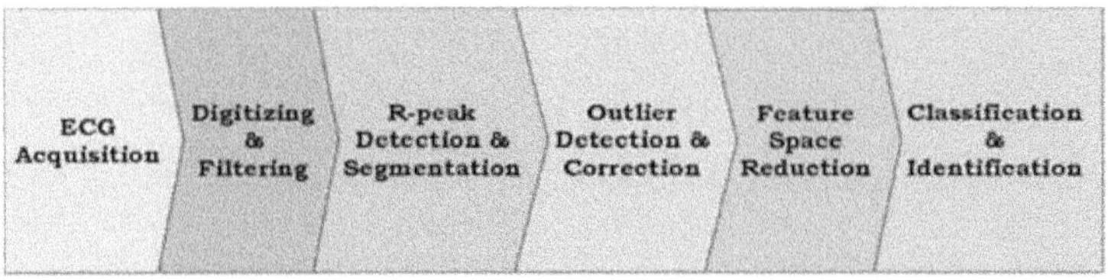


**Figure 5.10.** Generic structure of a human being identification system based on ECG.

### A. ECG Signal Processing Block

In this block, ECG signal acquisition takes place along with its digitization and de-noising. This signal is produced by electrical impulses originating in the brain and targeting our hart. Each of such impulses stimulates various parts of the heart muscle (myocardial) to perform the whole cycle (heartbeat). The process is being repeated cyclically. Furthermore, we can see the electrical field distribution over the whole body, and measuring the voltage differences in various parts of the body is the essence of ECG signal generation. The level of ECG signal is low, and this is why it is so much sense to different kinds of disturbances. To improve its quality, in the medical practice a multi-channel recording is being used with the patient laying on a bed. Such a way of the signal acquisition cannot be used in biometrical applications mentioned before. To improve the quality of the ECG-received signal, one can use well-known de-noising methods, e.g. Discrete Wavelet Transform for baseline drift correction, Adaptive Band Stop (Notch) Filter to eliminate the 50 Hz powerline frequency, Digital Low Pass Butterworth Filter to attenuate remaining noise.

### B. ECG Data Conditioning Block

The next block processing of the de-noised data is performed in two steps: a) Segmentation and normalization of isolated cycles (heartbeats), typically based on R-peak detection algorithm [40], b) Detection and correction of anomalies in the ECG segments (outlier detection and correction) [12]. The result of the first step of processing ECG data is the division of the whole electrocardiograph recording into isolated segments corresponding to one heartbeat. Normalization guarantees the best possible approximation for segments originating from various heart cycles.

$$K_{NORM} = \frac{\max[\tilde{x}(k,n)] - \min[\tilde{x}(k,n]}{2} \tag{5.2}$$

$$x(k,n) = \frac{\tilde{x}(k,n)}{K_{NORM}} \tag{5.3}$$

Here $\tilde{x}(k,n)$ are the de-noised and segmented ECG signal samples; x(k,n) are the de-noised, segmented, and normalized ECG signal samples, k is an index representing the number of the cycle (heartbeat); n is an index representing the number of samples for each cycle.

The need for the second processing step results from the presence of marginal distortions in the ECG data even after de-noising. It is impossible to diminish distortions deformations in which spectra lays in the range of the useful ECG signal to a satisfactory level using some conventional signal-processing methods. Therefore, it was proposed a new method [4], the major ideas of which is:

- Detect defective fragments in each ECG cycle, in which deviation of even one sample exceeds a defined threshold.
- Replace the defective fragments with equable values, acquired in the way of averaging all ECG recording cycles.

The way of computing the averaged recording and the threshold is expressed below:

$$\begin{bmatrix} x_{11} & \cdots & x_{1N} \\ \vdots & \ddots & \vdots \\ x_{K1} & \cdots & x_{KN} \end{bmatrix} \to [\bar{x}_1 \cdots \bar{x}_N] \to [s_1 \cdots s_N] \to S \tag{5.4}$$

x(k, n) is an element of the ECG record data matrix X(K, N); $k \in 1 \div K$ are the rows representing the number of heartbeats; $n \in 1 \div N$ are the columns representing the number of samples per beat. The average value for each sample $\bar{x}(n)$ and its standard deviation is calculated conventionally.

$$\bar{x}(n) = \frac{1}{K}\sum_{k=1}^{K} x(k,n) \tag{5.5}$$

$$s(n) = \frac{1}{K-1}\sqrt{\sum_{k=1}^{K}[x(k,n)-\bar{x}(n)]^2} \quad (5.6)$$

By S we denote the average value of standard deviation for matrix X (K, N), which is calculated as:

$$S = \frac{1}{N}\sum_{n=1}^{N} std(n) \quad (5.7)$$

The next equation describes the rule for the detection of deviating samples (exceeding the pain threshold) and marking their position using a binary matrix (the '1' values correspond to the positions of defective samples).

$$o(n,k) = \frac{|x(k,n)-x(n)|}{S} > gain \rightarrow \begin{bmatrix} o_{11} & \cdots & o_{1N} \\ \vdots & \ddots & \vdots \\ o_{K1} & \cdots & o_{KN} \end{bmatrix} \quad (5.8)$$

And finally, the key elements of the proposed method is the algorithm replacing the defective ECG cycle fragments from within the window of the length of L samples, which can be described by the bellow equation:

$$x(k, n \div L) = \begin{cases} \bar{x}(n \div L), & \text{if any in } o(k, n \div L) = 1 \\ x, (k, n \div L), & \text{if all in } o(k, n \div L) = 0 \end{cases} \quad (5.9)$$

The results of the proposed method for the detection and elimination of the anomalies are shown in Fig. 5.11.

The gain and L in the method are very important and justification of choosing values of gain=0.5 and L=5 as optimal are described in [12].

### C. The Classifier Block

In our research, we used ECG data coming from three different sources, such as Physionet ECG-ID, Physionet QT Database, and LBDS. Physionet ECG-ID database was created for testing the identification algorithms [40]. All recordings were acquired from 44 men and 46 women aged 13-75 years old. For some of the users, there are only a few records available, which means that they were recorded within one day. In the case of another user, there are 20 records available gathered throughout 6 months period. Physionet QT-Database was designed for the estimation of heartbeat segmentation algorithms in ECG [40]. This data set contains data coming not only from healthy people but also from patients with cardiological disorders. For the experiments only records with correct ECG rhythm were included (sel103, sel117, sel123, sel16265, sel16272, sel16273, sel16420, sel16483, sel16539, sel16773, sel16786, sel17152, sel17453, sel301, sel302, sel306, sel307, sel310, sele0111, sele0124, sele013 sele0210). Third database LBDS is being created by a group of researchers with Authors' contributions and at the time of working on this paper, it included 147 ECG records coming from 18 people. The minimal number of records per person was 3 [8].

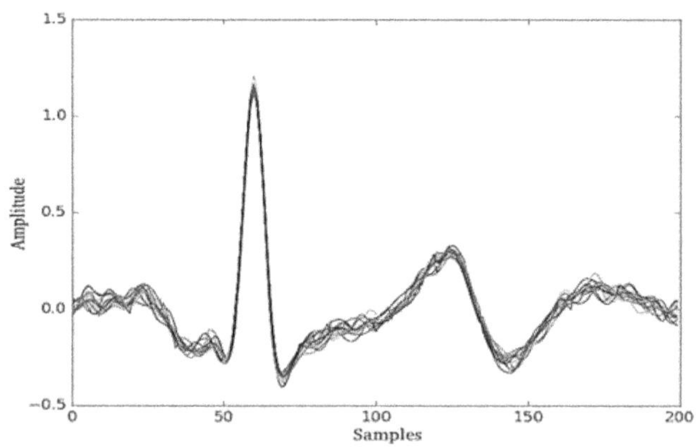


**Figure 5.11.** Generalized structure of the intelligent human identification system based on ECG.

All experiments were carried out twice. Only using LDBS, like in [8], and on a mix of all three databases. Because data have different sampling frequencies, all were reduced to one frequency of 250Hz. The research was conducted using Python 2.7 programming language. It was also used frameworks and libraries such as skflow, scipy, numpy, matplotlib, sci-kit learn, tensorflow. The source code can be found using URLs [14, 29].

As the essential architecture for the classifier, we selected the Feedforward Neural Network learning. The neural network comprises of the input layer (270 neurons corresponding to 270 pre-processed ECG signal samples), three hidden layers with, respectively, 70, 50, and

30 neurons and 18 neurons in the initial layer (corresponding to the number of users). As the transfer function for the hidden layers, we chose Rectified Linear Unit, for the output layer – softmax. As the learning algorithm – Adagrad, number of training cycles – 3000, learning rate – 0.05, regularization coefficient L1 – 0.001. Furthermore, for a more objective assessment of the efficiency of the artificial neural network, we used also Logistic Regression (LR) and KNN algorithms.

Results of the experiments demonstrating the identification accuracy obtained for the 3 artificial intelligence algorithms and the 2 databases were gathered in Table 5.6.

**Table 5.6.** Person identification accuracy using selected artificial intelligence algorithms.

| **Classification Algorithm** | **DataBases** | |
|---|---|---|
| | ***LBDS*** | ***LDBS+QT+ECG-ID*** |
| *ANN* | 0.9539 | 0.9282 |
| *LR* | 0.9417 | 0.7433 |
| *KNN (k=5)* | 0.9746 | 0.8785 |

As one can see in Table 5.6, the best results on the test set for the LBDS database were obtained using the KNN algorithm. The identification error was close to 2.5%. The ANN and LR algorithms resulted in slightly worse results and their error reached around 5%. Fundamentally, other results were obtained after learning the mentioned algorithms using extended data acquired through mixing the three ECG databases. Based on such a data set the ANN algorithm resulted in the best accuracy on the test with the error level reaching 7% while LR and KNN algorithms error level was, respectively, 12% and 25%.

## 5.2. Rehabilitation Virtual Cyber-Physical System Based on Heart Rate

Due to a certain gap between the current state of informational technologies, based on the use of smart sensors; on progressive technologies of normalization and processing of signals, including biosignals; on the uncertainty approach regarding the statistical processing of studies of persons of various professional and age groups - with the theoretical foundations of the majority of existing medical rehabilitation technologies and their practice applying means with a limited set of intelligent functions, there exists an opportunity to satisfy the social demand and real possibilities for further developments in the field of health-care, applied gerontology, and physical therapy.

Currently, Metrology 4.0 develops specific equipment for instance for the health-care CPS. The need for such equipment is perhaps the strongest here. CPS includes collaborating Smart Sensor(s) and Smart Actuator(s) that control the certain physical entity. Unlike the traditional embedded systems, a full-fledged CPS is typically designed as a network of interacting elements with physical input and output instead of as standalone devices [50]. The metrological problems of such systems are considered at the stage of their operation and maintenance.

### 5.2.1. Atherosclerosis of Vessels. The State of Art in Rehabilitation Techniques

The theme of the current issue relates to research in the field of health-care technology based on development and implementation of smart sensors and smart actuators as well as techniques of processing, normalizing and applying of information signals for the creation of means for adjusting the physiological state of the human body by electro stimulation agreed in real-time mode with cardiac rhythm.

Nowadays dangerous diseases concern the diseases of the cardiovascular system [51-52]. They rank high in the structure of the diseases, causing up to one-third part of all deaths and a fourth of the causes of disability. Often, the cause of such diseases is age-related changes due to non-future behavior and human habits: smoking, lack of physical activity, unhealthy eating, and excessive alcohol. Changing behavior, a person becomes able to reduce the risk of cardiovascular diseases. Obliterating atherosclerosis of the vessels of human limbs covers the detection rate of up to 2 % of the population. With age it happens more often, reaching 15-20 % for humans older than 60 years. Effective methods of treatment are mainly invasive. However, in diseases of the arteries of the lower limbs, rehabilitation, and motor exercises are significantly complicated (Fig. 5.12).

At the 1st stage of the disease, patients complain of pain in the muscles of the foot when walking about 1 km. The pain makes the patient limp, after rest, the lameness passes. This pain is associated with muscle ischemia and is caused by arterial blood difficult access to feet. At the stage 2A, pain occurs when walking a distance of 200 to 500 meters. Stage 2B - lameness occurs when walking less than 200 meters. At this stage, you can watch hypotrichosis, nail changes, malnutrition of the leg muscles, pale skin of the limbs. Pulse on the foot is usually absent. Many patients seek medical attention at this stage, as it significantly reduces their life quality.

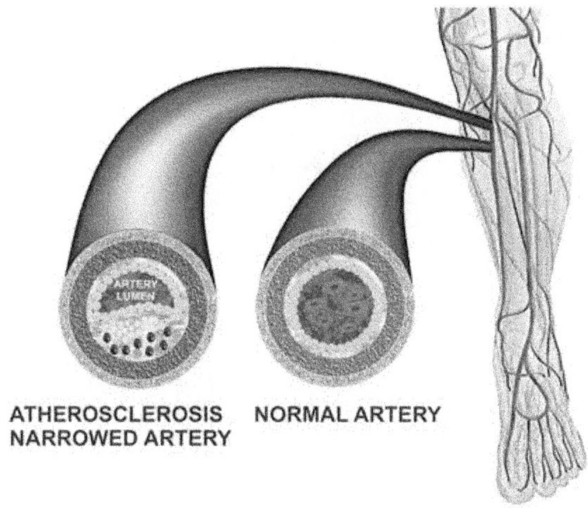


**Figure 5.12.** Vessels of the limbs in atherosclerosis.

Stage 3 is characterized by severe hemodynamic disorders in the limb, its main feature is the pain at the rest and while walking less than 50 meters. At stage 4 - the gangrene develops, leading to limb amputation (Fig. 5.13).

According to [53], peripheral artery diseases (obliterating atherosclerosis of the vessels of the limbs) cover the frequency of detection of up to 2 % of the population of different age groups. With age, it occurs more often, covering 15-20 % of the population over 60 years. This requires intensive treatment [54]. However, methods of treatment of sufficient effectiveness are still poorly developed, and methods of rehabilitation of patients are completely absent. This made it possible to substantiate the considered topic.

### 5.2.2. Virtual Device for Rehabilitation of Patients with Obliterating Atherosclerosis of Limb Vessels

Conducting research aimed at developing the foundations of the method of removing information from the human body, using the temperature and ultrasound [55], studying ECGs and determining certain categories of informative cardio signals [56] and features of arterial blood pressure, in particular, limbs; the processing of the received information, as one patient at different times, in different stages of the disease, as well as for different patients (sex, age, prehistory, family anamnesis), etc. - require a row of special equipment, MIs, means of creating virtual CPS based on various studies of biological, medical, recreational characters.

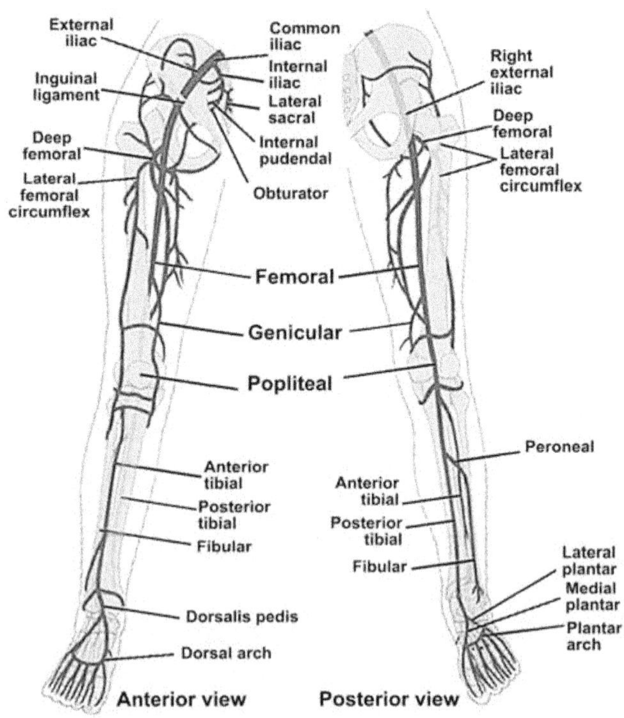


**Figure 5.13.** Pathology leading to amputation.

In particular, an ECG (to determine heart rate) and an EMS (to enhance arterial blood flow, taking into account the available frequency of contractions) were required to perform preliminary studies before the device design. Also, we consider it necessary - an ultrasound device (to assess the current condition of the patient and to study the changes in this condition), infrared thermometer with an adjustable emissivity of the measured object, i.e. the skin of the limb, changed due to disease [57], as well as several smart sensors.

The main goal of the study is the development of the instrumental method of rehabilitation of a person with obliteration of atherosclerosis of the limb vessels by in situ electrostimulation. Conducting research consists in obtaining the measuring information from the human body, its study, including of ECG information, and the peculiarities of the passage of the arterial blood pressure waves in the limbs; the strain gauges and thermistors methods of pulse determination, the processing of received information (in different phases of the disease) for different patients (sex, age, depth of disease), the temperature measurements, etc. Additionally, it requires special equipment, MIs as well as creating a virtual CPS based on a variety of different measured parameters of biological, medical, and recreational characters.

The main problem of current research is the development of the possibility of in-situ synchronizing each alternating wave of blood pressure measured by the smart sensor producing the start signal for the operation of the smart actuator. It seems that the recreating effect in the affected blood vessels is reached when the passage of the particular wave of the blood pressure along the vessels is combined with the physiological action of electromyograph electrodes tightly pressed to the skin of the limb. Under the latter, the blood vessels surrounded by muscles are located. So, these muscles are affected by an electrical signal from an electromyograph and its operation is synchronized in time with pressure wave passage by measuring the start of a certain cardiac contraction.

Previously we have involved the investigation of the abilities of the proposed technique of rehabilitation, basing on two combined devices (ECG and electromyograph). The pulse has been analyzed. The duration of blood pressure wave passage was determined; it was about some tenth of a second on the way from the heart to the foot. Then the inertia of electromyograph was studied (the lesser time). So, we have examined and found the possibility of conjugated operation of the mentioned two devices. Simultaneously there was studied the skin temperature of limbs by the micro pyrometer.

- **Experiment**

Currently, we are developing a device of stimulating blood movement in vessels of the body's limbs that can strengthen the patient physiological capabilities. The device combines the functions of the electrocardiograph (or smart sensor) and the electro myostimulator (or smart actuator) in-situ taking into account the frequency of heart rate. In general, in combination with other therapeutic and rehabilitation procedures, it promotes the achievement of the best result.

Development is realized based on the LabVIEW platform that is the system engineering software for applications that require test, measurement, and control with rapid access to hardware and data insights [58], a virtual mean for measuring the heart rate and a synchronized supply of the signal of electrostimulation is considered. The issue of applying the National Instruments USB 6008/6009 Bus-Powered Multifunction DAQ USB Devices for the creating the virtual mean of the heart rate accessing and synchronized electro stimulation signaling is examined below.

Having previously determined the needed parameters for the generated signals we formed a block diagram of the constructed device. For this purpose, we have thoroughly studied the capabilities of the high-mentioned platform. In particular, parameters of generated signals, limits of their values, and permissible uncertainties of the indicated quantities have been identified.

To determine the moment of passing the blood pressure wave, we have proposed studying the possibility of a thermistor that can feel quickly the rather small temperature changes being tightly pressed to the skin of the foot. The scheme of thermistor connection to the measuring and control unit of the general purpose of the NI USB-6009 type is presented in Fig. 5.14. To set the thermistor's operating current is recommended to use a DAC that is the component of the mentioned unit. The DAC output is connected to the thermistor via resistor R1. As a result, the last and the thermistor form the DAC output voltage divider, the factor of which depends on the thermistor temperature. The USB output of the NI USB-6009 is connected to the PC USB port by the appropriate cable. So, the proposed scheme of the measuring channel is quite simple. From the components of the elements included in the NI USB-6009 unit, only a thermistor and a resistor are used. As is shown below, the requirements of high accuracy, time stability, and low-temperature coefficient may be put to the resistor. Other requirements are the main for thermistor: low inertia (lesser the time of blood wave propagation from the place of its location on the body to the defined place on the foot, that is, lesser than 0.1 s.) and high sensitivity of feeling the temperature increase caused by the mentioned wave passing.

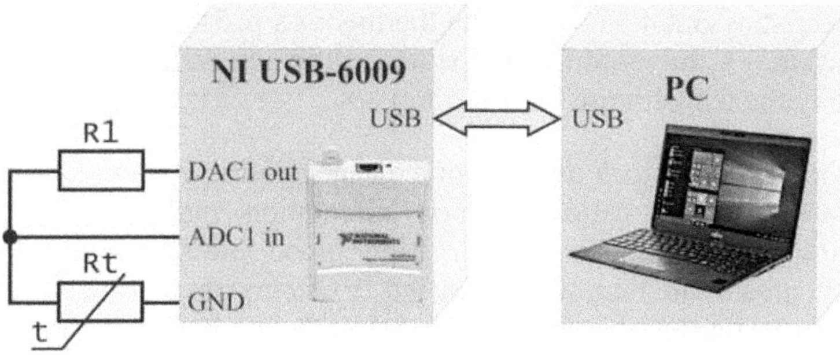


**Figure 5.14.** Scheme of connection of the thermistor to the measuring and control unit of NI USB-6009 block.

The transform function of the thermistor is an exponential [59]:

$$R_T = A \cdot e^{-\frac{B}{T}} \tag{5.10}$$

Here $R_T$ is the resistance of the thermistor at a given temperature T; A is the factor characterizing the change in thermistor's resistance under the temperature alteration. Studies [60] have established that the deviation from the exponential function does not exceed 0.01°C. Therefore, to determine the transform function of the certain thermistor in experimental studies, it suffices to calibrate it in two fixed temperatures. It has been noted [1] that the limit of the permissible deviation of the thermistor resistance at 293 K is ± 20 % of the nominal resistance. Also, the limit of the permissible deviation of the temperature coefficient of its resistance is ± 5%. The latter's value is ~4 % / K. Another considerable feature seems to be the response time. This one was studied previously in [61]. Because it takes longer for the temperature-sensing element to warm up, the response time is governed here by the time. For the response time to be short, the sensor must have good thermal-conductivity properties and a low mass. The thermistor meets these requirements and its response time is lesser than 0.1 s.

The next important feature consists of the following. The current studies are specific since the thermometric method with the thermistor application is planned to be involved for diagnosing the small (less the 0.1 K) temperature increase. Here the major of the high-mentioned characteristics is the temperature coefficient of thermistor's resistance. The latter defines the ability to start the actuator's operation. So, the output signal of the thermistor must be sufficient in electrical characteristics and 4 % / K for the thermistor's resistance ~ 20000 Ω at the current ~ 1 mA produces the voltage drop assessed in 80 mV while detecting the temperature increase in 0.1 K.

Development of a block diagram of a mean for measuring and analyzing peculiarities of a patient's blood circulation mainly consists of synchronizing in situ each alternating blood pressure wave in the problem area of the limb with produced an electro stimulation signal that results in increasing the throughput of the affected blood vessel. Taking into account the number of cardiac contractions during the 10-minute session, the patient is subjected to 600 repetitive actions, which eventually lead to a pronounced rehabilitation effect. It was evaluated by the contactless monitoring of the temperature changes of the same area of the affected limb.

To improve the effectiveness of the developed method, it is proposed to reconcile the presence of the mean control signal with the moment when the pulse from the heart arrives. Thus, even when differences between each successive heart attack and, accordingly, the arrival of the wave of blood flow have happened, the control signal automatically agrees with the change in rhythm. Also, when you change the location of the electro stimulator on the limb, the device automatically shifts the start moment for a smart actuator.

The conversion function of NI USB-6009 block [62] with the connected thermistor (Fig. 5.14) is not strictly known, especially for metrological tasks. To study it, we have simulated the measurement with the help of the thermistor by substituting it by an ambiguous measure of electrical resistance. We set alternately the following resistance values from 5 kΩ to 55 kΩ in steps of 5 kΩ, which was connected to inputs ADC1 and GND of the NI USB-6009 block. As a result, we have defined a significantly nonlinear conversion function of the mentioned block, arising up to 3 %, which, however, is insignificant for the operation of the created rehabilitation tool.

The temperature coefficient of resistance for metal oxide resistor R1 does not exceed 0.02 % per 1K. For comparison, the same coefficient of the thermistor is equal to 4 % per 1K. So, the impact of the used additional resistor is less than 5 % and can be neglected. The influence on the measurement result of the resistance the supply wires can also be neglected even for a two-wire connection circuit; copper wires of the cross-sectional area 0.2 $mm^2$ length can be even 50 m.

As a result of consideration, a thermistor is applied for revealing the temperature increment. For monitoring of limb's temperature is proposed the contactless method with the help of an IR pyrometer specially calibrated according to [63].

- **Virtual Device and its Software**

On the left side of the front panel (Fig. 5.15) there are indicators of average voltage on the thermistor "Voltage Therm", the maximum value of noise "max noise", the minimum value of noise "min noise", the average value of current through the thermistor "Current" and self-healing power caused by measuring current "Self-heating Power". Arrays of instantaneous voltage values on the thermistor "Instantaneous V", noise values "Noise"; a graph showing the values of the instantaneous voltages on the thermistor during the measurement. On the right, the block diagram shows a DAQ Assistant configured to measure the voltage from the analog input at a trial size of 96000 and a trial rate of 48000.

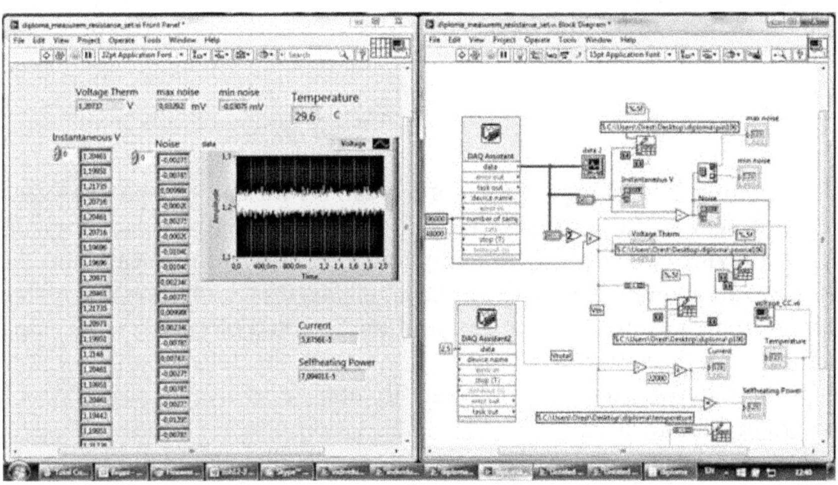

**Figure 5.15.** Virtual device for rehabilitation of patients with obliterating atherosclerosis of the limb vessels.

The measurement results are sent to the graph "data" and displayed on the front panel. Also, they are sent to the "Instantaneous V" indicator and displayed on the front panel, as well as recorded in a spreadsheet file. This data is also averaged and the obtained value is forwarded to the "Voltage Therm" indicator and converted into an array for writing into a tabular file. The determined differences between the array of measured values and the average value are sent to the indicator "Noise" for displaying on the front panel, as well as for recording in a spreadsheet file. Also, the maximal and minimal deviations from zero are selected from this array and displayed on the indicators "max noise" and "min noise", respectively. DAQ Assistant2 is configured to generate 2.5 V as the measuring current supply. Besides, the current through the thermistor and the self-heating thermistor are calculated. The program also contains a subroutine that converts the voltage on the thermistor to the temperature in degrees Celsius by using the conversion function:

$$t = (T - 273.15) = 21.359U_{THERM}^4 - 115.77U_{THERM}^3 + \\ +225.92U_{THERM}^2 - 228.49U_{THERM} \text{-} 138.63 \tag{5.11}$$

Besides, the block has produced the pulses of different shapes (rectangular, triangular, trapezoidal) for application to the electrodes glued to the skin of the problem area of the foot. Their amplitude did not exceed 5 V and duty cycle ~ 0.5 at a frequency corresponding to the pulse rate. The beginning of each successive cycle is a signal from the thermistor.

- **Operation**

Development of a block diagram of a mean for measuring and analyzing peculiarities of a patient's blood circulation mainly consists of synchronizing in situ each alternating blood pressure wave in the problem area of the limb with produced an electro stimulation signal that results in increasing the throughput of the affected blood vessel. Taking into account the frequency of cardiac contractions during the 10-minute session, the patient was subjected to 600 repetitive actions, which eventually led to a pronounced rehabilitation effect. It was evaluated, including the measured contactless temperature changes in the same area of the affected area. The temperature of the same point of the foot has increased at ~1 K after 10 minutes procedure or after 600-fold application of impulses to the tissues of the foot.

Currently, it is needed to determine the optimality of the procedure. That is the duration of the procedure, the number, and intensity of signals to achieve a tangible effect. There exist some peculiarities depending on the particular patient. So, we can recommend allocating traumatized young athletes into a separate rehabilitation group. The same group should be formed for the older generation with age-related changes. It is a must-see to work separately with smokers. For instance, the certain computerized program of self-assessing the state of the cardiovascular system, recommends to fix that doze from 100 cigarettes smoked in life would be a pronounced negative factor leading to a tangible negative consequence.

To improve the effectiveness of the developed method, it is proposed to reconcile the presence of the mean control signal with the moment when the pulse from the heart arrives. Thus, even when differences between each successive heart attack and, accordingly, the arrival of the wave of blood flow have happened, the control signal automatically agrees with the change in rhythm. Besides, when you change the location of the electro myostimulator on the limb, the device automatically shifts the start moment of the pulse. Currently, it is needed to determine the optimality of the procedure. That is the duration of the procedure, the number, and intensity of signals to achieve a tangible effect.

Promising in the direction of detecting the filling the arteries of the foot with blood is the known and proven method, which has received a new impetus in the development through the works of prof. T. Fröhlich [64]. The latter creatively combined modern approaches to the quantization of measurements with mass measurement, which allowed to improve the accuracy of mass and torque standards with high sensitivity. As a result, the technologies, including the above-mentioned rehabilitation technology can take full advantage of his team achievements.

## 5.3. Conclusions

1. Biometric identification is key for solving a lot of problems related to such areas as information security and access control, authorization, digital and online transactions, etc. ECG signal is a valid biometric marker quite robust to hardware variations, noise, and artifacts presence, stable over time, scalable over quite a solid number of users. It's hard to steal or mimic, easy to measure. A biometric system allows achieving high-operation speed, as just one heartbeat (≤ 1 s.) is enough for good classification (~90 %). The most promising algorithm for ECG identification is ANNs due to 2 reasons. First, deeply learning typically overperforms most of the other classification techniques in terms of accuracy, which can lead to more reliable and robust identification results. Second, deep learning allows performing identification while

using raw ECG data, without the preparation steps, required for most techniques.

2. It is developed and validated a new algorithm based on autoencoder neural networks which require at least five heartbeats for the detection and correction of ECG heartbeats outliers. Filtered, normalized, and reconstructed ECG-signal is compared with the input one, and RMS error is calculated. This error is used to detect and correct the samples of interest. The main advantages of the autoencoder based algorithm vs the based on standard deviation are fewer hyperparameters to be tuned and better variability.

3. Based on ECG studies there was proved the possibility of in-situ synchronizing each alternating wave of blood pressure measured by the smart sensor with starting the operation of the smart actuator (electromyostimulator). The rehabilitation effect in the affected blood vessels of the limb was reached when the passage of the particular wave of the blood pressure along the vessels was combined with the physiological action of electromyostimulator.

4. Design of virtual device for rehabilitation of patients with obliterating atherosclerosis of limb vessels was established basing on the NI USB-6009 block with the connected thermistor and the LabVIEW Programming Environment, which has built-in tools for both interactions with hardware (in particular, with standard data acquisition units), and advanced data processing tools. Available tools have been fully sufficient for realizing the channel of measuring temperature increment caused by the blood passage as well as for following generation of the voltage of the required shape and amplitude (1-7 V), supplying the electrodes tightly pressed to the sick area of the limb.

## References


[1]. A. Jain, P. Flynn, A. Ross, Handbook of Biometrics, *Springer,* 2008.

[2]. E. J. Kindt, Privacy and Data Protection Issues of Biometric Applications: A Comparative Legal Analysis, *Springer*, 2013.

[3]. A. Fratini, M. Sansone, P. Bifulco, M. Cesarel, Individual identification via electrocardiogram analysis, *BioMed Eng Online*, Vol.14, 78, 2015, pp. 1-23.

[4]. G. Kaur, D. Singh, S. Kaur, Electrocardiogram (ECG) as a Biometric Characteristic: A Review, *Int. Journ. Emerg. Res. in Man.t &Techn.,* Vol. 4, Issue 5, 2015, pp. 202-206.

[5]. A. Pal, Y. N. Singh, ECG Biometric Recognition, in *Proceedings of the 4th Int. Conf. on Communications in Computer and Information Science (ICMC 2018)*, Varanasi, India, January 9–11, 2018.

[6]. A. Pal, Yo. Singh, ECG Biometric Recognition, in *Proceedings of the 4th Int. Conf. in Math. and Comp.*, Varanasi, India, January 9-11, 2018, *Springer*, Singapore, 2018, vol. 834, pp. 61-73.

[7]. A. Matos, A. Lourenc, J. Nascimento, Embedded system for individual recognition based on ECG biometrics, in *Proceedings of the Conf. on Electronics, Telecomm. & Comp. (CETC' 13)*, Vilamoura, Algarve, Portugal, 19-21 Sept. 2013, pp. 265 – 272.

[8]. L. Wieclaw, Y. Khoma, P. Falat, D. Sabodashko, V. Herasymenko, Biometric identification from raw ECG signal using deep learning techniques, in *Proceedings of the 9th IEEE Int. Conf. Intel. Data Acquis. & Adv. Comp. Syst: Techn. & Appl.*, Bucharest, Romania, 21-23 September 2017, pp. 129-133.

[9]. Mawi band: stress and heart health monitor. Verification (https://mawi.band/).

[10]. SoftServe Biolock, Smart Identity Verification (https://demo.softserveinc.com/biolock/).

[11]. M. Bassiouni, W. Khalefa, E. S. A. El-Dahshanm, A. B. M Salemm, A study on the intelligent techniques of the ECG-based biometric systems, *Recent Adva. in El. Eng.,* 2015, pp. 26-31. (http://www.inase.org/library/2015/crete/bypaper/COCI/COCI-04.pdf).

[12]. V. Khoma, M. Pelc, Y. Khoma, D. Sabodashko, Outlier correction in ECG-based human identification, in Hunek W., Paszkiel S. (eds.), Biomedical Engineering and Neuroscience. BCI 2018. Advances in Intelligent Systems and Computing, Vol. 720, *Springer*, 2018, pp. 11-22.

[13]. M. Gertsch, The ECG: A Two-Step Approach to Diagnosis, *Springer: Verlag*, Berlin Heidelberg, 2004, pp.19-21.

[14]. Lviv Biometric Data Set. (https://github.com/YuriyKhoma/Lviv-Biometric-Data-Set/).

[15]. e-Health Sensor Platform V2.0 for Arduino and Raspberry Pi (https://www.cooking-hacks.com/documentation/tutorials/ehealth-biometric-sensor-platform-arduino-raspberry-pi-medical).

[16]. The ECG-ID Database (https://physionet.org/physiobank/database/ecgiddb/).

[17]. The QT Database (https://physionet.org/physiobank/database/qtdb/).

[18]. A. Goldberger, L. Amaral, L. Glass, J. Hausdorff, P. Ivanov, R. Mark, J. Mietus, G. Moody, C.-K. Peng, H. Stanley, PhysioBank, PhysioToolkit, and PhysioNet: Components of a New Research Resource for Complex Physiologic Signals, *Circulation,* 101, 23, 2000, pp. e215-e220.

[19]. C. M. Bishop, Pattern Recognition and Machine Learning, In Information Science and Statistics. *Springer,* Singapore, 2006.

[20]. F. Pedregosa et al., Scikit-learn: Machine Learning in Python, *Journ. Mach. Learn. Res.*, Issue 12, 2011, pp. 2825-2830.

[21]. M. AlMahamdy, B. Riley, Performance study of different denoising methods for ECG signals, *Procedia Computer Science,* Vol. 37, 2014, pp. 325 – 332.

[22]. A. Lourenco, H. Silva, D. Santos, A. Fred, Towards a finger-based ECG biometric system, in *Proceedings of the Int. Conf. on Bio-Inspired Systems and Sign. Proc. (BIOSIGNALS' 2011),* Rome, Italy, 26-29 January, 2011.

[23]. P. Khandait, N. Bawane, S. Limaye, Features extraction of ECG signal for detection of cardiac arrhythmias, in *Proceedings of the Nat. Conf. on Innov. Paradigms in Eng. & Techn. (NCIPET-2012),* 2012, pp. 6-10.

[24]. S. Karpagachelvi, M. Arthanari, M. Sivakumar, ECG feature extraction techniques - A survey approach, *International Journal of Computer Science and Information Security IJCSIS,* Vol. 8, No. 1, April 2010, pp. 76-80.

[25]. M. Varshney, C. Chandrakar, M. Sharma, A survey on feature extraction and classification of ECG signal, *Int. Journ. Adv. Res. in El., Electron. & Instr. Eng.,* Vol. 3, Issue 1, January 2014, pp.6572-6576.

[26]. N. Soorma, M. Tiwari, J. Singh, Feature extraction of ECG signal using HHT algorithm, *Int. Journ. Eng. Trends and Techn.,* Vol. 8, Issue 8, 2014, pp. 454-460.

[27]. Meet BioLock: Smart Biometrics for Tomorrow, 2016 (https://united.softserveinc.com/blogs/biolock-smart-identity-authentication/).

[28]. e-Health Sensor Platform V2.0 for Arduino and Raspberry Pi (https://www.cooking-hacks.com/documentation/tutorials/ehealth-biometric-sensor-platform-arduino-raspberry-pi-medical).

[29]. BioSPPy - Biosignal Processing in Python (https://github.com/PIA-Group/BioSPPy).

[30]. Arduino UNO & Genuino UNO (https://www.arduino.cc/en/Main/arduinoBoardUno).

[31]. The source code of the project, 2017 (https://github.com/YuriyKhoma/ecg-identification).

[32]. TensorFlow (https://www.tensorflow.org/).

[33]. D. Jenkins, S. Gerred, ECGs by Example, *Elsevier*, 2011.

[34]. J. Urigüen, B. Garcia-Zapirain, EEG artifact removal-state-of-the-art and guidelines, *Journ. Neural Eng*, Vol.12, Issue 3, 2015.

[35]. T. Shen, W. Tompkins, Y. Hu, One-lead ECG for identity verification, in *Proceedings of the 2nd Joint Conf. IEEE Eng. in Med. & Biology Soc. and Biomed. Eng. Soc.,* Vol. 1, 2002, pp. 62-63.

[36]. J. Cunha, B. Cunha, W. Xavier, N. Ferreira, A. Pereira, Vital-jacket: A wearable wireless vital signs monitor for patients' mobility, in *Proceedings of the Avantex Symp.,* 2007.

[37]. E. Aslanger, K. Yalin, Electromechanical association: a subtle electrocardiogram artifact, *Journ. Electrocardiology,* Vol. 45, Issue 1, 2012, pp. 15-17.

[38]. D. Tax, R. Duin, Outliers and data descriptions, in *Proceedings of the 7th Ann. Conf. Adv. School for Comp. and Imag. (ASCI)*, 2001.

[39]. V. Hodge, J. Austin, A survey of outlier detection methodologies, *Artificial Intelligence Review,* Issue 22, 2004, pp. 85–126.
[40]. T. Lugovaya, Biometric human identification based on electrocardiogram, Master's thesis, Fac. Comp. Techn. & Informatics, *Electrotechn. Univ. 'LETI', Saint-Petersburg*, RF, 2005.
[41]. A. Lourenco, H. Silva, C. Carreiras, A. Fred, Outlier Detection in Non-Intrusive ECG Biometric System, in Image Analysis and Recognition, *Springer Berlin Heidelberg*, 2013, pp. 43-52.
[42]. I. Jolliffe, Principal Component Analysis, *Springer Series in Statistics,* NY, 2002.
[43]. R. Duda, P. Hart, D. Stork, Pattern Classification, *Wiley Interscience*, 2000.
[44]. A. Dertat, Applied Deep Learning, part 3: Autoencoders (https://towardsdatascience. com/applied-deep-learning-part-3).
[45]. Hu Zhengbing, Su Jun, V. Jotsov, O. Kochan, M. Mykyichuk, R. Kochan, T. Sasiuk, Data science applications to improve accuracy of thermocouples, in *Proceedings of the IEEE 8$^{th}$ Int. Conf. on Intel. Syst,* Sofia, Bulgaria, 2016, pp. 180-188.
[46]. https://physionet.org/physiobank/database/ecgiddb/
[47]. I. Goodfellow, Y. Bengio, A. Courville, Deep learning, *The MIT Press*, 2016.
[48]. https://github.com/YuriyKhoma/ecg-identification
[49]. https://github.com/PIA-Group/BioSPPy
[50]. Cyber-Physical Systems. Metrological Issues, Ed. by S. Yatsyshyn, B. Stadnyk, *IFSA Publishing,* Barcelona, 2016.
[51]. Global, regional, and national life expectancy, all-cause mortality, and cause-specific mortality for 249 causes of death, 1980-2015: a systematic analysis for the Global Burden of Disease Study 2015, *Lancet,* 388, 10053, 2016, pp. 1459–1544.
[52]. Ruiz-Canela, M; Martínez-González, Lifestyle and dietary risk factors for peripheral artery disease, *Circulation Journal*, Vol. 78, Issue 3, 2014, pp. 553–559.
[53]. M. Papadakis, S. McPhee, M. Rabow, Current medical diagnosis & treatment, Series: Lange medical book, *McGraw Hill Medical*, N.Y., 2019.
[54]. G. Hankey, P. Norman, J. Eikelboom, Medical treatment of peripheral arterial disease, *JAMA*, Vol. 295, Issue 5, February 1, 2006, pp. 547–553.
[55]. S. Yatsyshyn, B. Stadnyk, Ya. Lutsyk, Handbook of Thermometry and Nanothermometry, *IFSA Publishing,* Barcelona, 2015.
[56]. V. Khoma, M. Pelc, Yu. Khoma, Artificial neural network capability for human being identification based on ECG, in *Proceedings of the 23$^{rd}$ Int. Conf. on Methods and Models in Automation and Robotics (MMAR),* Miedzyzdroje, 2018, pp. 479-482.
[57]. S. Yatsyshyn, B. Stadnyk, Method for determining the coefficient of emissivity of a material, *Pat. Ukraine 116684,* Bull. 8., 25.04.2018.
[58]. LabVIEW Environment Basics, National Instruments (https://www.ni.com/getting-started/labview-basics/environment).

[59]. Kochan, R., Kochan, O., Chyrka, M., Vasylkiv, N., Precision data acquisition (DAQ) module with remote reprogramming, in *Proceedings of the IEEE International Workshop on Intelligent Data Acquisition and Advanced Computing Systems: Technology and Applications*, Sofia, 2005, pp. 279-282.

[60]. S. Jun, O. Kochan, Common mode noise rejection in measuring channels, *Instruments and Experimental Techniques*, Vol. 58, Issue 1, 2015, pp. 86-89.

[61]. B. Stadnyk, S. Yatsyshyn, H. Pol'ova, Development of noise measurements, Part 7. Coriolis mass flowmeter and its errors, *Sensors and Transducers,* Vol. 158, Issue 11, November 2013, pp. 249-254.

[62]. User Guide NI USB-6008/6009 Bus-Powered Multifunction DAQ USB Device (http://www.ni.com/pdf/manuals/371303n.pdf )

[63]. M. Wozny, S. Prokhorenko, K. Mas, D. Ploch, Infrared radiation emitted due to scanning of a hot spot as a probe of hidden defects, *Infrared Physics & Technology,* Vol.76, May 2016, pp.574-579.

[64]. T. Fröhlich, N. Rogge, S. Vasilyan, *et al*, Neue wege zur kalibrierung von E2-massenormalen und darstellung von kräften bis 10 N, *Technisches Messen. Platform fur Methoden, Systeme und Anwendungen in der Messtechnik, Spec. Issue Das Erneuerte Internationale Einheitensystem (SI),* Vol. 87, Issue 4, 2020, pp. 280-293 (in German).

# Chapter 6

# Studies of Metrological Characteristics of Measuring Instruments

**Mykola Mykyichuk, Zenoviy Kolodiy, Serhiy Lazarenko and Ihor Likhnovsky**

## 6.1. Stability of Metrological Characteristics of the Measures of Electrical Quantities Based on Electric Noise Investigation

Often, when analyzing the results of the statistical measurements of the parameters of the same objects, it is advisable to ensure the conditions of the equality of measurements. The differences between the variances of the measurement results should be minimal. However, in practice, the differences between the variances can vary considerably. The question arises whether the condition of equality of measurements is observed.

Verification of the condition of measurement equality can be performed by statistical criteria, for example, by the Fisher criterion F, comparing the values of the variances of the measurement results with the critical value $F_{cr}$. For $F > F_{cr}$, it can be argued that the condition of the equality of measurements (null hypothesis) is violated and the measurements are non-equilibrium. However, statistical criteria do not answer the question of the cause of the inequality of measurements, and under what conditions the difference between the variances will be minimal. That is, in which cases the condition of the inequality of measurements will be fulfilled.

Unique, high-precision measures of electrical resistance - model coils of electrical resistance - have been selected as the object of research. They were manufactured in 1960. During storage at the Lviv Polytechnic Metrology Laboratory, electrical resistance measures were subject to natural aging. This makes it possible to assert that the mechanical stresses in their elements have been minimized, namely that the values of electrical resistance have been stabilized to the maximum and, accordingly, one can expect minimal values of the variance when re-measuring the values of their resistances. Similar physical considerations have been taken into account by Canadian orbital telescope specialists when using a lens that has been designed for more than 50 years with minimal, naturally eliminated mechanical stress, to reduce distortions of images of space objects.

The research aimed to use the high-mentioned unique means to establish the difference between the variances of the measurement results in the same conditions of electrical resistance of the same systems, to establish the energy spectrum of fluctuations of their nominal value as well as the energy spectrum in the frequency range in which the measurement process can be carried out in the conditions of quasi-equilibrium state. Then the measurement variances are minimal.

- **Model**

The main reasons for non-equilibrium measurements may be either a change in the measurement condition or the non-equilibrium state of the measurement object itself. Keeping in mind the conditions of measurement, the uniformity of which is not a problem, let us focus on the state of the measurement object. If the measurement object (hereinafter referred to as the test system) is in equilibrium, then the variance of the measurement results is not a function of time t, respectively, the standard deviation will not depend either on the time interval between the individual measurement results or on the time during which the number of n measurement results are obtained. However, the measurement results themselves can be represented as fluctuations in the values of the measured parameter $x_P(t)$ around the mean value $\overline{X_P}$ (Fig. 6.1).

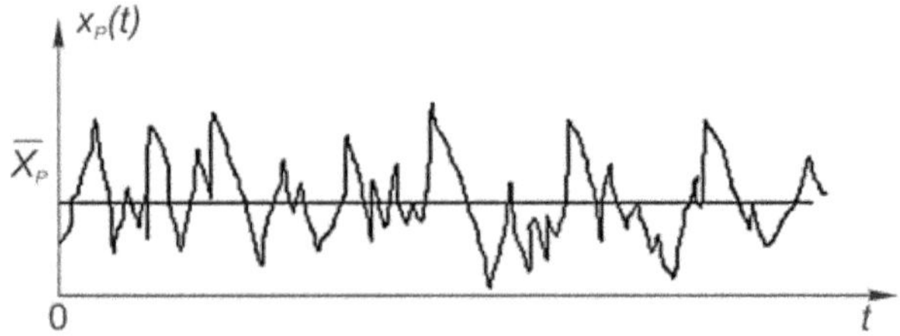


**Figure 6.1.** Changes of measured parameter $x_P$ of a balanced system in the time.

It is known that the energy spectrum of $S_{xp}(f)$ fluctuations of the parameter $x_P(t)$ of the equilibrium system is the same in the frequency range from $f \rightarrow 0$ up to ultrahigh frequencies (i.e., in the form of "white" noise) [1].

From Fig. 6.2 it is seen that the results of averaging 10 measurement results performed, for example, in 10 seconds (Δt) will be different from the average of 10 measurement results made in 10 days (ΔT). The energy spectrum of the fluctuations of the parameter $x_H(t)$ of the nonequilibrium (real) system $S_{xH}(f)$ is not the same in the entire frequency range and contains the flicker-component of the spectrum [2] (Fig. 6.3).

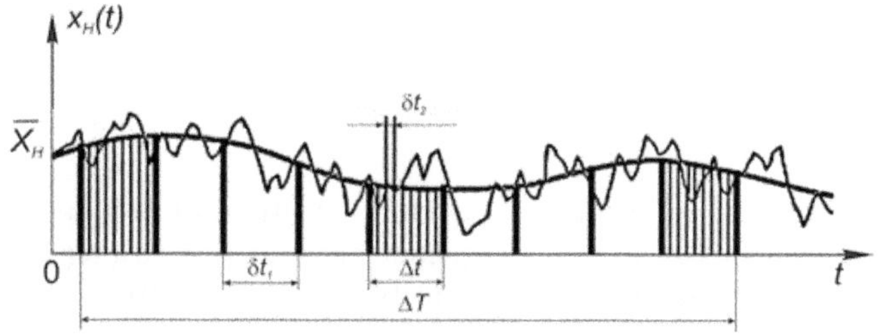


**Figure 6.2.** Time dependence of $x_H(t)$ parameter of the unbalanced system.

10 measurement results after 1 s in Fig. 6.11 evidence the parameter $x_H(t)$ of the certain system, which is quasi-equilibrium for a time Δt; 10 measurement results performed within 10 days characterize the system in the nonequilibrium state. Fig. 6.3 demonstrates the energy spectrum of the nonequilibrium system and the frequency range of 10 measurements through 1 s (a) of the parameter of system $x_H(t)$, which is in the nonequilibrium state, and frequency range of 10 measurements after 100 s (b).

When measuring variables, the ratio must be satisfied: $f_B \geq F_C$ , where $F_C$ is the upper frequency in the variable spectrum. From the analysis of Fig. 6.2-6.3 we can conclude that the variance of the results of the nonequilibrium system parameters depends not only on the number of measurements n but also on the time interval between measurements δt: the smaller is δt, the greater will be $f_B$ - frequency range the measurement is shifted to the band of high frequencies and the process of measuring the parameter of the system in the quasi-equilibrium state is realized (Fig. 6.3, $f_H$ = 0.1 Hz, $f_B$ = 1.0 Hz). As δt increases, $f_B$ decreases; the measurement frequency range shifts to the low frequencies, and the process of measuring the system parameter in the nonequilibrium state is realized (Fig. 6.3, $f_H = 0.001$ Hz, $f_B = 0.01$ Hz).

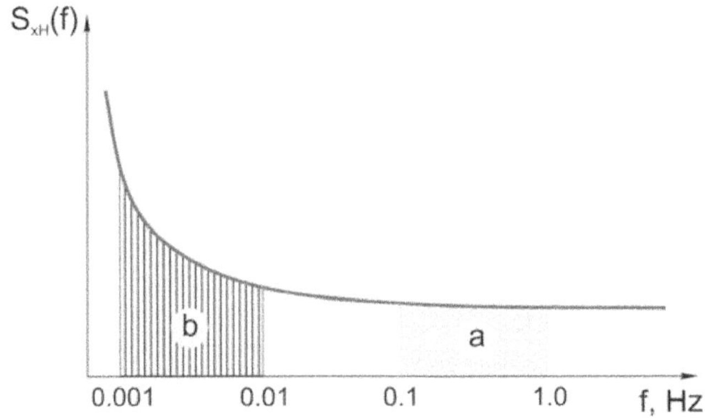


**Figure 6.3.** Energy spectrum $S_{xH}(f)$ of fluctuations of parameters of the unbalanced system and frequency range of 10 measurements of the system's parameter in 1s (a) and 100 s (b).

Based on the above, the condition of the equality of measurements while providing the same conditions for measurements can be determined by analyzing the energy spectrum of the system under study. According to the known energy spectrum, it is possible to determine the frequency range $\Delta f = f_B - f_H$ for which the measurement process will be carried out under quasi-equilibrium conditions of the system. For the known $f_H$ and given $\delta t$, it is possible to determine the number of measurements n for which the difference between the variances will be minimal and, accordingly, the condition of the equivalence of the measurements will be provided.

It is possible to determine whether the system under study is in equilibrium or non-equilibrium state based on the experimentally determined energy spectrum S(f). it is established in [2] that the energy spectrum of real systems is in the form of

$$S_H(f) = \frac{\exp(f \cdot \tau)}{\exp(f \cdot \tau) - 1} S_0 , \tag{6.1}$$

where $S_0$ is the value of the power spectral density at the middle and high frequencies; $\tau$ is the relaxation time (the time by which value of $S_H(f)$

changes to $S_H(f) \cong 1.58 \cdot S_0$, where $f_0 = \frac{1}{\tau}$). The definition of the value of $\tau$ is given in Fig. 6.4. The criterion for the state of the system is the value of $\tau$: as $\tau \to \infty$, $S_H(f) \cong S_0$ - the system is in equilibrium state; at $\tau < \infty$ the graph of the energy spectrum of the system is as in Fig. 6.4, or the system is in a non-equilibrium state. Depending on the value of $\tau$, the energy spectrum of $S_H(f)$ (6.1) will have one or another appearance (Fig. 6.4).

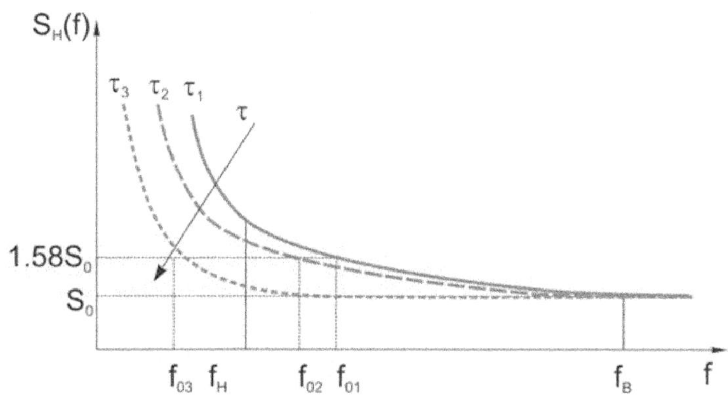


**Figure 6.4.** The energy spectrum of nonequilibrium systems.

Therefrom, it is seen that in the frequency range from $f_H$ to $f_B$, the $S_H(f)$ for the system with $\tau_3$ changes the least, and the $S_H(f)$ for the system with $\tau_1$ changes the largest. So, depending on the state of the system under study (what is the value of $\tau$), measuring its parameters in the mentioned frequency range will occur in quasi-equilibrium (for $\tau_3$) or nonequilibrium (for $\tau_1$, $\tau_2$) states at the same time interval between measurements $\delta t$ ( $f_B = \frac{1}{\delta t}$ ) and with the same number of measurements n ( $f_H = \frac{1}{n \cdot \delta t}$). Therefore, the variance of results under different measurement conditions differs. For example, under the same conditions of measuring the resistance of resistors in one batch, the different value of the variance may not be an error of the experiment, but a consequence of the different state of the tested resistors (the inner structure of some of them is quite different from the others). To verify this statement, electrical resistance measurements of two 1 Ω resistance coils were carried out. Measurements were made under the same conditions (with the same methodological errors and with the same external factors - temperature, humidity, etc.) using a DC bridge according to a four-wire scheme. One coil was conventionally designated Measure A, the other was Measure B. 100 electrical resistance measurements were performed under the same conditions for each coil (Fig. 6.5).

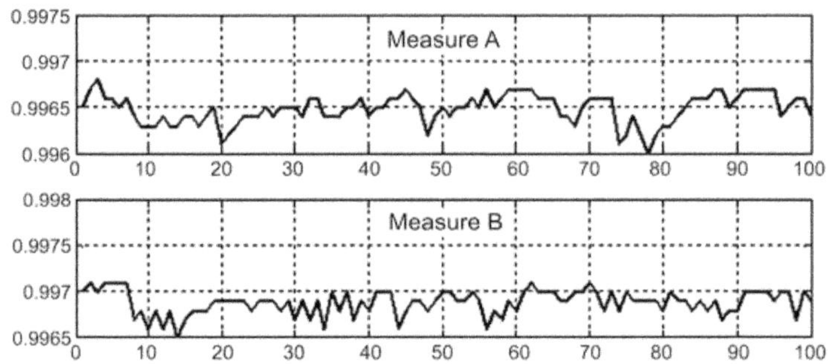


**Figure 6.5.** Resistance measurements of Measure A and Measure B.

As results, we have obtained the mean resistance $\overline{R}$, the standard deviation σ and the correlation coefficient κ between the results of Measure A and Measure B given below in Table 6.1. The correlation coefficient κ was determined to establish the influence of external factors on the measurement results. The small value of κ suggests that the measurement results are independent and that the impact of external factors is minimal.

**Table 6.1.** Results of measurements of Measure A and Measure B.

| | $\overline{R}$, Ω | σ, Ω | $\overline{R}$±3σ, Ω | κ |
|---|---|---|---|---|
| Measure A | 0.9967 | $1.8524 \cdot 10^{-4}$ | $0.9967 \pm 5.5572 \cdot 10^{-4}$ | 0.2271 |
| Measure B | 0.9969 | $1.2978 \cdot 10^{-4}$ | $0.9969 \pm 3.8934 \cdot 10^{-4}$ | |

It can be seen that the difference between the standard deviations σ of Measure A and Measure B is ≈ 40 %. It was assumed that such a difference between the variances is a consequence of the nonequilibrium state of Measure A and Measure B. The EFS of their resistances were determined based on measurements (Fig. 6.6). Since the measurement results were determined after δt = 25 seconds, the sampling rate $f_D$ ($f_D = f_B$) was 40 mHz. Figure 6.6 demonstrates the EFS plot within the frequency range from zero to $f_D/2$. Here, the EFS of Measure A and

Measure B raise while decreasing frequency and is similar to the EFS of flicker-noise for which $S_{FN} \sim f^{-1}$.

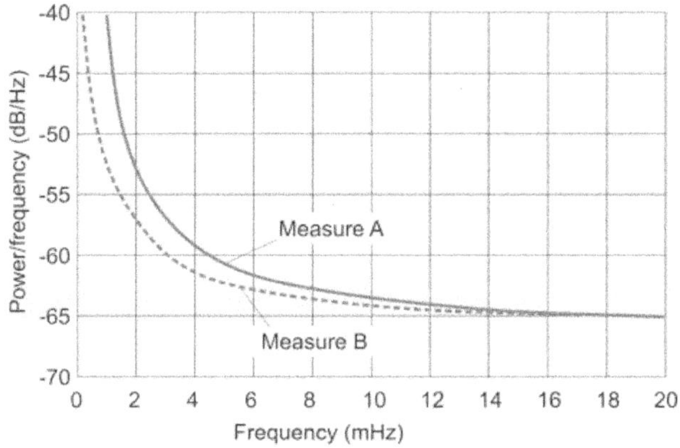


**Figure 6.6.** EFS of Measure A and Measure B.

For determined EFS values, the lower frequency $f_H'$ ( $f_H' < f_H = \frac{1}{n \cdot \delta t}$ ) of the range from $f_H'$ to $f_B$, in which Measure A and Measure B are in the quasi-equilibrium state: $f_H' = 10 \cdot 10^{-3}$ Hz; $f_B = \frac{1}{\delta t} = 40 \cdot 10^{-3}$ Hz. That's, at $f_H' = 10 \cdot 10^{-3}$ Hz and value of the time interval between the readouts equal to $\delta t = 25$ s, the number of results $n'$ for which measurements made at the quasi-equilibrium state of the object under study and, accordingly, the difference between the variances is minimal, the condition of equality of measurements will be satisfied. This number is determined as $n' = 1/f_H' \cdot \delta t = 4$. Such a conclusion was verified by determining the variances for any $n' = 4$ consecutive measured resistance of Measure A and Measure B took at the interval $\delta t = 25$ s: mean of standard deviations of Resistor A $\overline{\sigma_A} = 1.0210 \cdot 10^{-4}$ Ω, and of Resistor B $\overline{\sigma_B} = 1.0087 \cdot 10^{-4}$ Ω ($\left| \frac{\overline{\sigma_A}}{\overline{\sigma_B}} \cdot 100\% - 100\% \right| \approx 1$ % ). For comparison, the difference between the standard deviations of the resistance of Measure A and Measure B coils is determined for a larger number of measurements ($n'' = 25$) taken after $\delta t = 25$ s: $\left| \frac{\overline{\sigma_{A1}}}{\overline{\sigma_{B1}}} \cdot 100\% - 100\% \right| \approx 8$ %. That is, for a higher quantity of measurements at the same time interval $\delta t = 25$ s, the difference between the variances increases since the measuring is carried out for the nonequilibrium state: $f_H'' = 1/n'' \cdot \delta t = 1.6 \cdot 10^{-3}$ Hz (see Fig. 6.15).

So, in equilibrium measurements of the parameters of the same-type systems being in the quasi-equilibrium state, the difference between the variances is minimal, and for the systems in the nonequilibrium state, the difference between the variances rises substantially reaching the maximum.

To check the influence of the internal structure of the studied systems on their energy spectrum and, accordingly, on the nonequilibrium or quasi-equilibrium state, EFS resistances of metal film resistor (Resistor Ca) and the same resistor with damaged structure (Resistor Cb) were studied (Fig. 6.7).

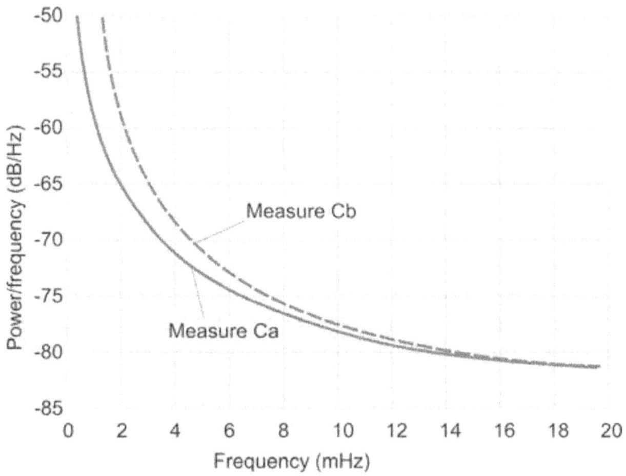


**Figure 6.7.** Energy spectra of resistance fluctuations of Measure Ca and Measure Cb.

The structure was damaged by slight mechanic cutting of the upper conductive layer of the resistor. The average resistance value of the Resistor Ca $\overline{R}_{Ca}$ was 1.7545 Ω, the average resistance value of the Resistor Cb $\overline{R}_{Cb}$ was 1.8605 Ω ( $\left|\frac{\overline{R}_{Ca}}{\overline{R}_{Cb}} \cdot 100\% - 100\%\right|$ ≈6 %). Their variances were counted and accordingly were equal to $\sigma_{Ca}$=3.8895·10$^{-5}$ Ω and $\sigma_{Cb}$ = 5.0574·10$^{-5}$ Ω ($\left|\frac{\sigma_{Ca}}{\sigma_{Cb}} \cdot 100\% - 100\%\right|$ ≈ 30 %).

EFS of the resistance Measures Ca and Cb in Fig. 6.8 confirm the effect of structural defects of the metal-film resistor on its energy spectrum. As noted above, the measurement conditions (number of measurements n and time interval δt between individual measurements) determine the frequency range $f_H$ and $f_B$ (Fig. 6.4), within which the EFS either changes (nonequilibrium state) or is constant (equilibrium state). Therefore, depending on the selected n and δt, the performed measurements are fulfilled on the object under study in equilibrium or non-equilibrium state.

Therefore, the significant difference between the variances of the results under the same measurement conditions (see higher) can be a consequence of the different degree of disequilibrium of the considered systems (different τ - Fig. 6.4). In general, the possible reasons for the differences between the variances of the equivalent measurements received results for the same-type systems may be the non-identity of their inner structure. The latter directly affects the energy spectrum.

The results of this work can be used in the processing of statistical equilibrium measurements, as well as the detection of the extraneous factors impact. If the conditions of measurements - the number of measurements n and the time interval between measurements δt - are such that satisfy the condition of the quasi-equilibrium state of the system under study, then the influence of external factors is the reason for the difference invariances.

## 6.2. Dependence of Results Dispersion on Duration of Measurement Including Intervals Between Them

Precise statistical measurements of the physical quantity X (t) demands the presentation of the received results in the form of temporal series $x_1$, $x_2$, ... $x_N$. The graphs of such series are similar to values that fluctuated around the mean of the measured value. When processing such results, the average value $\overline{X}$ and variance $D_x$ of the measured value X(t) are first and foremost determined. This raises the question: how long should be the temporal series of the measurement study, or what should be the total time of measurement T, or more exactly what should be the number N of measurement results performed at equal intervals Δt during the total duration of the experiment) to minimize the variance.

If the measurement results are independent, then the variance should decrease with an increasing number of measurements N as $D_x = \frac{\sigma_x^2}{N}$, where $\sigma_x$ is the standard deviation of the measurement results. That is if the measurement results are independent, a random measurement error characterized by a variance $D_x$ or standard deviation $\sigma_{cp} = \sqrt{D_x} = \frac{\sigma_x}{\sqrt{N}}$, can be reduced to any small value by increasing the number of measurements N. However, an increase in N is associated with the duration of the measurements, and the condition of independence of the measurement results does not uniquely account for the change in the measurement value itself over time. At large measurement intervals, the real physical quantities alter over time, and thus the statistical stability of the average measurement value is broken. So, it seems quite interesting to study the dependence of the variance of the electrical resistance on the duration of multiple measurements (more exactly, on the number of measurements N fulfilled at the same intervals $\Delta t$ within the total duration of measurement T).

### 6.2.1. Impact of the System State on the Errors of Measuring Its Parameters

Until recently, real statistics have been considered to be inherent in the convergence properties. Therefore, known probabilistic models that assume statistics convergence may be used at large measurement intervals. However, numerous studies of the phenomenon of statistical stability have proved that the possibilities of classical probabilistic models are limited [3]. At the limited number N of measurements, the mean value, variance, and standard deviation tend to stabilize. However, with increasing N, the trend of statistics convergence to fixed values is not confirmed.

Violation of the statistical stability of the mean and the increase in variance may be due to the nonequilibrium state of the system. Otherwise, it can be concluded unambiguously that all real systems are in a nonequilibrium state. However, when measuring, the effect of the nonequilibrium state on the measurement results is manifested with a comparability of the measurement time T and the relaxation time $\tau$ of the system under study: $T \approx \tau$. In other cases, $T \gg \tau$ and $\ll \tau$, the effect of the nonequilibrium state of this system on the measurement results is lesser. One way to find out whether the considered system is in equilibrium or non-equilibrium state is to analyze the fluctuations of its statistical results of measuring the system's parameters, for example, the temporal series.

From [1], the energy spectrum of the fluctuations S(f) of the equilibrium system is immutable in the range from 0 Hz to the ultrahigh frequencies:

S(f) = const. Therefore, if the energy spectrum of the fluctuations is changeable throughout the frequency range, the system is not in equilibrium. In particular, the presence of a flicker component in the energy spectrum of fluctuations evidence of system imbalance. In [4], an expression for the dispersion of measurement results is given if the energy spectrum of the S(f) fluctuations of the measurement results is in the form of "white" noise $S_x(2\pi f) = S_0 = \text{const}$:

$$D_x = \frac{2}{N^2}\int_0^\infty \left[\frac{\sin(\pi f \Delta t N)}{\sin(\pi f \Delta t)}\right]^2 \cdot S(f) \cdot df \approx$$
$$\approx 2S_0 \int_0^\infty \left[\frac{\sin(\pi f \Delta t N)}{N\pi f \Delta t}\right]^2 \cdot df = \frac{S_0 f_\infty}{N} = \frac{\sigma_x^2}{N} \quad (6.2)$$

That is, if the system under study is in equilibrium, then the random error can be reduced $\sqrt{N}$ times by averaging N measurement results. An expression is provided for the dispersion of measurement results if their EFS contains a flicker-component which occurs when the considered system is nonequilibrium [4]:

$$D_x = \frac{2}{N^2}\int_{f_H}^\infty \left[\frac{\sin(\pi f \Delta t N)}{\sin(\pi f \Delta t)}\right]^2 \cdot aS_0 \cdot \frac{1}{f} df \approx aS_0 \frac{1}{x^2}(1 - \cos^2(x) +$$
$$2x\sin(x)\cos(x) - 2x^2 \mathrm{Ci}(2x)) \quad (6.3)$$

Here $f_B$ is the lower cutoff frequency of flicker component of fluctuations; $= \pi f_H \Delta t N$ ; $\mathrm{Ci}(x) = \gamma + \ln(x) + \int_0^x \frac{\cos(\phi)-1}{\phi} d\phi$ is an integral cosine; $\gamma = 0.577\ldots$ is Euler's constant. From (6.3), if the system under study is nonequilibrium, then for the number of measurement results $N^I$, after averaging of which the random error is the same as after accessing based on (6.2). This value is more significant compared with the same for an equilibrium system: $N^I > N$. In the work [5] is given an equation for the measurement time $T_{measmax}$ necessary to fix a nonequilibrium system in one of its possible states $M_{max}$:

$$T_{measmax} = \frac{-\tau}{\ln(1-\exp(-\ln M_{max}))}. \quad (6.4)$$

Here $\tau$ is the relaxation time of the system under study. The fixation of one state of the system means the minimum probability of fluctuations of its parameters and, consequently, the minimum value of random error. From (6.4) follows that $T_{measmax} \approx M_{max}\tau$ . That is, $T_{measmax}$ depends on how much the system under study (its relaxation time is $\tau$) deviated from the equilibrium; to ensure the minimal random error, the measurement time of the nonequilibrium (real) system parameters should be in $M_{max}$ times greater than τ. If the interval between individual measurements is Δt, the number of $N_H$ measurement results, for which the random error of measurements for the nonequilibrium system is minimal, is defined as: $N_H = \frac{T_{measmax}}{\Delta t \frac{\tau}{\Delta t_{max}}}$.

### 6.2.2. Study of the Standard Deviation of the Measurements of Electrical Resistance on the Number of Measurements

Three series of measurements of the electrical resistance of metal oxide film resistors were carried with the help of BR2820 LCR Meter. The time between the measurement series was four days. The time interval between individual readouts was Δt = 4 s. The average value of the measured resistance $\overline{R}$ and standard deviation $\sigma_R$, as well as its dependence $\sigma_R(N)$ on the number of measurements, were determined. Fig. 6.8 demonstrates the statistical results of three series of measurements in the form of graphs a, c, e, and the corresponding dependencies $\sigma_R(N)$ showed in Fig. 6.8 b, d, f.

The average values of the electrical resistance of the three measurement series were respectively: 1.0385, 1.0382, and 1.0382 Ω. Analysis of the dependences of the standard deviation on the number of readings $\sigma_R(N)$ (Fig. 6.8 b, d, f) envisages an increase in $\sigma_R$ with N. It indicates that the measurement results are not independent since their fluctuations are not the "white" noise. To check the influence of MI on the measurement results, the fluctuations of readouts (Fig. 6.9, a; c), as well as the standard deviations (Fig. 6.9, b; d), were fixed at short-circuiting of the device input after the 2nd and 3rd series of resistance measurements.

Analyzing obtained results demonstrates the following. The average value of the MI readouts is: $\overline{R}$ = 3.9573 mΩ (Fig. 6.9a), $\overline{R}$ = 4.0008 mΩ (Fig. 6.9c). Dependence of the standard deviation of the fluctuations on the number of readings $\sigma_R(N)$ in the 1st case (Fig. 6.9b) shows a slight increase with N, and in the 2nd case, stabilization is observed. That is, the MI impact on the variance of the measurement results is negligible. Thus, studies confirm the conclusions of [3] that with increasing number N and duration T of measurement, the dispersion of obtained results ceases to decrease.

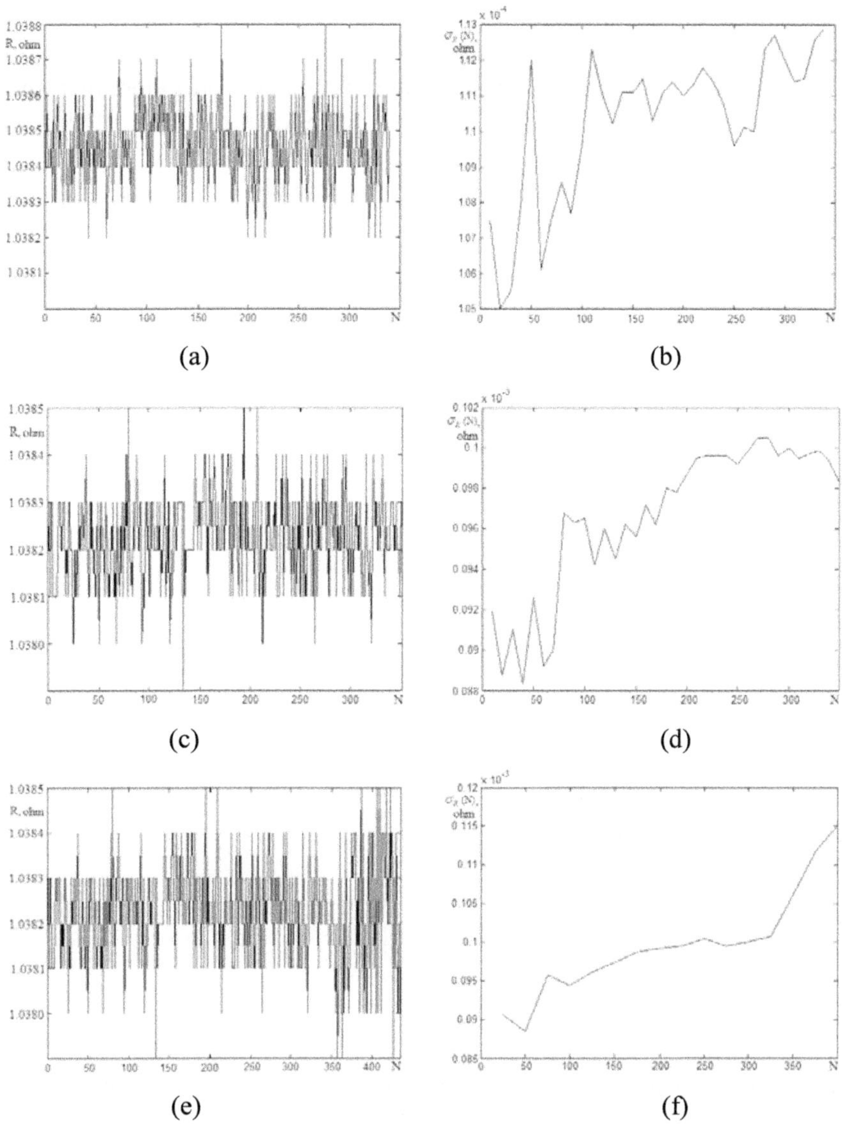


**Figure 6.8.** Results of measurements of the electrical resistance R and the dependence of the standard deviation of the measurement results on the number of measurements $\sigma_R(N)$.

Fig. 6.10 shows the EFS of the measured resistance. This spectrum is not the same throughout the frequency band and is similar to flicker noise as is increasing while decreasing the frequency. The investigated resistor is not in thermodynamic equilibrium, and therefore the variance $\sigma_R$ does not change as provided by formula (6.2). Figure 6.11 envisages the dependence of the normalized standard deviation of the results $\sigma_m/\sigma_m(N = 1)$ on the number of measurements, provided that the energy spectrum of the measurement results S(f) in the form of flicker noise FN (2). For comparison, the same dependence is provided that S(f) for the “white” noise WN (equilibrium system) (6.2) and the combined “white” noise and flicker noise WN + FN.

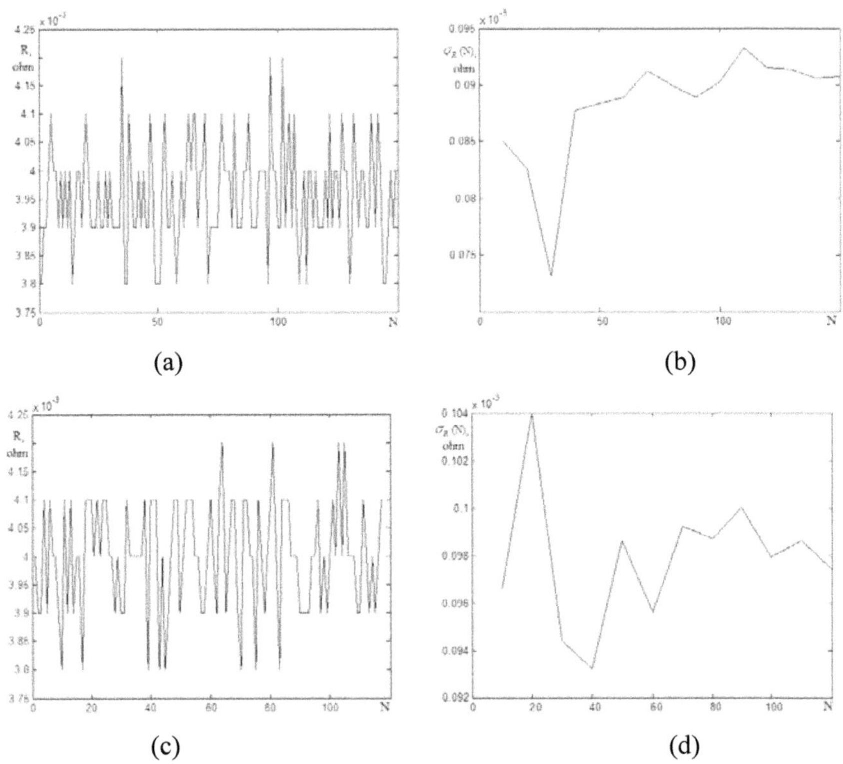


**Figure 6.9.** Fluctuations of readouts (a; c) as well as the standard deviations $\sigma_R(N)$ (b; d) of BR2820 type of MI.

It can be seen from Fig. 6.11 that the averaging of 100 measurements provides reducing the standard deviation of the measured value by an order of magnitude if the EFS is in the form of “white” noise; reducing the standard deviation by 1.4 times if the EFS is similar to flicker noise, and ≈ 1.7 times if the EFS looks like “white” noise and flicker component noise together.

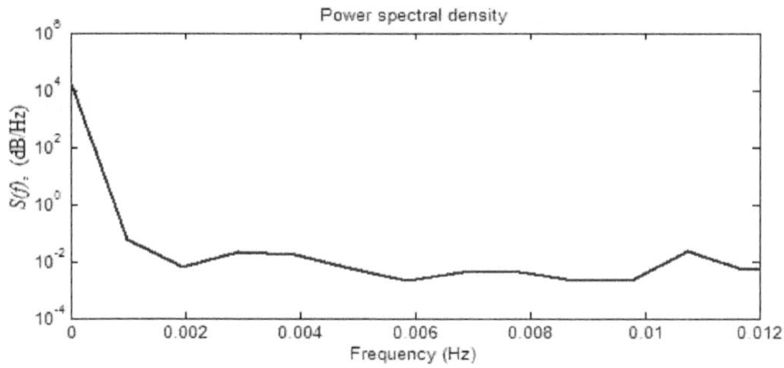


**Figure 6.10.** EFS of measured resistance.

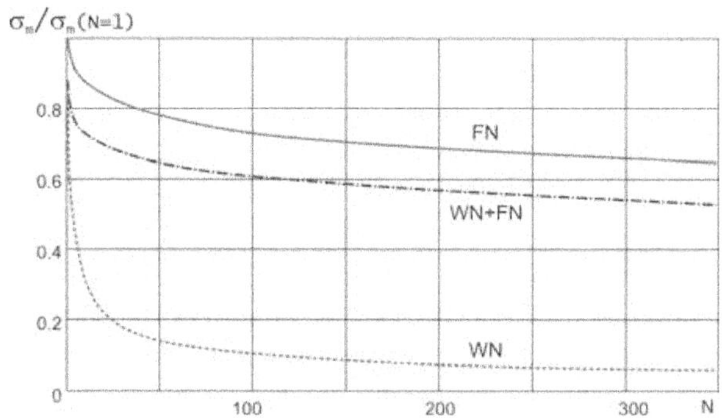


**Figure 6.11.** The dependence of the normalized standard deviation of results on the number of measurements (WN is the "white" noise, FN is flicker noise, WN + FN is combined "white" noise and flicker noise).

## 6.3. Operational Monitoring of Metrological Characteristics of Measuring Instruments

Requirements for Measurement Processes and Measuring Equipment of Measurement Management Systems are set by [6] and for the Competence of Testing and Calibration Laboratories by [7]. They are continuously developing especially due to the permanent implementation of dispersed energy-generating and energy-consumption objects. Such objects are conjugated in their cooperation. Therefore, ensuring the measurement efficiency of dispersed measuring subsystems is the nowadays challenge [8].

Depending on the place of application in the measuring process, methods of improving the accuracy of measurements can be divided into two groups. They are the methods of improving the MI accuracy and the methods of enhancing the accuracy of measurement results [10-11]. The fundamental difference between them is as follows. The object of the 1st group is the stabilization of the MI transform function. The object of the 2nd group of methods is the stabilization of the MI output signal. Hence two types of impact on the MI transform function: the 1st group of methods involves the internal influence on the function, that is, the correction; the 2nd group involves the impact on the MI readouts by the introduction of an amendment [9].

The correction is implemented by various methods. Namely, by the design and technological methods that provide better protection against interferences and the application of the advanced element base. Structural-algorithmic methods consist of the structural stabilization of the MI transform function and the inner processing of measuring information. The latter is realized by various metrological tests of MIs aimed at establishing the real metrological properties of a particular MI. Basing on the obtained results of these tests, amendments may be introduced into the results of measurements. The first way is directed at improving the MI internal perfection, and the result of this way is to improve the quality of the MI. The second way is aimed at ensuring external perfection, and its result is the improvement of the MI operation quality.

To improve the quality, determine, and eliminate the systematic error, the most effective way is to increase the number of MI calibration for mutual checking with a more accurate MI or calibrator. The analysis of the influence of the calibration frequency on the MI metrological reliability showed that the most effective are two ways of checking implementation: in each cycle of measurements, when it needs the built-in reliance element or periodically with help of an external measure.

Besides it, the establishment of optimal interverification or intercalibration intervals is recommended. Moreover, the generally accepted approach is based on the principle of assigning one intercalibration interval for all MIs of the same type by solving the problem of its definition based on cost minimization. The latter consists of the cost of MI checking and forced downtime (encouraging to increase the interval); and the cost of receiving and applying the unreliable measurement results (encouraging to reduce the interval). Often the assignment of intercalibration intervals by comparison with the intercalibration intervals of similar MIs of other manufacturers leads to the establishment of unbelievably long intercalibration intervals for MIs. The noted shortcomings due to the uncertainty of the initial, significantly limit the scope of a fairly developed, albeit extremely complex, the theory of metrological reliability both in the appointment of the primary intercalibration interval and in its adjustment during operation.

To predict the MI metrological serviceability at the stage of operation, it is advisable to use metrological inspection protocols. The information that can be obtained from them is differs depending on the verification methods. There exist three ways to verify:

- The calibration, that is the establishment of actual values of physical quantities assigned to the MI at points in the range of measurements being tested;
- Determination of suitability for use according to stability standards, followed by MI calibration recognized as fit,
- Determination of suitability for use with the rejection of those MIs, whose errors exceed the permissible limit.

Metrological stability is indirectly checked only in the 2$^{nd}$ method of verification since the MI suitability for application is established primarily by comparing its instability obtained from the inspection with the accepted norms, and only then the unit size of the considered physical quantity is performed from the measure. However, this kind of verification applies only to working standards.

The 1$^{st}$ method of verification is also used mainly for working standards, and its implementation is possible in two ways. In the first case, the actual values of the physical quantity assigned to MI are set (f.i. for unambiguous measure - a resistance coil, and so on). This makes it possible to obtain the dependence of the actual values of the physical quantity assigned to MI on the time basing on the results of several inspections, and, consequently, to assess the MI instability and the correctness of the intercalibration interval assignment. It is impossible to assess MI instability if the test only transferring the unit size of the physical quantity from the standard (f.i., in the case of weight transfer from mass measures to the balance, or from the unambiguous voltage measure to the calibrator) without establishing the actual value of the physical quantity at the moment of inspection. Then, it becomes impossible to obtain reliable information about the current level of metrological serviceability of a particular MI.

The 3$^{rd}$ method of verification controls the alternative feature (suitable - not suitable) without quantifying. Of course, that reduces the probability of metrological serviceability defined for the set of MIs on the results of its assessment.

Therefore, the optimal way to improve the quality of industrial measurements is to develop methods and algorithms for operational control of MIs metrological characteristics. The implementation of operational control of metrological characteristics dramatically reduces the share of metrologically defective MIs and significantly increases the reliability of measurements.

Requirements for the methods of operational control of MIs' performance can be formulated as follows: efficiency, that is, the ability to control the metrological characteristics at any moment; objectivity, as the assessment of MI metrological characteristic to correspond the established level of probability; integration, that is, insurance of the compatibility of all components of the operating system to control in the particular operation process; reliability, as an ability of the operating system to function for an appropriate time without unacceptable loss of its metrological properties.

Summarizing the methods of operational control of metrological characteristics of the MIs and methods of their implementation, can be classified as it is shown in Table 6.2.

**Table 6.2.** Classification of methods of operational monitoring of MI metrological characteristics.

| Feature | Method of operational monitoring |
|---|---|
| By the method of control | Auxiliary measurement method |
| | Method of reference signals |
| By control time | Continuous monitoring |
| | Periodic monitoring |
| By influence on the technological process | With technological interruption for MI extraction aiming the checking and monitoring |
| | Without technological interruption for MI extraction aiming the checking and disassembly monitoring |

Studies envisage that for the implementation of operational control, it is advisable to apply the following methods of MI monitoring of errors (Fig. 6.12).

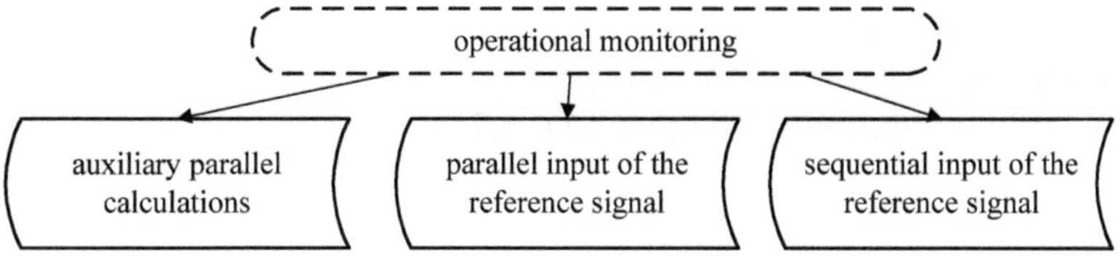


**Figure 6.12.** Implementation of methods of MI operational monitoring.

Let us consider in more detail the specifics of the implementation of methods of operational error monitoring in industrial MIs as components of the local information- measurement systems. Non-dismantling operational monitoring of errors of industrial MI is expedient to carry out by the following methods: method of verification of homogeneous MIs using a common signal source; method of circular comparisons in groups of homogeneous MIs; method of mutual verification regarding the stable MI; method of MI verification using an industrial calibrator.

The method of MIs' verifications using a common signal source can be performed as follows. In the case of compact placement of homogeneous MIs, for example on the control panel, it is necessary to implement a scheme of periodic disconnection of their inputs from the sources of measuring signals and parallel supply to all inputs simultaneously the stable signal (see Fig. 6.13). As a result, we obtain some readouts of this set of homogeneous MIs - {X1, X2,..., Xm}. Using the known methods of mathematical statistics, you can find the mean of the readings of the set, that is the $\overline{X}$ which is taken as a reference value:

$$\overline{X} = \frac{1}{m}\sum_{i=1}^{m} X_i \tag{6.5}$$

Then, determining the errors of MIs - {Δ1, Δ2,...,Δm} and comparing their values with the allowable limits for each one, we can select from the set those of them for which the errors exceeded the mentioned limits. As a result, they must be removed from the technological process and be calibrated.

Thus, the verification by using a common signal source should be used: in the case of compact displacement of homogeneous MIs; if short-term disconnections of the MI from the controlled parameter circuit are permitted while operating. If the technological process is organized in such a way that there are periodic breakouts, and it is impractical to carry out preliminary verification, it is possible to make them circular in the groups of homogeneous MIs (see Fig. 6.14).

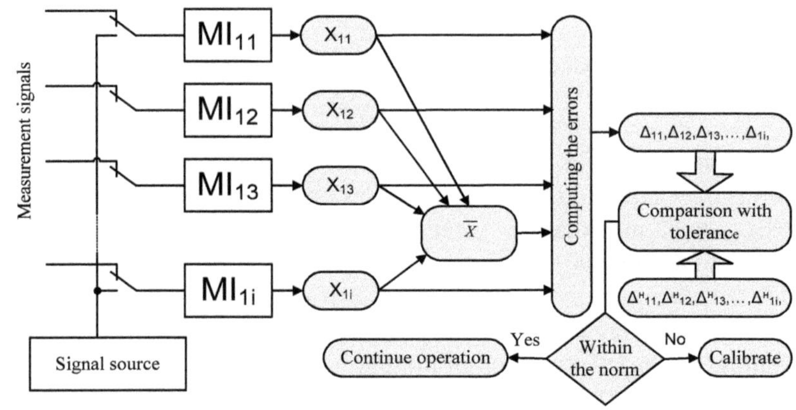


**Figure 6.13.** Method of verification of homogeneous MIs using a common signal source.

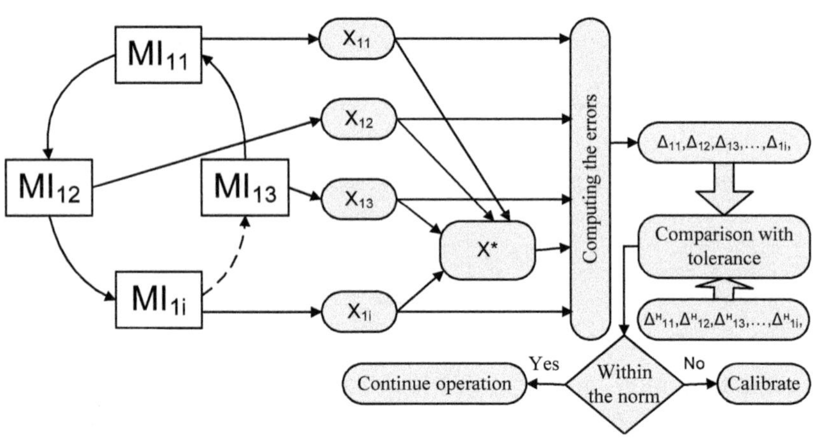


**Figure 6.14.** The method of circular verifications in groups of homogeneous MIs with the definition of the most plausible assessment of the result.

For this option, the reference value is the most plausible estimation of the result of the verification:

$$X^* = \arg\max\left\{p(X_1,...X_j / X)\right\}, \quad (6.6)$$

where X is the actual value of the unit of the physical quantity measured during the verification. Further processing of the results is carried out as it was done for the previous verification.

Therefore, the method of circular verifications in groups of homogeneous MIs should be applied: while applying such groups is the technological process; when the breaks are allowed for the particular TP long enough to disconnect the MIs; when there are no possibilities to use traditional methods of MIs' verification.

When the company exploits several groups of homogeneous MIs or homogeneous MIs are located at different points of the TP, it is advisable to use the method of mutual verifications that is based on the stable MI. The aforesaid method is shown below (Fig. 6.15).

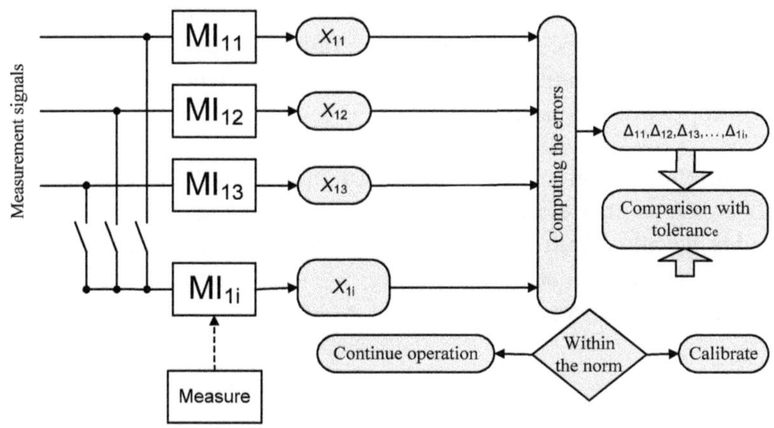


**Figure 6.15.** Method of mutual verification concerning stable MI.

It is realized as follows. From each group of homogeneous MIs {MI11, MI12... MI1i}, basing on the results of their previous operation and calibrations, is necessary to choose the MI with the lowest error and the highest stability. To ensure the traceability of measurements within the local information-measuring network with units of physical quantities reproduced by state standards, it is sufficient to calibrate the selected most stable MI1i. Then the procedure of mutual verification consists of the organization of parallel measurements by stable MI1i and each MI from this set of homogeneous MIs.

The possibility of operational monitoring over the readouts of the stable MI is confirmed by [6], proving that a reliable assessment of the metrological serviceability of MIs can be fulfilled by ensuring the ratio between the error of verified and standard MIs as 1: 1.05. So, the method of mutual verification based on the readouts of the stable MI should be applied: in case of technological application of several groups of homogeneous MIs; if technological breaks are allowed in the operation, long enough to disconnect the MI; if it is impossible to apply the calibrators.

The most effective method of operational monitoring the MI's errors is the method based on the application of an industrial calibrator. As the latter may be an industrial calibrator or the MI that is used in metrological confirmation processes for the production chain. Requirements for industrial calibration can be formulated as follows: reproduction of sample signals required for calibration of all MIs used in the technological process; ensuring the required accuracy of transfer of units of physical quantities in this process; invariance to the impact of the MI application conditions in it.

The implementation of the method of operational monitoring of MI errors by an industrial calibrator is presented in Fig. 6.16. This scheme provides the maximum degree of uniformity of measurements in local networks of information-measuring systems, and hence the high probability of their results.

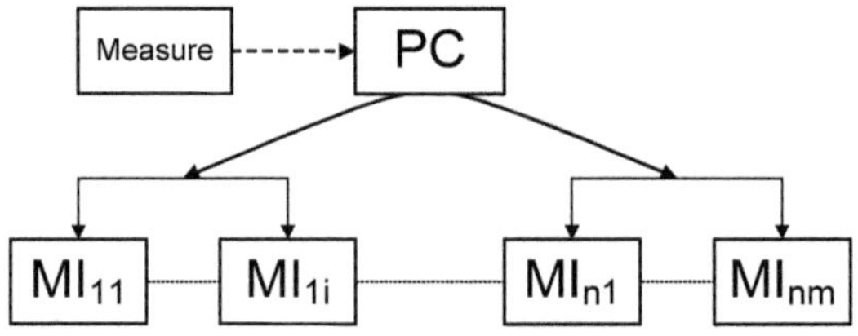


**Figure 6.16.** Method of monitoring with the help of an industrial calibrator.

It should be noted that the MIs used for the verification as it is shown in Fig. 6.15, are the working MIs, however, their high individual metrological stability allows applying them periodically to enhance the reliability of measurements of the whole set of MIs which are yielded as a part of the local network.

The considered methods of operational monitoring of the metrological characteristics of industrial MIs increase the accuracy of reproduction, storage, and transfer of several units of physical quantities in terms of autonomous metrological maintenance of MIs as well as reduce costs in their operation in the industry. Their application on a place of operation allows increasing the probability of monitoring the parameters of technological processes and indicators of final product quality.

### 6.3.1. Operational Monitoring of the Errors of Industrial Measuring Instruments by the Auxiliary Parallel Measurements

The method of auxiliary parallel measurements is often used in the implementation of classical metrological inspections (reference device method). To implement it, the MIch, which is checked, the reference MIref and the source of the input signal SS are needed (Fig. 6.17). The main advantage of the considered method consists in the absence of the need to break the measuring circuit. The disadvantage is the possibility of distortion of the input signal, which can reduce the accuracy.

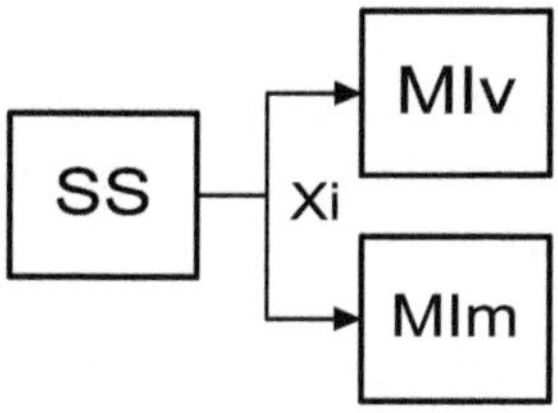


**Figure 6.17.** The method of auxiliary parallel measurements.

Since industrial MIs measure the output signals of primary transducers and unified DC voltage and current signals, we consider below the features of the implementation of the method of auxiliary parallel measurements to perform the mentioned measurements. Fig. 6.18 presents the implementation of this method for MIs, which measure the voltage and output signals of thermoelectric transducers. Here an algorithm of operational monitoring consists of simultaneous measurements of RPS and MI and definition with the help of microcontroller MCU of the value of MI error. The main requirement for RPS is the absence of influence of the U/N meter on the measuring circuit, in particular, due to its input currents.

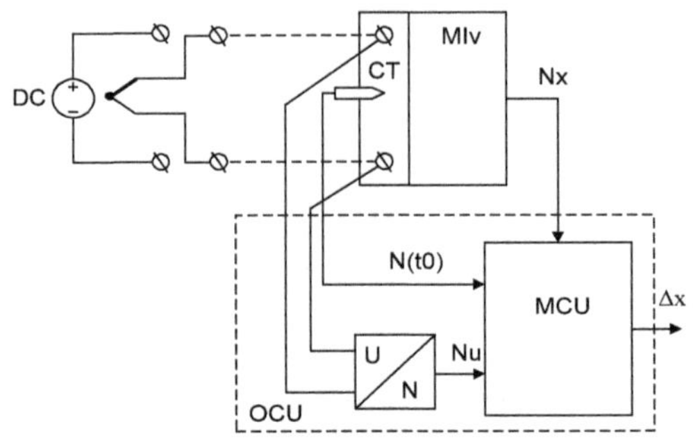


**Figure 6.18.** The method of auxiliary parallel measurements for MIs which measure the voltage and output signals of thermocouples.

Providing the condition of neglecting the error of the means of operational monitoring, the error of RPS is determined by the dependence:

$$\Delta_x = U(N_X) - U(N_U), \quad (6.7)$$

where $U(N_X)$, $\Delta_X$ are the voltage measured by RPS and its error accordingly; $U(N_U)$ is the voltage measured utilizing operational monitoring. Fig. 6.19 presents the implementation of the method of auxiliary parallel measurements for MIs that measure the output signals of resistance thermometers.

The realization of RPS for this scheme is more complex than the previous one. This is due to the need to measure another parameter, that is the current. Then, it needs to break the circuit of the measuring current from the MI while monitoring and pass it through the resistance reference measure $R_0$ which is the RPS's unit.

The algorithm of operative fulfillment for the scheme of Fig. 6.19 consists of simultaneous measurement of resistance of the thermometer by the checked MI and measurement utilizing operational monitoring of voltage on Rx and the current passing through the same resistor. The next step is to calculate the value of the resistance of the measured RPS $R(N_U, N_I)$ and defining the MI error $\Delta_X$by the formula:

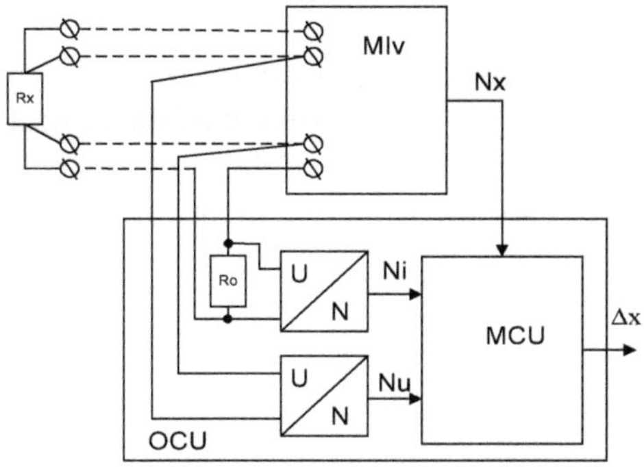


**Figure 6.19.** The method of auxiliary parallel measurements for MIs which measure the output signals of resistance thermometers.

$$\Delta_X = R(N_X) - R(N_U, N_I) \qquad (6.8)$$

Here $R(N_X)$, $\Delta_X$are the resistance measured by RPS and its error; $R(N_U, N_I)$ is the resistance value measured by the mean of monitoring. A certain inconvenience while using the considered method is the need to break the circuit of the current measurement to switch up a measure of resistance $R_0$. However, if frequent monitoring is required, this problem is simply solved by the permanent switching of R0. in the measuring circuit of MI.

Fig. 6.20 demonstrates the realization of the method of auxiliary parallel measurements for MIs by measuring the output signals of rheochord transducers and strain gauges.

The peculiarity of strain gage bridges is that to ensure thermal stability it is included in the measuring circuit with the help of a six-wire line, and the circuit is supplied with a persistent voltage. Most often, the excitation of such converters is carried out in a given voltage mode and the strain gauge is connected to the secondary MI using a six-wire line, two conductors of which are used to supply an excitation voltage to the bridge, two others - to transmit voltage to the bridge diagonal and the next two - to transfer the bridge imbalance voltage to the secondary device.

Measuring device $MI_V$ that is verified, measures two voltages, determines their ratio, and performs a functional transformation inverse to the transformation of the primary strain gauge. The algorithm of operational monitoring for the scheme presented in Fig. 6.20 consists of simultaneous measurement of imbalance of a strain gage by the verified $MI_V$ - K (Nx), and measurement utilizing operational monitoring of supply voltage of a strain gage and voltage of imbalance of a strain gage. The next step is computing the value of the conversion factor of the strain gage RPS - $K(N_1, N_2)$ and defining the error of $MI_V$ -$\Delta_X$:

$$\Delta_X = K(N_X) - K(N_1, N_2) \tag{6.9}$$

Here $K(N_X)$, $\Delta_X$ are accordingly the value of imbalance of a strain gage bridge measured by $MI_V$ and its error; $K(N_1, N_2)$ is the value of imbalance of a strain gauge bridge measured by the mean for monitoring.

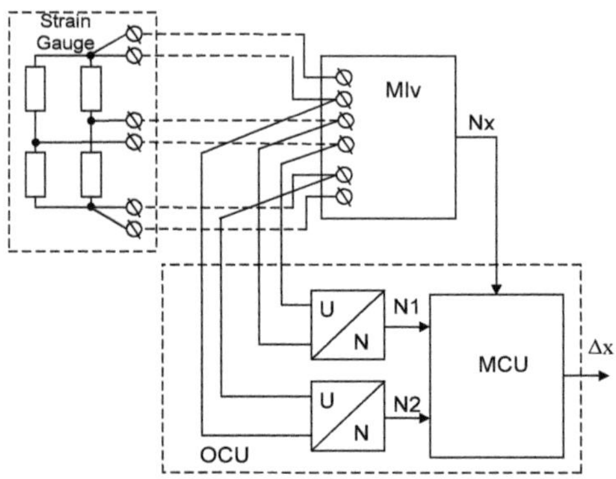


**Figure 6.20.** Method of auxiliary parallel measurements for MIs that are measuring the output signals of the strain gauge.

The method of auxiliary parallel measurements can be used not only for the classic strain gage bridge scheme (Fig. 6.20) but also for the half-bridge scheme as well as for measuring the output signals of rheochord transducers. In technological processes often use measuring transducers whose output signal is a direct current. Fig. 6.21 presents the implementation of the method of auxiliary parallel measurements for MIs, which measure the output signals of DC.

The algorithm of operational monitoring for the scheme of Fig. 6.21 consists of simultaneous measurement of current by the monitored $MI_V$, and measurement utilizing operational monitoring of the voltage caused by the current passing through R0. At the next step the MIver error $\Delta_X$ is determined by the formula:

$$\Delta_X = I(N_X) - I(N_I) \qquad (6.10)$$

Here $I(N_X)$, $\Delta_X$ are accordingly the value of current measured by MIVer and its error; $I(N_I)$ is the value of current measured by the mean of operational monitoring. A certain inconvenience while using the considered method is the need to break the circuit of the current measurement to switch up a measure of resistance R0. However, if frequent monitoring is required, this problem is simply solved by the permanent switching of R0. in the measuring circuit of MI.

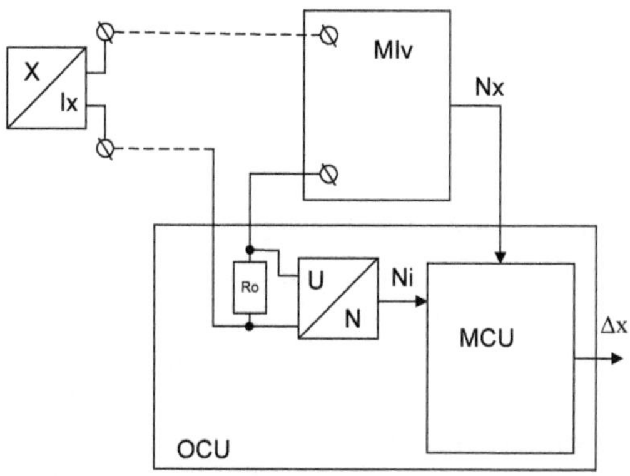


**Figure 6.21.** Method of auxiliary parallel measurements for MIs that are measuring the output signals of the direct current.

Thus, the considered schemes of realization of a method of auxiliary measurements permit to realize the function of control of errors of industrial MIs without their dismantling from an operation point. It increases the efficiency and reliability of metrological compliance processes of industrial MIs as well as reduces the cost of their metrological assurance.

### 6.3.2. Operational Monitoring of the Errors of Industrial Measuring Instruments by Introducing the Reference Signals

The above methods of operational monitoring implemented by auxiliary measurements become an effective tool for controlling the errors of industrial MIs but can be applied episodically due to the possible distortion of the measured signal. Therefore, it is advisable to use the method of input of the reference signal since it more efficiently reproduces the measurement conditions. The latter predicts two options of implementation: parallel and sequential inputs of the reference signal. A block diagram of the implementation of the considered method is demonstrated in Fig. 6.22.

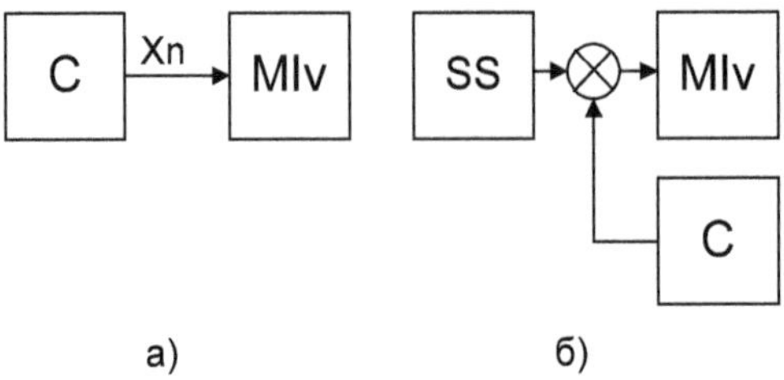


**Figure 6.22.** Method of input of reference signals.

Methods of a parallel input of reference signals seem to be well-known [7] and apply to carry out the procedures of methods of verification and calibration. They are introduced applying to the MI input that is tested, the reference signal from the calibrator of the corresponding physical quantity. However, this method has not yet become widespread for non-dismantling operational monitoring of MI errors due to the lack of calibrators for industry purposes with sufficient appropriate metrological and functional capabilities.

Less studied is the implementation of the method of sequential input of reference signals. Let's consider the features of the last regarding the peculiarities of the source of the measured signal. Fig. 6.23 presents the implementation of the method of sequential input of reference signals for MIs, aiming at the measurements of the voltage and output signals of thermocouples. Its implementation involves the supply of the reference voltage value from the output of the converter N/U in series with the source of the measured input voltage.

Since the algorithm of operation consists of supplying the reference voltage in series with the MI voltage, 2 measurements must be performed to determine the value of the MI error. The 1$^{st}$ one is fulfilled at a reference value of the RPS voltage equal to 0, and the 2$^{nd}$ measurement - at a non-zero voltage, f. e. equal to the first measurement result. The obtained two measurement results are stored in the RPS controller, which has to compute the MI's error by the equation:

$$\Delta_X = \frac{U_X^2 - U_X^1 - U_0(N_0)}{2} \tag{6.11}$$

Here $U_X^2$, $U_X^1$, $\Delta_X$are accordingly the results of the 1st and the 2nd measurements and the defined error; $U_0(N_0)$ is the value of the reference voltage is reproduced utilizing operational monitoring. The existing problem of the implementation of the considered method consists of sequential introducing electrical resistance into the measuring circuit. It is simplified when the resistance calibrator is built based on an active resistance simulator.

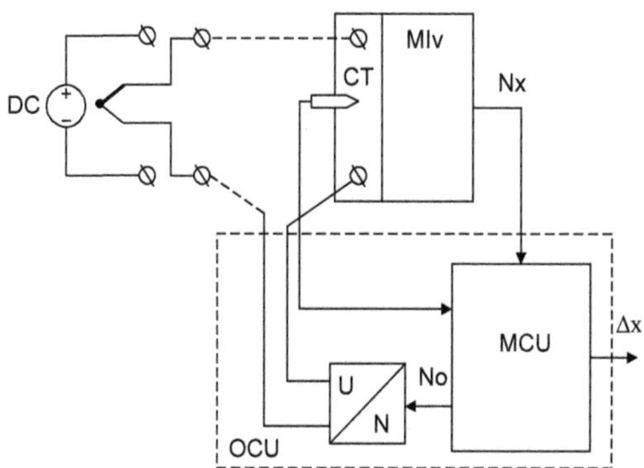


**Figure 6.23.** Implementation of the method of sequential input of reference signals for MIs, measuring voltage, and output signals of thermocouples.

Fig. 6.24 presents the implementation of the method of sequential input of reference signals for MIs, which measure the output signals of the resistance thermometers. Its algorithm of operation consists of supplying from the reference electrical resistance switched sequentially with the measured resistance and simultaneous measuring the sum of voltages on these resistances by the verified MI. To determine its error is necessary to fulfill 2 measurements: the first one at $R_0^1(N_0^1) = 0$ and the second one at $R_0^2(N_0^2)$ other than 0.

The results of measurements of MI - $R_X^1(N_X^1)$, $R_X^2(N_X^2)$ are sent to the controller RPS where MI error $\Delta_X$ is computed by the formula:

$$\Delta_X = \frac{R_X^2(N_X^2) - R_X^1(N_X^1) - R_0^2(N_0^2)}{2} \tag{6.12}$$

Here $R_X^1(N_X^1)$, $R_X^2(N_X^2)$ are accordingly the 1st and the 2nd results of measurements as also an error of considered MI; $R_0^2(N_0^2)$ is the reference resistance value reproduced utilizing operational monitoring. Fig. 6.25 presents the implementation of the method of input of reference signals for monitoring MI that measures the output signals of strain gauges.

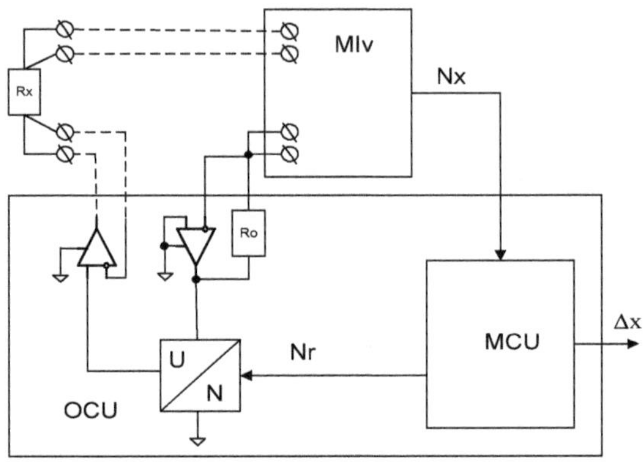


**Figure 6.24.** Implementation of the method of sequential input of reference signals for MIs, that measure the output signals of the resistance thermometers.

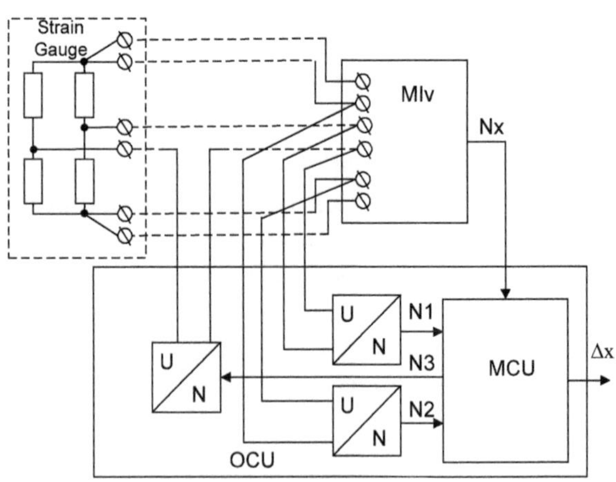


**Figure 6.25.** Method of input of reference signals for monitoring MI that measures the output signals of strain gauges.

The algorithm of operational monitoring for the scheme of Fig. 6.25 consists of supplying the reference value of the voltage sequentially to the voltage from the strain gage and measuring the difference of these voltages by the MI and RPS. Also, it is necessary to realize the function of measuring the RPS voltage supplying the strain gage bridge. To determine the MI error, it is recommended to perform 2 measurements: the 1st one at $U_0^1(N_1^1) = 0$, and the 2nd one at $U_0^2(N_1^2)$. The results of these measurements $U_X^1(N_X^1)$, $U_X^2(N_X^2)$ are fed to the controller in which MI error $\Delta_X$ is defined by:

$$\Delta_X = \frac{U_X^2(N_X^2) - U_X^1(N_X^1) - U_0(N_3)}{2} \tag{6.13}$$

Here $U_X^2(N_X^2)$, $U_X^1(N_X^1)$, $\Delta_X$ are accordingly the 1st and the 2nd results of measurements as also an error of considered MI; $U_0(N_3)$ is the reference voltage value reproduced utilizing operational monitoring. Fig. 6.26 presents the implementation of the method of input of reference signals for MI that measures the DC signals.

The algorithm of operational monitoring for the scheme of Fig. 6.26 consists of supplying the reference value of current $I_0(N_I)$ in parallel with the current from the measuring transducer $I_X$ and sequential measurement of currents by the monitored MI. To determine the MI error, it is necessary to fulfill 2 measurements: the 1st one at $I_0(N_I) = 0$, and the 2nd one at a certain value of $I_0(N_I)$.

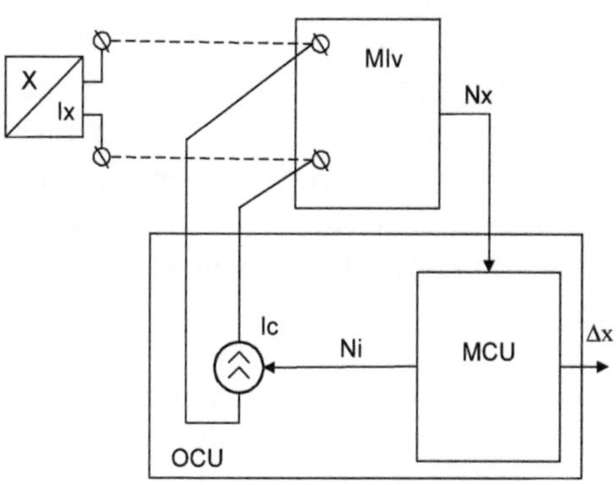


**Figure 6.26.** Method of input of reference signals for MI that measures the DC signals.

The results of 2 measurements $I_X^1(N_X^1)$ and $I_X^2(N_X^2)$ pass to the controller where the MI error is computed:

$$\Delta_X = \frac{I_X^2(N_X^2) - I_X^1(N_X^1) - I_0(N_I)}{2} \tag{6.14}$$

Here $I_X^2(N_X^2)$, $I_X^1(N_X^1)$, $\Delta_X$ are accordingly the 1st and the 2nd results of measurements as also an error of considered MI; $I_0(N_I)$ is the reference current value reproduced utilizing operational monitoring.

As a result of the analysis of the high-mentioned realizations of methods of operative monitoring the errors of industrial MIs, the following generalizations are determined:

1) For effective non-dismantling operational monitoring the MI errors it is necessary to apply the following means: two ADC of the voltage, DACs of the voltage, the current and the electrical resistance, controller and software adapted to the local measurement network;
2) Workability of considered methods is provided due to the rational design of the scheme of re-switching of MI's measuring circuits;
3) In the case when it is necessary to verify the MI metrological serviceability, it is sufficient to determine its error at such point of the measurement range which corresponds to the value of the controlled technological parameter; therefore, it is more appropriate to propose the method of parallel measurements for this goal;
4) For MIs that measure technological parameters in a wide range, it is advisable to apply the method of reference signals;
5) The input circuit of the RPS should be constructed in such a way that does not affect the input circuit of the monitored MI (high ADC input resistance, dynamic characteristics and noise protection have to be much higher regarding the monitored MI);
6) Re-switching of the measuring circuit of the monitored MI should not cause the inadmissible changes in the modes of the technological process.

To select the optimal option for operational monitoring, it is advisable to use the appropriate criterion for effectiveness. The above analysis envisages that the mentioned criterion should take into account the maximum number of impact factors [12]. Therefore, as a criterion, it is expedient to use the maximum estimation of the weighted mean value (taking into account productivity and accuracy) of the share of measurements for the given local measurement network which is carried out by metrologically serviceable MIs. This approach can be represented by the following relationship:

$$E_{ОП} = \frac{\sum_{i=1}^{M} (1\text{-}K_{Mi} + P_{Hi}) \cdot \frac{N_i \cdot П_i}{\Delta_{P_i}^2}}{\sum_{i=1}^{M} \frac{N_i \cdot П_i}{\Delta_{P_i}^2}} \qquad (6.15)$$

Here M is the number of MIs under the monitoring which are included in the local network; $N_i$ is the number of MIs in an i-th group; $П_i$ is the productivity (number of measurement per 1 s.) of MIs of i-th group; $\Delta_{Pi}$ is the permissible error of MIs of i-th group; $K_{Mi}$ is the metrological reserve factor of MIs of i-th group; $P_{Hi}$ is the maximal probability of error of the $2^{nd}$ kind for MIs of i-th group in the implementation of the particular type of operational monitoring.

## 6.4. Metrological Support for Gamma-Ray Measuring Instruments

Independently on the direction of future nuclear technique and energetics, it seems quite important is to study ways to improve the reference base in the field of dosimetric measurements, to increase its accuracy based on the analysis of the state of metrological support for measuring the characteristics of gamma radiation. Despite the intensive development of nuclear measuring equipment, a slow replacement and modernization of the reference equipment for the metrological support of measuring instruments for ionizing radiation are noticed. The main issue consists of the metrological characteristics and accuracy of the standards that are used by metrologists, the conditions, and methods of creating a standard radiation source for means' verification/calibration. Particularly difficult is the reference setting for gamma radiation dosimetry. It needs to ensure the safety in operation with the IRS, especially high-activity gamma radiation sources.

Also, the gamma-ray sensors are commonly used for measuring the fluid levels in the industry. Typically, the $^{60}Co$ or $^{137}Cs$ isotopes as the radiation source are applied. For instance, in the USA, such detectors are beginning to be used as part of the Container Security Initiative [13]. Problems of their metrological assurance are becoming more general. They also require the development of metrological support. So, the conducted researches of calibration, metrological verification, and determination of the sensitivity of gamma-ray dosimeters, based on the joint application of installations equipped with ionizing radiation sources and of X-ray installations may be the time.

According to the series of standards ISO 4037 [14-16], the requirements for the characteristics of standard X-ray and gamma radiation as well as for the methods for their production are set. For the latter, we need to use IRS with radionuclides, which are listed in Table 6.3.

**Table 6.3.** Radionuclides.

| No | Radionuclide | Radiation energy, keV | Half-life, days |
|---|---|---|---|
| 1. | $^{60}Co$ | 1173.3<br>1332.5 | 1925.5 |
| 2. | $^{137}Cs$ | 661.6 | 11050 |
| 3. | $^{241}Am$ | 59.54 | 157788 |

It is important to use radioactive material with sufficient activity per unit mass in the radiation sources and to provide conditions in which the power of the dose rate in the air due to impurities does not exceed 1% of the power in the air. Installations of this type are particularly suitable for the use of IRSs with $^{60}Co$ and $^{137}Cs$ radionuclides since the contribution of scattered radiation is less than 5%. Generally, they have to meet two basic requirements: the contribution of the external scattered radiation to the radiation from the source in the capsule should not be greater than 5% of the total power of dose rate in the air; the latter must be proportional within 5% to the inverse of the square of the distance from the center of the source to the center of the detector.

Non-collimated geometry installations must be located in rooms with a minimum size of 4m × 4m × 3m. The scheme of an example of the installation of the collimated geometry is presented in Fig. 6.27.

The metrological basis for checking or calibrating gamma-ray dosimeters in Ukraine is provided by reference UPGD and UPD-Inter installation (Table 6.4).

They meet the .requirements of [14]. Consider the spectrum (Fig. 6.28) obtained as a result of the CsI gamma-ray detector calibration at the UPGD-2 installation. It is quite similar to the $^{137}Cs$ spectrum from a point source; the effect of scattered radiation is negligible, and the peak/comp ratio is satisfactory.

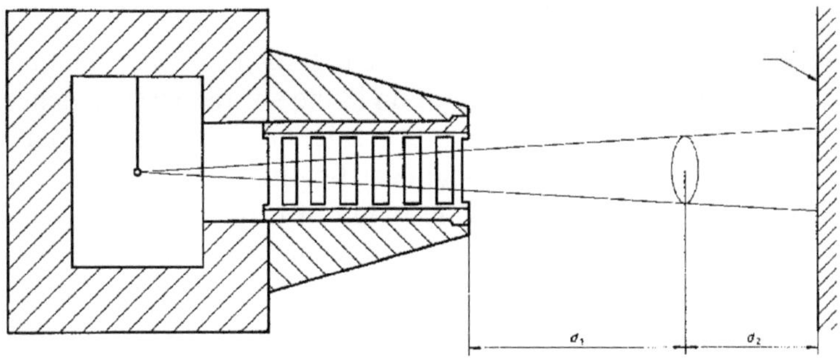


**Figure 6.27.** Installation of the collimated geometry.

**Table 6.4.** The main characteristics of the reference installations UPGD and UPD-Inter.

| No | Main characteristics | UPGD | UPD-Inter |
|---|---|---|---|
| 1. | Source of ionizing radiation | $^{60}Co$, $^{137}Cs$, $^{241}Am$ | $^{137}Cs$ |
| 2. | Measuring range, Gy / h | until $5\times10^{-2}$ | until 3 |
| 3. | Range of working distances, m | 0.5 – 5.0 | 0.2 – 1.5 |

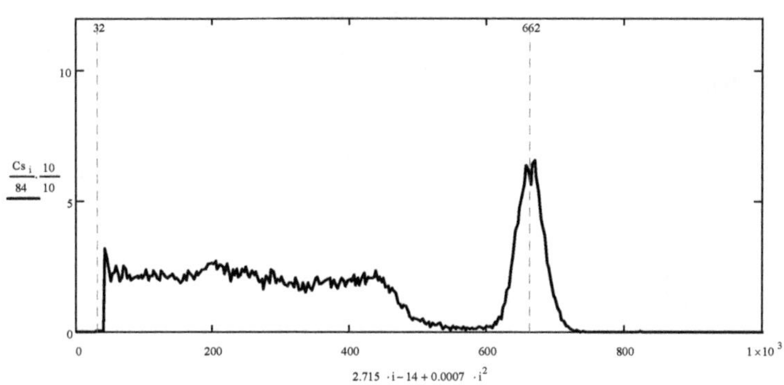


**Figure 6.28.** Spectrum response of the CsI gamma-ray detector from the $^{137}Cs$ IRS in a thin steel capsule (UPGD-2 installation).

While calibrating gamma-ray dosimeters, several (usually 3-4) IRSs of maximum activity of up to 5 Curie are used. This ensures the dosimeters' studying in the power range up to 50 mGy/h (at a distance of 1 m from the source), whereas the measuring range for many devices expands up to 10 Gy/h. Individual dosimeters for current and emergency monitoring used on radiation-hazardous objects also exploit a measurement range of up to 10 Sv (10 Gy); and for calibration in the entire measurement range, they must be irradiated for 160 hours at UPGD installation. The latter (Intertype) provides metrological verification of gamma radiation dosimeters in Ukraine over a wide range of dose rates. Structurally, it consists of an irradiation chamber and a radiation head with attenuators that provide the stepped dose rate change. The radiation head containing one $^{137}Cs$ source is made of tungsten alloy and provides a radiation beam. The dependence of the size of the uniform field of the installation from the distance to the IRS is shown in Table 6.5.

**Table 6.5.** Dependence of the size of the uniform field of the installation from the distance to the IRS.

| **Distance to the source, mm** | 20 | 40 | 1000 | 1200 | 1650 |
|---|---|---|---|---|---|
| **Size of the field, mm** | 85×68 | 155×125 | 370×290 | 435×345 | 600×470 |

Fig. 6.29 demonstrates an apparatus response of the spectrum for a CsI gamma detector from $^{137}Cs$ point source. The spectrum clearly shows $^{137}Cs$ peak with an energy of 662 keV and an X-ray peak with an energy of 32 keV, while the contribution of scattered radiation is insignificant. This spectrum can be taken as a standard.

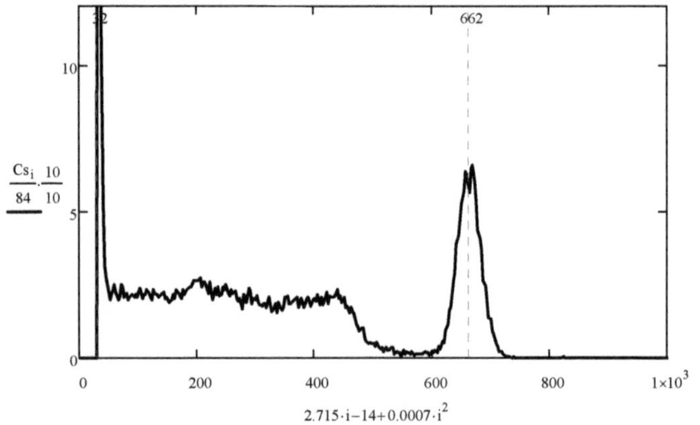


**Figure 6.29.** Apparatus response of the spectrum for CsI gamma detector from $^{137}Cs$ point source.

Figs. 6.30-6.31 show the responses of the spectra for the CsI gamma detector installed in UPD - Inter with $^{137}Cs$ IRS.

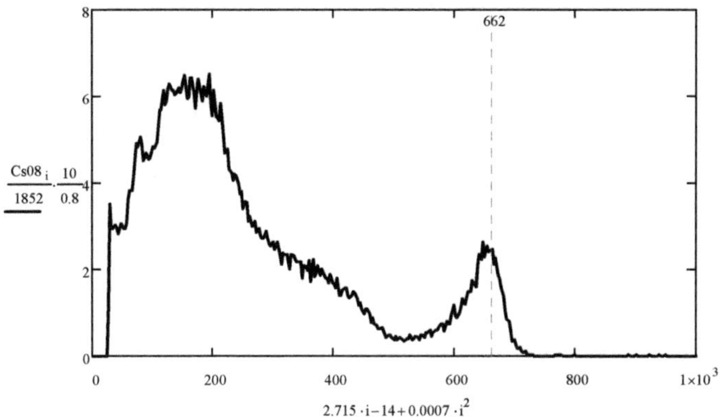


**Figure 6.30.** Apparatus response of the spectrum for CsI gamma detector from $^{137}Cs$ point source (installation UPD – Inter with a few attenuators made from the tungsten alloy).

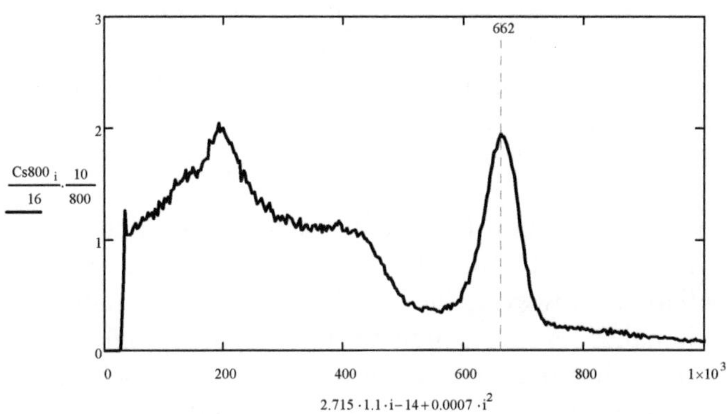


**Figure 6.31.** Apparatus response of the spectrum for CsI gamma detector from $^{137}Cs$ point source (installation UPD – Inter without attenuators).

The spectrum (Fig. 6.30) was obtained on the UPD - Inter installation using attenuators of different thicknesses. Here, the contribution of scattered radiation is significant (noticeably large contribution of the Compton distribution - the ratio "peak / Compton" is significantly small).

The spectrum of Fig. 6.31 was received without attenuator's application. Therefore, the contribution of scattered radiation is lower, but there arises a peak of backscattering (~ at 184 keV), which also affects the result. As can be seen, these spectra are characterized by different shapes under non-similar calibration conditions. However, due to the use of attenuators for weakening the gamma radiation beam, the distance from the IRS to the effective center of the instrument detection units as the gamma-ray spectrum becomes unknown while the scattered radiation is growing. This causes difficulties in determining the gamma radiation spectrum, and the contribution of scattered radiation becomes significant.

Exactly different energy dependence of the measuring instruments and different influence of the accompanying radiations can lead to considerable differences in the readings of dosimeters. Therefore, the unit size transfer must be carried out by the group comparator method. Moreover, only the particular types of dosimeters can be checked with a specific installation for which it was calibrated previously.

Hence, each installation is inherent in own advantages and disadvantages: on UPGD-2 installations it is possible to carry out metrological verification (calibration) of practically all types of dosimeters in a wide energy range, but not in the whole measuring range; on the UPS-Inter installation, it is possible to carry out metrological verification in a wider measurement range (due to a highly active source of radiation), but the requirements for the characteristics of standard radiation are not fulfilled.

The most obvious solution is to create standard installations that meet the requirements of ISO 4037-1 : 1996, with highly active sources of gamma radiation. However, the problems arise with radiation safety. Therefore, it makes sense to conduct studies on metrological verification of dosimeters with different types of detectors (gas-discharge counters, scintillators, ionization chambers, and others) on the UPD-Inter installations, if refined energy spectra; selected the different methods of calibrating the installations and dosimeters, etc.

It also considered the possibility and conducted preliminary studies of the use of an X-ray unit in combination with installations equipped with IRS, for calibration, verification and determination of gamma radiation dosimeters sensitivity over a wide range of dose rates. X-ray units are characterized by some advantages compared with powerful gamma installations: much easier to ensure radiation safety during transport and operation. They are capable of overlapping up to three orders of magnitude of dose rate providing easier and cheaper radiation protection of the room. The essence of the method [17] consists in the determination of the sensitivity to the $^{137}Cs$ gamma radiation source at one starting point of the range, then the linearity of the calibration characteristic over the entire measurement range is studied at the X-ray unit, and finally, whole calibration characteristic is determined.

Detection units of dosimeters are inherent in the nonlinear calibration characteristic; for example, the dosimeter based on gas-discharge meters. At high dose rates, due to the presence of time zone meters ("dead-time zone"), the multiplicative component of the error increases. For a lot of dosimetric MIs, the upper limit of measurement is equal to 10 Sv/h (10 Gy / h). Linearization of the calibration characteristic is carried out using the microprocessor by introducing an adjustment according to the formula:

$$X = K N (1 - \tau N)^{-1} \quad (6.16)$$

Here X is the measured value, f. e., the air-kerma rate of gamma radiation, Gr / s.; K is the transforming factor (sensitivity) of the detection unit; N is the pulse counting speed, p/s; τ - "dead time", s.

Therefore, for the calibration of gamma-radiation dosimeters, to determine the mentioned component of the error, IRS is needed, which ensures the reproduction of the dose rate at the level of 6-8 Sv / h. Preferably, only the sensitive subband (or detector) of the devices is checked on the UPGD and UPD-Inter installations; and for the rough subband of dosimeters, only the detector operability at the beginning of the subrange is checked. In advance, the sensitivity to the $^{137}Cs$ source of gamma radiation at one point of the range is determined at the high-mentioned installation. Next, at the X-ray installation, the linearity of the calibration characteristic in the entire range of measurements is studied aiming the restricting radiation exposures in the workplace and applying current occupational radiation protection techniques in the world [18].

## 6.5. Conclusions

1. The key task of metrology is the assurance of the unity of measurements, that is, the state of measurements, in which their results

are expressed in measurement units established by law, and the characteristics of errors or uncertainty of measurements are known with a certain probability and do not exceed the established limits.

2. It has been studied the method, positive feature of which is the possibility of obtaining the random estimation parameters of objects close to the optimal under conditions of the absence of a priori information about their distribution. From theory, the proposed method of measurement observations processing is based on their comparison with the set of sequences of the reference observations that are equal to the expectation of order statistics corresponded to the assigned. In practice, the considered method is realized by a quite simple procedure: first, the sorting of input observations and then averaging them by the set of previously calculated weight coefficients, corresponded to the assigned models of distribution. If the number of the input observations is $n \geq 10 \ldots 15$, then the proposed algorithm provides a smaller value of the measurement result standard uncertainty in comparison with the standard uncertainty of the mean value recommended by the Guide to the Expression of Uncertainty in Measurement. In 50…70 % cases, the considered method detects the true model of distribution correctly and in approximately 90…95% and more cases the true distribution or nearest to it (after the contra kurtosis) is detected. The effectiveness of the proposed algorithm of the processing of random observations grows with the amplitude of the deviations of the distribution of observations. Quality of method can be improved by the large numbers of reference samples corresponding to the reference density distributions that are closely located in the contra kurtosis axle as well as by using the more complicated procedure of RMS analysis.

3. There were performed the studies of naturally obsolete metrological objects, namely unique, high-precision measures of electrical resistance - model coils of electrical resistance - manufactured in 1960 and stored at the Lviv Polytechnic Metrology Laboratory. The mechanical stresses in these measures were minimal and the values of electrical resistance were stabilized; accordingly, one can expect minimal values of the variance when re-measuring the values of their resistances. Suggesting that the difference between the RMS of studied units is formed as a result of their nonequilibrium thermodynamic states, there were determined their energy spectra of resistance fluctuations and the frequency range within which the energy spectrum changes insignificantly. For a certain frequency range, you can determine the measurement conditions - the time interval between the following measurements $\delta t$ and the number of

results n, for which the difference between the variances of the results of equilibrium measurements becomes minimal.

4. Determined by the results averaging the standard deviation and, therefore, the random error of the measurement can be reduced to a negligible value only if the spectrum of altering the measurand is immutable within the band from 0 to high frequencies. Under real conditions, the mentioned spectrum is similar to flicker noise; so, a random error cannot be reduced to any small value by averaging a certain number of measurement results. Only if the relation $T_{measmax} \approx M_{max}\tau$ ( $T_{measmax}$ is the time of measuring the fluctuation parameters of the real system, $\tau$ is the relaxation time of the latter, $M_{max}$ is the number of possible ways within which the equilibrium state of the system can be realized) is pursued, a random error can reach a minimum.

5. As a result of the conducted researches the possibility of providing calibration, verification, and determination of the sensitivity of gamma-ray dosimeters, has been proved based on joint applying of a) installations equipped with ionizing radiation sources and b) X-ray installations. However, the main problems limiting the use of X-ray installations in the metrological verification of dosimeters are related to the instability of the X-ray tube and the extent of X-ray spectra. If the first problem is solved with the help of a reference dosimeter or a witness camera, the second problem is much more complicated.

## References


[1]. H. Nyquist, Thermal agitation of electric charge in conductors, *Phys. Rev.*, Vol. 32, Issue 1, 1928, pp. 110-113.

[2]. Z. Kolodiy, Flicker-noise of electronic equipment: Sources, ways of reduction and application, *Radioelectronics and Com. Syst.*, Vol. 53, Issue 8, 2010, pp. 412-417.

[3]. I. Gorban, Evaluation of statistically unpredictable changes in physical quantities over large observation intervals, *Journ. Techn. Phys.*, Vol. 88, Issue 12, 2018, pp. 1779-1786.

[4]. Z. Kolodiy, B. Stadnyk, S. Yatsyshyn, Development of Noise Measurements. Part 2. Random Error, *Sensors & Transducers*, Vol. 151, Issue 4, April 2013, pp. 107-112.

[5]. Z. Kolodiy, S. Yatsyshyn, Entropy of Measurements of Electric and Non-electric Systems Fluctuating Parameters, *Journ. Electronic Res. and Appl.*, Vol. 2, Issue 3, 2018, pp. 26-30.

[6]. ISO 10012:2003, Measurement Management Systems: Requirements for Measurement Processes and Measuring Equipment, 2003.

[7]. ISO/IEC 17025:2017, General Requirements for the Competence of Testing and Calibration Laboratories, 2017.
[8]. V. Yatsuk, M. Mykyjchuk, and T. Bubela, Ensuring the Measurement Efficiency in Dispersed Measuring Systems for Energy Objects, in Sustainable Production: Novel Trends, in Energy, Environment and Material Systems (Królczyk G., Wzorek M., Król A., Kochan O., Su J., Kacprzyk J. (Eds.)), *Springer, Cham,* part 9, 2020, pp. 131-149.
[9]. M. Mykyichuk, Metrological Support of Production, Lviv, Ukraine, *Lviv Polytechnic Publishing House*, 2017 (in Ukrainian).
[10]. T. Cooke, Calibration vs. Verification, *Cal-Tek Company, Inc.,* 2017, (https://www.repaircalibration.com/wp-content/uploads/public_docs/Calibration-vs-Verification.pdf).
[11]. R. Northrop, Introduction to Instrumentation and Measurements, *CRC Press Taylor & Francis Group*, Boca Raton, 2014.
[12]. V. Yatsuk, T. Bubela, Y. Pokhodylo, Y. Yatsuk, Improvement of data acquisition systems for the measurement of physical-chemical environmental properties, in *Proceedings of the 9th IEEE Int. Conf. on Intel. Data Acq. and Adv. Comp. Systems: Technology and Applications (IDAACS)*, 2017, pp. 41-46.
[13]. Container Security Initiative ([https://www.cbp.gov/border-security/ports-entry/cargo-security/csi/csi-brief).
[14]. ISO 4037-1:1996, IDT. Radiological protection -- X and gamma reference radiation for calibrating dosemeters and dose-rate meters and for determining their response as a function of photon energy -- Part 1: Radiation characteristics and production methods (https://standartgost.ru/g/ISO_4037-1:2019).
[15]. ISO 4037-2:2019, IDT. Radiological protection – X and gamma reference radiation for calibrating dosemeters and dose-rate meters and for determining their response as a function of photon energy – Part 2: Dosimetry for radiation protection over the energy ranges from 8 keV to 1,3 MeV and 4 MeV to 9 MeV (https://www.iso.org/standard/66873.html).
[16]. ISO 4037-3:1999, IDT. BS ISO 4037-3:2019. Radiological protection. X and gamma reference radiation for calibrating dosemeters and dose-rate meters and for determining their response as a function of photon energy – Part 3 : Calibration of the area and personal dosemeters and the measurement of their response as a function of energy and angle of incidence (https://www.en-standard.eu/bs-iso-4037-3-2019-radiological-protection-x-and-gamma-reference-radiation-for-calibrating-dosemeters-and-doserate-meters-and-for-determining-their-response-as-a-function-of-photon-energy-calibration-of-area-and-personal-dosemeters-and-the-measurement-of-).
[17]. L. Sinnikov, E. Andrievski, Investigation of the Possibility of calibrating the Photon Radiation Detection Units Using X-ray Installation, *ASRI,* Vol. 80, Issue 1, 2015, pp. 48-51.
[18]. Calibration or radiation protection monitoring instruments, STI/PUB/1017, Safety reports, Series №16, Internat, *Atomic Agency,*

*Vienna*, 2000 (https://www-pub.iaea.org/MTCD/Publications/PDF/P074_scr.pdf).

# Index

## M



## N



## O



## P



## Q



## R



## S

## W

www.ingramcontent.com/pod-product-compliance
Ingram Content Group UK Ltd.
Pitfield, Milton Keynes, MK11 3LW, UK
UKHW020220290926
15470UKWH00001B/300

9 788409 268993